T0275817

LONDON MATHEMATICAL SOCIETY LECTURE NOTE SERIES

The books in the series listed below are available from booksellers, or, in case of difficulty, from Cambridge University Press.

34 Representation theory of Lie groups, M.F. ATIYAH *et al*
36 Homological group theory, C.T.C. WALL (ed)
39 Affine sets and affine groups, D.G. NORTHCOTT
40 Introduction to H_p spaces, P.J. KOOSIS
43 Graphs, codes and designs, P.J. CAMERON & J.H. VAN LINT
45 Recursion theory: its generalisations and applications, F.R. DRAKE & S.S. WAINER (eds)
46 p-adic analysis: a short course on recent work, N. KOBLITZ
49 Finite geometries and designs, P. CAMERON, J.W.P. HIRSCHFELD & D.R. HUGHES (eds)
50 Commutator calculus and groups of homotopy classes, H.J. BAUES
51 Synthetic differential geometry, A. KOCK
57 Techniques of geometric topology, R.A. FENN
59 Applicable differential geometry, M. CRAMPIN & F.A.E. PIRANI
62 Economics for mathematicians, J.W.S. CASSELS
66 Several complex variables and complex manifolds II, M.J. FIELD
69 Representation theory, I.M. GELFAND *et al*
74 Symmetric designs: an algebraic approach, E.S. LANDER
76 Spectral theory of linear differential operators and comparison algebras, H.O. CORDES
77 Isolated singular points on complete intersections, E.J.N. LOOIJENGA
78 A primer on Riemann surfaces, A.F. BEARDON
79 Probability, statistics and analysis, J.F.C. KINGMAN & G.E.H. REUTER (eds)
80 Introduction to the representation theory of compact and locally compact groups, A. ROBERT
81 Skew fields, P.K. DRAXL
82 Surveys in combinatorics, E.K. LLOYD (ed)
83 Homogeneous structures on Riemannian manifolds, F. TRICERRI & L. VANHECKE
86 Topological topics, I.M. JAMES (ed)
87 Surveys in set theory, A.R.D. MATHIAS (ed)
88 FPF ring theory, C. FAITH & S. PAGE
89 An F-space sampler, N.J. KALTON, N.T. PECK & J.W. ROBERTS
90 Polytopes and symmetry, S.A. ROBERTSON
91 Classgroups of group rings, M.J. TAYLOR
92 Representation of rings over skew fields, A.H. SCHOFIELD
93 Aspects of topology, I.M. JAMES & E.H. KRONHEIMER (eds)
94 Representations of general linear groups, G.D. JAMES
95 Low-dimensional topology 1982, R.A. FENN (ed)
96 Diophantine equations over function fields, R.C. MASON
97 Varieties of constructive mathematics, D.S. BRIDGES & F. RICHMAN
98 Localization in Noetherian rings, A.V. JATEGAONKAR
99 Methods of differential geometry in algebraic topology, M. KAROUBI & C. LERUSTE
100 Stopping time techniques for analysts and probabilists, L. EGGHE

101 Groups and geometry, ROGER C. LYNDON
103 Surveys in combinatorics 1985, I. ANDERSON (ed)
104 Elliptic structures on 3-manifolds, C.B. THOMAS
105 A local spectral theory for closed operators, I. ERDELYI & WANG SHENGWANG
106 Syzygies, E.G. EVANS & P. GRIFFITH
107 Compactification of Siegel moduli schemes, C-L. CHAI
108 Some topics in graph theory, H.P. YAP
109 Diophantine Analysis, J. LOXTON & A. VAN DER POORTEN (eds)
110 An introduction to surreal numbers, H. GONSHOR
111 Analytical and geometric aspects of hyperbolic space, D.B.A.EPSTEIN (ed)
112 Low-dimensional topology and Kleinian groups, D.B.A. EPSTEIN (ed)
113 Lectures on the asymptotic theory of ideals, D. REES
114 Lectures on Bochner-Riesz means, K.M. DAVIS & Y-C. CHANG
115 An introduction to independence for analysts, H.G. DALES & W.H. WOODIN
116 Representations of algebras, P.J. WEBB (ed)
117 Homotopy theory, E. REES & J.D.S. JONES (eds)
118 Skew linear groups, M. SHIRVANI & B. WEHRFRITZ
119 Triangulated categories in the representation theory of finite-dimensional algebras, D. HAPPEL
121 Proceedings of *Groups - St Andrews 1985*, E. ROBERTSON & C. CAMPBELL (eds)
122 Non-classical continuum mechanics, R.J. KNOPS & A.A. LACEY (eds)
124 Lie groupoids and Lie algebroids in differential geometry, K. MACKENZIE
125 Commutator theory for congruence modular varieties, R. FREESE & R. MCKENZIE
126 Van der Corput's method for exponential sums, S.W. GRAHAM & G. KOLESNIK
127 New directions in dynamical systems, T.J. BEDFORD & J.W. SWIFT (eds)
128 Descriptive set theory and the structure of sets of uniqueness, A.S. KECHRIS & A. LOUVEAU
129 The subgroup structure of the finite classical groups, P.B. KLEIDMAN & M.W.LIEBECK
130 Model theory and modules, M. PREST
131 Algebraic, extremal & metric combinatorics, M-M. DEZA, P. FRANKL & I.G. ROSENBERG (eds)
132 Whitehead groups of finite groups, ROBERT OLIVER
133 Linear algebraic monoids, MOHAN S. PUTCHA
134 Number theory and dynamical systems, M. DODSON & J. VICKERS (eds)
135 Operator algebras and applications, 1, D. EVANS & M. TAKESAKI (eds)
136 Operator algebras and applications, 2, D. EVANS & M. TAKESAKI (eds)
137 Analysis at Urbana, I, E. BERKSON, T. PECK, & J. UHL (eds)
138 Analysis at Urbana, II, E. BERKSON, T. PECK, & J. UHL (eds)
139 Advances in homotopy theory, S. SALAMON, B. STEER & W. SUTHERLAND (eds)
140 Geometric aspects of Banach spaces, E.M. PEINADOR and A. RODES (eds)
141 Surveys in combinatorics 1989, J. SIEMONS (ed)
142 The geometry of jet bundles, D.J. SAUNDERS
143 The ergodic theory of discrete groups, PETER J. NICHOLLS
144 Introduction to uniform spaces, I.M. JAMES
145 Homological questions in local algebra, JAN R. STROOKER
146 Cohen-Macaulay modules over Cohen-Macaulay rings, Y. YOSHINO
147 Continuous and discrete modules, S.H. MOHAMMED & B.J. MÜLLER
148 Helices and vector bundles, A.N. RUDAKOV et al
149 Solitons, nonlinear evolution equations and inverse scattering, M.A. ABLOWITZ & P.A. CLARKSON
150 Geometry of low-dimensional manifolds 1, S.K. DONALDSON & C.B. THOMAS (eds)
151 Geometry of low-dimensional manifolds 2, S.K. DONALDSON & C.B. THOMAS (eds)
152 Oligomorphic permutation groups, P.J. CAMERON
153 L-functions in arithmetic, J. COATES & M.J. TAYLOR (eds)
154 Number theory and cryptography, J. LOXTON (ed)
155 Classification theories of polarized varieties, TAKAO FUJITA
156 Twistors in mathematics and physics, T.N. BAILEY & R.J. BASTON (eds)
158 Geometry of Banach spaces, P.F.X. MÜLLER & W. SCHACHERMAYER (eds)

London Mathematical Society Lecture Note Series. 158

Geometry of Banach Spaces

Proceedings of the Conference held in Strobl, Austria, 1989

Edited by
P.F.X. Müller
W. Schachermayer
Johannes Kepler Universität, Linz

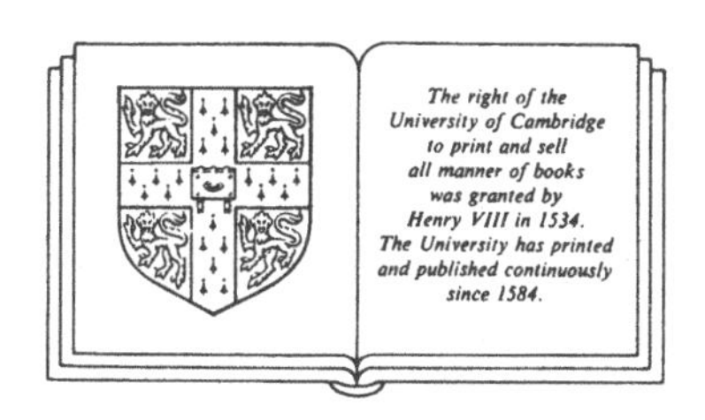

CAMBRIDGE UNIVERSITY PRESS
Cambridge
New York Port Chester Melbourne Sydney

Published by the Press Syndicate of the University of Cambridge
The Pitt Building, Trumpington Street, Cambridge CB2 1RP
40 West 20th Street, New York, NY 10011, USA
10, Stamford Road, Oakleigh, Melbourne 3166, Australia

First published 1990

Library of Congress cataloguing in publication data available

British Library cataloguing in publication data available

ISBN 0 521 40850 4

Transferred to digital printing 2002

CONTENTS

FOREWORD

In the week from June 12th to June 17th, 1989, the functional analysis group of the Mathematics Department, Johannes-Kepler-Universität at Linz played host to an international conference on the geometry of Banach spaces which was sponsored by the International Mathematical Union. This conference united 119 participants from 27 countries, including most of the leading specialists in the field, for a week of intensive study and discussion on recent progress. The scientific programme consisted of 23 plenary lectures which were held by a number of experts invited by the members of the organizing committee. In addition, there were a series of half-hour talks held in parallel sessions.

It is the pleasant duty of the editors of the proceedings of the above conference to express their gratitude to the following persons and organisations whose support, financial or otherwise, played an important role in its successful realisation:
The members of the organizing committee

Jean Bourgain, IHES, Bures-sur-Yvette
Nassif Ghoussoub, University of British Columbia, Vancouver
William Johnson, Texas A&M University
Hermann König, University of Kiel
Joram Lindenstrauss, The Hebrew University, Jerusalem
Bernard Maurey, Université de Paris VII
Aleksander Pelczynski, Academy of Sciences, Warsaw
Gilles Pisier, Université de Paris VI and Texas A&M University
Haskell Rosenthal, University of Texas, Austin

for their help in arranging of the details of the scientific programme.

The following organizations for their financial assistance:
- International Mathematical Union
- Österreichisches Bundesministerium für Wissenschaft und Forschung
- Linzer Hochschulfonds
- Land Oberösterreich
- Handelskammer Oberösterreich
- Verband der Versicherungsunternehmen Österreichs
- Wiener Städtische Versicherung.

The conference was held at the Bundesinstitut für Erwachsenenbildung in Strobl, in the Austrian Salzkammergut and the organisers would like to express their appreciation to its Director, Dr. Klaus Gattol, and his staff for their efficient and friendly assistance in the organisation of the scientific aspects of the conference.

Paul Müller Walter Schachermayer
(Editors)

List of Main Lectures

1. J. Bourgain: *Homogeneous polynomials on the complex ball, polynomial bases and the ball algebra.*
2. A.V. Bukhvalov: *Nonlinear majorisation of linear operators.*
3. H.G. Feichtinger: *Characteriasation of Banach spaces of distributions through non orthogonal expansions.*
4. T. Figiel: *Singular linear operators in function spaces.*
5. N. Ghoussoub: *Critical points in variational methods.*
6. E. Gluskin: *Estimates of the volume of some convex bodies.*
7. P. Gruber: *The space of convex bodies.*
8. M.I. Kadets: *The latest results on conditionally convergent series.*
9. N. Kalton: *Commutators and complex interpolation.*
10. B.S. Kashin: *Uniform convergence of orthogonal series.*
11. J. Lindenstrauss: *Almost isometric Euclidean sections.*
12. B. Maurey: *Some results and problems in infinite dimensional theory.*
13. V. Milman: *Convex perestroika.*
14. T. Odell: *On distorted norms and nonseparable duals.*
15. V.V. Peller: *Hankel operators and continuity poinst of best approximation operators.*
16. G. Pisier: *Factorisation of operator-valued functions on* H^p.
17. D. Preiss: *Determination on measures and differentiability.*
18. C. Read: *Different forms of the approximation property.*
19. H.P. Rosenthal: *Non-distorted norms and the fine subspace structure of general Banach spaces.*
20. G. Schechtmann: *Complemented subspaces of* ℓ_p^n.

List of Participants

Javier Alonso
Badajoz, Spain

Abdul Latif Alsiddiqui
Manamma, Bahrain

Dale Alspach
Stillwater, Oklahoma, U.S.A.

Dan Amir
Tel Aviv, Israel

Spiros Argyros
Heraklion, Greece

Alvaro Arias
College Station, Texas, U.S.A.

Keith Ball
Cambridge, England

Wojciech Banaszczyk
Lodz, Poland

Jesus Bastero
Zaragoza, Spain

Erhard Behrends
Berlin, BRD

Steven F. Bellenot
Talahassee, Florida, U.S.A.

Yoav Benyamini
Haifa, Israel

Oscar Blasco
Zaragoza, Spain

Gordon Blower
Oxford, England

Bela Bollobas
Cambridge, England

Jean Bourgain
Bures-sur-Yvette, France

A.V. Bukhvalov
Leningrad, USSR

Peter Casazza
Columbia, Missouri, U.S.A.

Bill Davis
Columbus, Ohio, U.S.A.

Andreas Defant
Oldenburg, BRD

Martin Defant
Kiel, BRD

Irini Deligianni
Heraklion, Greece

Robert Deville
Besancon, France

Steven Dilworth
Columbia, South Carolina, U.S.A.

Marian Fabian
Prague, Czechoslovakia

Jeff Farmer
College Station, Texas, U.S.A.

Hans-Georg Feichtinger
Wien, Österreich

Helga Fetter
Guanajuato, Mexico

Tadeusz Figiel
Sopot, Poland

Catherine Finet
Mons, Belgium

Berta Gamboa
Guanadjuato, Mexico

Ted Gamelin
Los Angeles, U.S.A.

D.J.H. Garling
Cambridge, England

Nassif Ghoussoub
Vancouver, Canada

Maria Girardi
Urbana, Illinois, U.S.A.

E.D. Gluskin
Leningrad, USSR

Gilles Godefroy
Paris, France

Yehoram Gordon
Haifa, Israel

Tim Gowers
Cambridge, England

Michael Grosser
Wien, Österreich

Peter Gruber
Wien, Österreich

Jim Hagler
Denver, Colorado, U.S.A.

Peter Harmand
Oldenburg, BRD

Stefan Heinrich
Berlin, DDR

W. Hensgen
Regensburg, BRD

Hans Jarchow
Zürich, Switzerland

Kamil John
Prague, Czechoslovakia

Marius Junge
Kiel, BRD

Michail I. Kadets
Charkow, USSR

Nigel Kalton
Columbia, Missouri, U.S.A.

B.S. Kashin
Moscow, USSR

Hermann König
Kiel, BRD

Ryszard Komorowski
Edmonton Alberta, Canada

Omran Kouba
Paris, France

Elton Lacey
College Station, Texas, U.S.A.

Miguel Lacruz
Madrid, Spain

Imre Leader
Cambridge, England

Michel Ledoux
Strasbourg, France

Aswald Lima
Vollebekk, Norway

Bor Luh Lin
Iowa City, Iowa, U.S.A.

Pei-Kee Lin
Memphis, Tennessee, U.S.A.

Werner Linde
Jena, DDR

Joram Lindenstrauss
Jerusalem, Israel

Francoise Lust-Piquard
Paris-Orsay, France

Vania D. Mascioni
Zürich, Switzerland

David McLaughlin
Edmonton, Alberta, Canada

Ted Odell
Austin, Texas, U.S.A.

Alain Pajor
Paris, France

P.L. Papini
Bologna, Italy

Aleksander Pelczynksi
Warszawa, Poland

Minos Petrakis
Heraklion, Greece

Gilles Pisier
Paris, France

Gerhard Racher
Salzburg, Östrreich

Frank Räbiger
Tübingen, BRD

Charles Read
Cambridge, England

Wolfgang Ruess
Essen, BRD

Georg Schlüchtermann
München, BRD

Abderrazzak Sersouri
Urbana. Illinois, U.S.A.

Gun-Marie Lövblom
Stockholm, Sweden

Pjotr Mankiewicz
Warszawa, Poland

Bernard Maurey
Paris, France

Vitali Milman
Tel Aviv, Israel

Jose Orihuela
Murcia, Spain

Antonio Pallares
Murcia, Spain

Rafael Paya
Granada Spain

V.V. Peller
Leningrad, USSR

Bob Phelps
Seattle, Washington, U.S.A.

David Preiss
Prague, Czechoslovakia

Jamie Radcliffe
Cambridge, England

Yves Raynaud
Paris, France

Haskell P. Rosenthal
Austin, Texas, U.S.A.

Gideon Schechtman
Rehovot, Israel

Thomas Schlumprecht
Austin, Texas, U.S.A.

Stanislaw Szarek
Cleveland, Ohio, U.S.A.

Michel Talagrand
Paris, France

G. Terenzi
Milano, Italy

Nicole Tomczak-Jägermann
Edmonton, Alberta, Canada

Andrew Tonge
Kent, Ohio, U.S.A.

S.L. Troyanski
Sofia, Bulgaria

Barry Turett
Rochester, U.S.A.

Hans-Olav Tylli
Helsinki, Finland

Lior Tzafiri
Jerusalem, Israel

Ulf Uttersrud
Oslo, Norway

Manuel Valdivia
Burjasot, Spain

Hermann-Josef Wanning
Essen, BRD

Lutz Weis
Baton Rouge, Louisiana, U.S.A.

Dirk Werner
Berlin, BRD

Elisabeth Werner
Stillwater, Oklahoma, U.S.A.

Wend Werner
Paderborn, BRD

Tomasz Wolniewicz
Torun, Poland

Quanhua Xu
Paris, France

David Yost
Berlin, BRD

James Bell Cooper
Paul F.X. Müller
Reinhard Riedl
Walter Schachermayer
Michel Schmuckenschläger
Charles Stegall
Bernhard Stöger

Institut für Mathematik, Universität Linz, Österreich

A NOTE ON H. ISHIHARA AND W. TAKAHASHI MODULUS OF CONVEXITY

Javier Alonso and Antonio Ullán
Departamento de Matemáticas, Universidad de Extremadura
06071–BADAJOZ (SPAIN)

AMS Subject Class. 46B20

1. Introduction:

The classical modulus of convexity of a normed linear space E, introduced by J.A. Clarkson[2] in 1936, is defined as

$$\delta_E(\epsilon)=\inf\{1-\left\|\frac{x+y}{2}\right\| : \|x\|\le 1,\ \|y\|\le 1,\ \|x-y\|\ge\epsilon\} \qquad (0\le\epsilon\le 2)$$

and it is at the origin of a great number of moduli defined since then by several authors[1,3,...,11] up to the present date. One of these is the modulus of convexity of a convex subset C, defined by H. Ishihara and W. Takahashi[7] as

$$\delta(C,\epsilon)=\inf\{1-\frac{1}{r}\left\|z-\frac{x+y}{2}\right\| : x,y,z\in C,\ 0<r\le d(C),\ \|z-x\|\le r,\ \|z-y\|\le r,\ \|x-y\|\ge r\epsilon\}$$

where $0\le\epsilon\le 2$ and d(C) denotes the diameter of C.

We prove in this paper that this modulus does not depend on C but on the affine linear subspace aff(C), spaned by C. Moreover, if $0\in$aff(C) then this modulus coincides with the Clarkson modulus of the linear subspace $\langle C\rangle$, spaned by C.

Throught the paper we shall refer only to convex subsets with more than one point.

2. Results:

It is easy to prove the following properties for every $0\le\epsilon\le 2$:

a) If B_E denotes the unit ball of E then

$$\delta(B_E,\epsilon)=\delta(E,\epsilon)=\delta_E(\epsilon).$$

b) If C and D are convex subsets of E and $D\subset C$ then

$$\delta(C,\epsilon)\le\delta(D,\epsilon).$$

c) For every convex subset $C\subset E$, $x\in E$ and $\lambda\in\mathbb{R}\setminus\{0\}$ it holds

$$\delta(C+x,\epsilon)=\delta(\lambda C,\epsilon)=\delta(C,\epsilon).$$

Lemma. Let C be a convex subset of E and let $x,y,z \in aff(C)$. Then there exist $\lambda>0$ and $u \in C$ such that $x,y,z \in \lambda(C-u)+u$.

Proof. Let $x,y,z \in aff(C)$. Then there exist $u_1, \ldots, u_n \in C$ such that $x,y,z \in C^*=aff(u_1, \ldots, u_n)$. Let u be an interior point of the convex hull of $\{u_1,\ldots,u_n\}$, $Co(u_1,\ldots,u_n)$, with the relative topology of C^*, and let $B(u,s)$ be an open ball of center u and radious s such that $B(u,s)\cap C^* \subset Co(u_1,\ldots,u_n) \subset C$. Then $B(0,s)\cap(C^*-u) \subset C-u$ and $x-u$, $y-u$, $z-u$ are in C^*-u. Let $\lambda>0$ be such that $x-u$, $y-u$, $z-u$ are in $B(0,\lambda s)$. Then $x-u$, $y-u$, $z-u$, are in $\lambda(C-u)$ and therefore, $x,y,z \in \lambda(C-u)+u$.

Proposition. Let C be a convex subset of E. Then $\delta(C,\epsilon)=\delta(aff(C),\epsilon)$, for every $0\leq\epsilon\leq 2$.

Proof. Let $0\leq\epsilon\leq 2$. From Property (b) it follows $\delta(C,\epsilon)\geq\delta(aff(C),\epsilon)$. Conversely, let $x,y,z \in aff(C)$ and $r>0$ be such that $\|z-x\|\leq r$, $\|z-y\|\leq r$ and $\|x-y\|\geq \epsilon r$. From the above lemma there exist $\lambda>0$ and $u\in C$ such that $x,y,z \in \lambda(C-u)+u$. We can also suppose that $\lambda\geq r/d(C)$. So, we have $0<r\leq\lambda d(C)=d(\lambda(C-u)+u)$ and, by Property (c)

$$1-\frac{1}{r}\left\|z-\frac{x+y}{2}\right\| \geq \delta(\lambda(C-u)+u,\epsilon)=\delta(\lambda(C-u),\epsilon)=\delta(C-u,\epsilon)=\delta(C,\epsilon).$$

Therefore, $\delta(aff(C),\epsilon)\geq\delta(C,\epsilon)$.

Corollary 1. Let C be a convex subset of E such that $0\in aff(C)$. Then $\delta(C,\epsilon)=\delta_{<C>}(\epsilon)$, for every $0\leq\epsilon\leq 2$.

Proof. It follows easily from Property (a) and the fact that $aff(C)=<C>$.

It is also obvious the next corollary.

Corollary 2. If C and D are two convex subsets of E such that either $aff(C\cap D)= aff(C)$ or $aff(C\cap D)=aff(D)$ then $\delta(D\cap C,\epsilon)=\max\{\delta(C,\epsilon), \delta(D,\epsilon)\}$ for every $0\leq\epsilon\leq 2$.

In [8] the above result is stated without the assumption "$aff(C\cap D)=aff(C)$ or $aff(C\cap D)=aff(D)$". Unfortunately in this case it is not true. Let, for example, E be the linear space $\mathbb{R}^3$ endowed with a norm whose unit ball is the set $B_E=\{(x,y,z)/ x^2+y^2+z^2\leq 1, |z|\leq \frac{1}{2}\}$ and let $a=(0,0,\frac{1}{2})$, $C=B_E+a$ and $D=B_E-a$. Then $\delta_E(1)=0$. On the other hand, the space $<C\cap D>$ is the usual inner product space $(\mathbb{R}^2,\|\ \|_2)$, and and so $\delta_{<C\cap D>}(\epsilon)=1-\sqrt{1-\frac{\epsilon^2}{4}}$ for every $0\leq\epsilon\leq 2$. Therefore,

$$\delta(C,1)=\delta(D,1)=\delta_E(1)=0<1-\sqrt{1-\tfrac{1}{4}}=\delta_{<C\cap D>}(1)=\delta(C\cap D,1).$$

3. References:

1. J. BANAS, *On moduli of smoothness of Banach spaces*, Bull. Polish Acad. Sci. Math. **34**, (1986), no. 5–6, 287–293.

2. J.A. CLARKSON, *Uniformly convex spaces*, Trans. Amer. Math. Soc. **40**, (1936), 394–414.

3. D.F. CUDIA, *On the localization and directionalization of uniform convexity*, Bull. Amer. Math. Soc. **69**, (1963), 265–267.

4. C. FINET, *Uniform convexity properties of norms on a superreflexive Banach space*, Israel J. Math. **53**, (1986), no. 1, 81–92.

5. K. GOEBEL and T. SEKOWSKI, *The modulus of noncompact convexity*, Ann. Univ. Mariae Curie–Sklodowska **38**, 3,sect. A, (1984), 41–48.

6. V.I. GURARII, *On moduli of convexity and flattening of Banach spaces*, Soviet Math. Dokl. **6**, (1965), 535–539.

7. H. ISHIHARA and W. TAKAHASHI, *Modulus of convexity, characteristic of convexity and fixed points theorems*, Kodai Math. J. **10**, (1987), 197–208.

8. A.R. LOVAGLIA, Locally uniformly convex Banach spaces, Trans. Amer. Math. Soc. **78**, (1955), 225–238.

9. V.D. MILMAN, *Infinite–dimensional geometry of the unit sphere of a Banach space*, Soviet Math. Dokl. **8**, (1967), 1440–1444.

10. V.L. SMULIAN, *Sur la derivavilite de la norme dans l'espace de Banach*, Dokl. Akad. Nauk SSSR (N.S.) **27**, (1940), 643–648.

11. F. SULLIVAN, *A generalization of uniformly rotund Banach spaces*, Canadian J. Math. **31**, (1979), 628–636.

A PROPERTY OF NON-STRONGLY REGULAR OPERATORS

by

S. ARGYROS - M. PETRAKIS

Department of Mathematics
University of Crete
Heracleion - Grete

Introduction:

Let m be the Lebesgue measure on [0,1]. By L^1 we denote the Banach space of all Lebesgue integrable functions on [0,1] equipped with the norm $\| f \|_1 = \int |f| \, dm$. Let X be a Banach space. In [B_1] J. Bourgain proved that if $T : L^1 \to X$ is a non Dunford - Pettis operator then there exists a Dunford - Pettis operator $D : L^1 \to L^1$ such that the operator $T \circ D$ is not Bochner representable. Recall that on operator from L^1 into a Banach space X is called Dunford - Pettis if it maps weakly compact subsets of L^1 into norm compact subsets of X. (see [B_1], [D-U] for details and undefined notions).

In this paper we prove the following:

Theorem 1: *Let* $T : L^1 \to X$ *be any non strongly regular operator. Then there exists a Dunford - Pettis operator* $D : L^1 \to L^1$ *such that the operator* $T \circ D$ *is not Bochner representable.*

Recall (see [G-G-M-S] Theorem IV. 10) that an operator $T : L^1 \to X$ is strongly regular iff for each $A \subset [0,1]$, $m(A)>0$ and $\varepsilon>0$ there is a relatively weakly open subset W of the set $F_A = \{f \in L^1 : f \geq 0, \int f dm = 1, \text{supp } f \subset A\}$ such that diam $(T(W)) < \varepsilon$. It is known ([G-G-M], [Wes]) that any strongly regular operator $T : L^1 \to X$ is Dunford - Pettis.

The proof of Theorem 1 strongly depents on Lemma 2 which is probably of some independent interest. In Lemma 2 we prove an "uncoditional" version of the clasical Theorem of Mazur in the case of a bounded subset of the positive cone of L^1. Lemma 1 is used in the proof of Lemma 2. A consequence of Theorem 1 is following result (Corollary 2): Let K be a closed convex subset of a Banach space X. Suppose that every closed bounded subset of K with the PCP has the RNP. Then if $T: L^1 \to X$ is any non representable "K valued" operator (i.e. $T(F) \subset K$ wheve by F we denote the set of densities in L^1) there exists a Dunfort - Pettis operator $D: L^1 \to L^1$ such that the operator $T \cdot D$ is non representable. Finally we present two examples. Example 1 shows that the D - P operator D in the statement of Theorem 1 can not in general be replaced by a D - P convolution operator.

In Example 2 we construct a non - dentable strongly regular subset K of c_0 with the additional property that for every "K valued" operator $T: L^1 \to c_0$ and for every D-P operator $D: L^1 \to L^1$ the operator $T \cdot D$ is representable.

Remark 1: As noted in [P-S] it follows easily from Proposition VII.4 of [G-G-M-S] that the class of strongly regular operators and the class of representable operators from L^1 to L^1 coincide.

The proof of Corollary 2 shows that the above is true if we replace the target space L^1 by any space X with the property that the PCP and the RNP are equivalent on the subsets of X. (For a simple proof that any strongly regular convolution operator from $L^1(\mathbb{T})$ to $L^1(\mathbb{T})$ is representable see also [P-S]).

Hence it follows from Theorem 1 that for any non representable operator $T: L^1(\mathbb{T}) \to L^1$ there exists a D-P operator $D: L^1(\mathbb{T}) \to L^1(\mathbb{T})$ such that the operator $T \cdot D$ is not representable. This result has been proved in [K - P - R - U] and [P]. Actually it is proved there that the operator D can be taken to be a convolution D - P operator.

Remark 2. (i) One of the first examples of non representable operators from L^1 into $C[0,1]$ is the integral operator $V: L^1 \to C[0,1]$ defined by

$Tf(s) = \int_0^s f\, dm$, $f \in L^1$. Bourgain has proved in [B1] that if $D: L^1 \to L^1$ is any D-P operator then the operator $V \cdot D$ is representable.

Hence it follows from Theorem 1 that the operator V is strongly regular. This can also be proved using Theorem IV. 10 of [G-G-M-S].

(ii) In view of Schachermayer's result ([S]) Theorem 1 reduces the KMP versus RNP problem to the following question: Suppose K is a closed convex bounded subset of a Banach space X. Suppose also that K has the KMP. Does this imply that every "K valued" operator $T: L^1 \to X$ has the property that the operator $T \cdot D$ is representable for every D - P operator $D: L^1 \to L^1$?

Lemma 1 : *Let* $G_n \subset L^2$, $n=1,2,\ldots$ *be a sequence of norm compact sets. Suppose that* $\forall g_n \in G_n$, $g_n \to 0$ *weakly. Then there is an increasing* *sequence of positive integers* $n_1 < n_2 < \ldots < n_k < \ldots$ *and a positive constant* M' *such that* $\left\| \sum_{l=1}^{k} \varepsilon_l g_{n_l} \right\|_2 \leq M'\sqrt{k}$ *for all* g_{n_l} *in* G_{n_l} *and for all choices of signs* $\varepsilon_l = \pm 1$, $l = 1,2,\ldots k$.

Proof: Let $M = \sup\{ \|g\|_2, g \in \bigcup_{n \in \mathbb{N}} G_n \}$ Notice that $M < \infty$ otherwise $\forall k \in \mathbb{N}$ $\exists n_k$ such that $\| g_{n_k} \|_2 > k$. This is impossible since $g_{n_k} \to 0$ weakly in L_2.

Let h_k, $k=1,2,\ldots$ be an orthonormal basis in L^2. If $g = \sum_{i=1}^{\infty} a_i h_i$ and $k \in \mathbb{N}$ we denote by $g|[1,2,\ldots k]$ the function $\sum_{i=1}^{k} a_i h_i$ and by $g|[k,k+1, \ldots \infty]$ the function $g = \sum_{i=k}^{\infty} a_i h_i$. For $l<m$ $g|[l, m]$ denotes the function $g = \sum_{i=l}^{m} a_i h_i$.

Fix $k \in \mathbb{N}$. Let $a_n = \sup\{\|g|[1, \ldots k]\|_2, g \in G_n\}$. Notice that $a_n \to 0$ since $g_n|[1, \ldots k] \to 0$ for all choices of g_n in G_n.

Using induction we can construct sequences of positive integers $n_1 < n_2 < \ldots < n_s < \ldots$ and $k_1 < k_2 < \ldots < k_s < \ldots$ such that

(i) $\sup\{ \| g|[1, 2, \ldots k_{i-1}] \|_2 , g \in G_{n_i} \} < \frac{M}{2^i}$

and

(ii) $\sup \{ \| g|[k_i+1, \ldots \infty] \|_2 , g \in G_{n_1} \cup \ldots \cup G_{n_i} \} < \frac{M}{i\,2^i}$

Notice that if $g_{n_l} \in G_{n_l}$, $\varepsilon_l = \pm 1$

$$\Big\| \sum_{l=1}^{k} \varepsilon_l g_{n_l} \Big\|_2 \le \Big\| \sum_{l=1}^{k} \varepsilon_l (g_{n_l} \mid [k_{l-1}+1, \dots, k_l]) \Big\|_2 + \Big\| \sum_{l=1}^{k} \sum_{\substack{j=1 \\ j\neq l}}^{k} (g_{n_j} \mid [k_{l-1}+1, \dots, k_l]) \Big\|_2$$

The functions $g_{n_l}|[k_{l-1}+1, \dots K_l]$ are disjointly supported and by Bessel's inequality we have that

$$\Big\| \sum_{l=1}^{k} \varepsilon_l (g_{n_l}|[k_{l-1}+1, \dots, k_l]) \Big\|_2 \le M\sqrt{k}$$

On the other hand $\displaystyle\sum_{\substack{j=1 \\ j\neq l}}^{k} \| g_{n_j} \mid [k_{l-1}+1, \dots, k_l] \|_2 =$

$$\sum_{j=1}^{l-1} \| g_{n_j} [k_{l-1}+1, \dots, k_l] \|_2 + \sum_{j=l+1}^{k} \| g_{n_j} \mid [k_{l-1}+1, k_l] \|_2 \le \frac{lM}{12^l} + \frac{M}{2^l}$$

and hence $\displaystyle\Big\| \sum_{l=1}^{k} \sum_{\substack{j=1 \\ j\neq l}}^{k} g_{n_j} \mid [k_{l-1}+1, \dots k_l] \Big\|_2 \le C$ for some constant C. It is clear now that $\| \sum_{l=1}^{k} \varepsilon_l g_l \|_2 \le M\sqrt{k} + C \le M'\sqrt{k}$ for some constant $M' > 0$, for all choises of signs $\varepsilon_l = \pm 1$, $l = 1, 2, \dots k$.

Lemma 2 : *Let* K *be a bounded subset of the positive cone of* L^1. *Let* f *be a weak limit point of* K. *Then for every* $\varepsilon > 0$ *there exist a finite set* $F = \{f_1, f_2, \dots f_d\} \subset K$ *and real numbers* a_i, $i = 1, 2, \dots d$, $a_i \ge 0$, $\sum_{i=1}^{d} a_i = 1$ *such that* $\| \sum_{i=1}^{d} \varepsilon_i a_i (f_i - f) \|_1 \le 5\varepsilon$ *for all choices of signs* $\varepsilon_i = \pm 1$, $i = 1, 2, \dots, d$.

Proof : Let M be large enough such that $\int_A f\,dm < \varepsilon$ where $A = \{x \in [0,1] : f(x) > M\}$. Let (h_k) $k = 1, 2, \dots$ be a biorthogonal system in $L^\infty(A^c)$ such that $\| h_k \|_\infty = \| h_k \|_2 = 1$, $\forall k \in \mathbb{N}$. For each $n \in \mathbb{N}$ there exists a finite subset F_n of K, $F_n = \{f_1^n, \ldots, f_{k_n}^n\}$ and real numbers $a_i^n \geq 0$

$i = 1,2, \ldots k_n$, $\sum_{i=1}^{k_n} a_i^n = 1$ such that

$$(1) \quad \sup_{i=1,2,\ldots k_n} \left| \int (f_i^n - f) h_k \, dm \right| < \frac{1}{n} \quad \text{if } k < n.$$

and

$$(2) \quad \left\| \sum_{i=1}^{k_n} a_i^n f_i^n - f \right\| < \frac{1}{n}$$

For $n > \frac{1}{\varepsilon}$ let G_n be the set of all functions g that can be written in the form $g = \sum_{i\in S} a_i^n (f_i^n - f)$ where $\sum_{i\in S} a_i^n \geq \frac{1}{2}$.

We claim that the set $\bigcup_{n>\frac{1}{\varepsilon}} G_n$ is uniformly integrable: If not, there exists a $\vartheta > 0$ such that for all $\delta > 0$ we can find a set $B \subset [0,1]$ with $m(B) < \delta$ and a function $g \in \bigcup_{n>\frac{1}{\varepsilon}} G_n$ such that $\int_B g \, dm > \vartheta$. Choose $\delta_0 > 0$ such that $\int_B f \, dm > \frac{\vartheta}{2}$ for all $B \subset [0,1]$ with $m(B) < \delta_0$.

Assume that the function g is of the form $\sum_{i\in S} a_i^n (f_i^n - f)$ where $\sum_{i\in S} a_i^n \geq \frac{1}{2}$

We have that $\vartheta < \int_B g \, dm = \int_B \sum_{i\in S} a_i^n (f_i^n - f) = \int_B \sum_{i\in S} a_i^n f_i^n - \int_B \sum_{i\in S} a_i^n f \leq$

$\int_B \sum_{i\in S} a_i^n f_i^n \leq \int_B \sum_{i=1}^{k_n} a_i^n f_i^n$. On the other hand for $n > \frac{4}{9}$ we have that

$$\left| \int_B \sum_{i=1}^{k_n} a_i^n f_i^n - f \right| \leq \left\| \sum_{i=1}^{k_n} a_i^n f_i - f \right\| < \frac{1}{n} < \frac{9}{4} \qquad (2)$$

This implies that $\int_B \sum_{i=1}^{k_n} a_i^n f_i^n < \int_B f + \frac{9}{4} < \frac{9}{2} + \frac{9}{4} = \frac{39}{4}$. This is a contradiction since $\int_B \sum_{i=1}^{k_n} a_i^n f_i^n \geq 9$.

Notice that if for each $n \in \mathbb{N}$ we select a function $g_n \in G_n$ then the sequence $(g_n | A^c)_{n\in\mathbb{N}}$ converges weakly to zero.

Now let $g \in G_n$, and suppose that g can be written in the form $g = \sum_{i\in S} a_i^n (f_i^n - f)$ where $\sum_{i\in S} a_i^n \geq \frac{1}{2}$. Consider the set

$$A_g = \{ x \in [0,1] : \sum_{i\in S} a_i^n (f_i^n(x) > 10M \}$$

We claim that $\int_{A_g \setminus A} |g| < \frac{2}{n}$

To prove the claim notice that $\int_{A_g \setminus A} \left| \sum_{i\in S} a_i^n (f_i^n - f) + \sum_{i\in S^c} a_i^n (f_i^n - f) \right| dm < \frac{1}{n}$

Let $(A_g \setminus A)^- = \{ x \in A_g \setminus A : \sum_{i\in S^c} a_i^n (f_i^n - f)(x) < 0 \}$

and $(A_g \setminus A)^+ = \{ x \in A_g \setminus A : \sum_{i\in S^c} a_i^n (f_i^n - f)(x) \geq 0 \}$.

Notice that for $x \in (A_g \setminus A)^-$ we have $\frac{1}{2} M > \sum_{i \in S^c} a_i^n f(x) > \sum_{i \in S^c} a_i^n f_i(x)$

Therefore $\left| \sum_{i \in S^c} a_i^n (f_i^n(x) - f(x)) \right| < M$ and hence

$\frac{1}{9} \sum_{i \in S} a_i^n (f_i^n(x) - f(x)) > \left| \sum_{i \in S^c} a_i^n (f_i^n(x) - f(x) \right|$, for $x \in (A_g \setminus A)^-$. We can see that

$$\frac{1}{n} > \int_{(A_g \setminus A)^+} \left| \sum_{i \in S} a_i^n (f_i^n(x) - f(x)) + \sum_{i \in S^c} a_i^n (f_i^n(x) - f(x)) \right| dm$$

$$+ \int_{(A_g \setminus A)^-} \left| \sum_{i \in S} a_i^n (f_i^n(x) - f(x)) + \sum_{i \in S^c} a_i^n (f_i^n(x) - f(x)) \right| dm$$

$$\geq \int_{(A_g \setminus A)^+} \sum_{i \in S} a_i^n (f_i^n(x) - f(x))\, dm + \int_{(A_g \setminus A)^+} \sum_{i \in S^c} a_i^n (f_i^n(x) - f(x))$$

$$- \frac{1}{9} \int_{(A_g \setminus A)^-} \sum_{i \in S} a_i^n (f_i^n(x) - f(x))\, dm + \int_{(A_g \setminus A)^-} \sum_{i \in S} a_i^n (f_i^n(x) - f(x)) \geq$$

$$\frac{8}{9} \int \sum_{i \in S} a_i^n (f_i^n(x) - f(x))\, dm.$$ Therefore

$$\int_{A_g \setminus A} g\, dm = \int_{A_g \setminus A} \sum_{i \in S} a_i^n (f_i^n(x) - f(x)) < \frac{9}{8} \frac{1}{n} < \frac{2}{n}$$

This proves the claim.

Now for $g \in G_n$ define

$$K_g(x) = \begin{cases} g(x) & \text{for} \quad x \in A \cup (A^c \setminus A_g) \\ 0 & \text{for} \quad x \in A_g \setminus A \end{cases}$$

It is clear that $\| K_g - g \|_1 < \frac{2}{n}$ and that $\| K_g | A^c \|_\infty \leq 11M$.

The family $\{ K_g | A^c : g \in \bigcup_{n > \frac{1}{\varepsilon}} G_n \}$

is uniformly bounded in $L^2(A^c)$ and moreover if $g \in G_n$ then

$| \int K_g h_k | < \frac{3}{n}$, for $k = 1,2, \dots, n$.

Consider the sequence $\{ G_n \}_{n \in \mathbb{N}, n > \frac{1}{\varepsilon}}$

By Lemma 1 there is a subsequence $\{ G_{n_l} \}_{l \in \mathbb{N}}$ and $M' > 0$ such that

$\| \sum_{l \in H} \varepsilon_l K_{g_l} \|_2 \leq M' \sqrt{\# H}$ for all $g_l \in G_{n_l}$ and for all choices of signs $\varepsilon_l = \pm 1$.

By taking $\# H$ large enough we have that $\frac{1}{\# H} \| \sum_{l \in H} \varepsilon_l K_{g_l} \|_1 < \varepsilon$

Since every g_l is ε-close in L^1- norm to K_{g_l} we have that $\frac{1}{\# H} \| \sum_{l \in H} \varepsilon_l g_l \|_1 < 2\varepsilon$ for all choices of signs $\varepsilon_l = \pm 1$.

For $l \in H$ consider the set $F_{n_l} = \{ f_1^{n_l}, f_2^{n_l}, \dots, f_{k_{n_l}}^{n_l} \}$. We know that

$$\| \sum_{i=1}^{k_{n_l}} a_i^{n_l} (f_i^{n_l} - f) \|_1 < \frac{1}{n_l}$$

We want to prove that for all choices of signs $\varepsilon_i = \pm 1$, $i=1,2,\dots,k_l$ we have that $\frac{1}{\# H} \| \sum_{l \in H} \sum_{i=1}^{k_{n_l}} \varepsilon_i a_i^{n_l}(f_i^{n_l} - f) \| \leq 5\varepsilon$. Notice that for every choise of signs $\varepsilon_l = \pm 1$, $\frac{1}{\# H} \| \sum_{l \in H} \varepsilon_l \sum_{i=1}^{k_{n_l}} a_i^{n_l}(f_i^{n_l} - f) \| \leq \varepsilon$ since $n_l > \frac{1}{\varepsilon}$.

For each $l \in H$ there exists $\bar{\varepsilon}_l$ such that the set $S_l = \{ i : \varepsilon_i = \bar{\varepsilon}_l \}$ has the property that $\sum_{i \in S_l} a_i^{n_l} \geq \frac{1}{2}$. For these $\bar{\varepsilon}_l$'s we have that

$$\frac{1}{\# H} \| \sum_{l \in H} \bar{\varepsilon}_l \sum_{i \in S_l} a_i^{n_l}(f_i^{n_l} - f) \| = \frac{1}{\# H} \| \sum_{l \in H} \bar{\varepsilon}_l g_l \| < 2\varepsilon .$$

Therefore $\frac{1}{\# H} \| (\sum_{l \in H} \bar{\varepsilon}_l \sum_{i=1}^{k_{n_l}} a_i^{n_l}(f_i^{n_l} - f) - \sum_{l \in H} \bar{\varepsilon}_l \sum_{i \in S_l} a_i^{n_l}(f_i^{n_l} - f)) \| \leq 3\varepsilon$.

This implies that $\frac{1}{\# H} \| (\sum_{l \in H} \sum_{i=1}^{k_{n_l}} \varepsilon_i a_i^{n_l}(f_i^{n_l} - f) \| \leq 3\varepsilon + 2\varepsilon = 5\varepsilon$

This proves Lemma 2.

Remark 3. The above result suggests the following general question: Let K be a bounded subset of a Banach space X and let x be a weak limit point of K and $\varepsilon > 0$. Does there exist a finite set $F = \{ x_1, x_2, \dots x_d \} \subset K$ and real numbers a_i, $i = 1, 2, \dots d$, $a_i \geq 0$, $\sum_{i=1}^{d} \alpha_i = 1$ such that $\| \sum_{i=1}^{d} \varepsilon_i \alpha_i (x_i - x) \| < \varepsilon$ for all choices of signs $\varepsilon_i = \pm 1$, $i = 1, 2, \dots d$?

In the case x is the weak limit of a sequence of elements of K the answer is positive.

The following notations are taken from [R]. See also [Bo], [D-U].

Let $\mathbb{N}^{(\mathbb{N})}$ be the set of all finite sequences of positive integers. For $a = (\alpha_1, \alpha_2, \dots, \alpha_n) \in \mathbb{N}^{(\mathbb{N})}$, we set $|a| = n$. For $a = (\alpha_1, \alpha_2, \dots, \alpha_n)$, $\beta = (\beta_1, \beta_2, \dots, \beta_m)$ elements of $\mathbb{N}^{(\mathbb{N})}$ we say that $a < b$ if $n \leq m$ and $\alpha_k = \beta_k$ $\forall\, k = 1,2,\dots,n$.

A non empty subset $\mathbf{A}$ of $\mathbb{N}^{(\mathbb{N})}$ is called a tree if for $\gamma \in \mathbf{A}$ and $\alpha \in \mathbb{N}^{(\mathbb{N})}$

with $\alpha < \gamma$ then $\alpha \in \mathbf{A}$. For any tree $\mathbf{A}$ in $\mathbb{N}^{(\mathbb{N})}$ and $\alpha \in \mathbf{A}$ we set $S_\alpha = \{ \beta \in A : \alpha < \beta \text{ and } |\beta| = |\alpha| + 1 \}$. A tree $\mathbf{A}$ is called finitely branching if $2 < \# S_\alpha < \infty$ $\forall\, \alpha \in \mathbf{A}$. Call $B \subset \mathbf{A}$ a branch if B is a maximal well-ordered subset of $\mathbf{A}$.

Let X be a Banach space and K a bounded subset of X Let $\varepsilon_n \geq 0$ be a sequence such that $\sum \varepsilon_n < \infty$. Let $\mathbf{A}$ be a finite branching tree. and $\delta > 0$.

A δ-approximate bush $(x_\alpha)_{\alpha \in \mathbf{A}}$ $x_\alpha \in K$, is characterised by the following two properties.

(i) $\forall \alpha \in \mathbf{A}$ $\exists \lambda_\beta \geq 0$, $\beta \in S_\alpha$ such that $\sum_{\beta \in S_\alpha} \lambda_\beta = 1$ and $\| x_\alpha - \sum_{\beta \in S_\alpha} \lambda_\beta x_\beta \| < \varepsilon_{|\alpha|}$.

(ii) $\| x_\alpha - x_\beta \| > \delta$ $\forall \alpha \in \mathbf{A}$ $\forall \beta \in S_\alpha$.

It is well known (see [K - R]) that any δ-approximate bush $(x_\alpha)_{\alpha \in \mathbf{A}}$ can be associated with a vector valued quasi - martingale (ξ_n, Σ_n) as follows : Let Ω be the set of branches of $\mathbf{A}$. For $\alpha \in \mathbf{A}$, let $U_\alpha = \{ B \in \Omega, \alpha \in B \}$. Let Σ denote the σ-algebra of subsets of Ω generated by $\{ U_\alpha : \alpha \in \Omega \}$. Let $\Sigma_n = \sigma(\{ U_\alpha : |\alpha| = n \})$. We define a probability measure P on (Ω, Σ) : Set $P(\Omega) = 1$. If P is defined on Σ_n and $\beta \in S_\alpha$ with $|\alpha| = n$ then $P(U_\beta) = \lambda_\beta P(U_\alpha)$

where λ_β is the coefficient of x_β in the convex combination $\sum_{\beta \in S_\alpha} \lambda_\beta x_\beta$ which is $\varepsilon_{|\alpha|}$ close to x_α.

Now $\xi_n = \sum_{|\alpha|=n} x_\alpha \chi_{U_\alpha}$ where χ_{U_α} is the characteristic function of U_α.

It is easy to see that $\| E(\xi_{n+1} \mid \Sigma_n) - \xi_n \| \leq \varepsilon_n$ and $(\xi_n, \Sigma_n)_{n \in \mathbb{N}}$ is a quasi martingale.

We are now ready to give the

Proof of Theorem 1 :

Let $T : L^1 \to X$ be any non strongly regular operator. There is a set $A \subset [0,1]$, $m(A) > 0$ and a $\delta > 0$ such that if W is any non-empty weakly open subset of the set $F_A = \{ f \in L^1 : f \geq 0, \int f\, dm = 1, \operatorname{supp} f \subset A \}$ then $\operatorname{diam}(T(W)) > 2\delta$. Let $f_0 \in F_A$. We can find a net $(f_i)_{i \in I}$ in F_A such that $f_i \to f$ weakly and $\|Tf_i - Tf_0\| > \delta$. Notice that $\|f_i - f_0\|_1 > \frac{\delta}{\|T\|}$. By Lemma 2 we may select a finite number $f_1, f_2, \ldots, f_d$ of the f_i's such that there are positive scalars λ_i , $i = 1,2, \ldots, d$, $\sum_{i=1}^{d} \lambda_i = 1$ with the property that

$$\| \sum_{i=1}^{d} \varepsilon_i \lambda_i (f_i - f_0) \| < \frac{1}{2^0} \quad \text{for all choises of signs } \varepsilon_i = \pm 1 , i = 1,2, \ldots, d.$$

Now set $x_0 = f_0$ and $x_{01} = f_1$, $x_{02} = f_2$, ..., $x_{0d} = f_d$. Clearly

$$\| \sum_{\alpha \in S_0} \varepsilon_\alpha \lambda_\alpha (x_\alpha - x_0) \| < \frac{1}{2^0} \quad \text{for all choises of signs } \varepsilon_\alpha = \pm 1 , \alpha \in S_0.$$

Similarly for everyone of the x_α's , $\alpha \in S_0$ we can find a finite number of elements of F_A say $x_{\alpha 1}, x_{\alpha 2}, \ldots, x_{\alpha k}$ such that

$$\| \sum_{\beta \in S_\alpha} \varepsilon_\beta \lambda_\beta (x_\beta - x_\alpha) \| < \frac{1}{2^{|\alpha|}} \cdot \frac{1}{\#\{\gamma : |\gamma| = |\alpha|\}} ,$$ for some positive scalars λ_β, $\beta \in S_\alpha$, $\sum_{\beta \in S_\alpha} \lambda_\beta = 1$ and for all choises of signs $\varepsilon_\alpha = \pm 1$. Also we may choose the x_β, $\beta \in S_\alpha$ such that $\|x_\alpha - x_\beta\| > \frac{\delta}{\|T\|}$ for all $\beta \in S_\alpha$. In this way

we construct an approximate $\frac{\delta}{\|T\|}$ bush in F_A. Let $(\xi_n, \Sigma_n)_{n\in\mathbb{N}}$ be the corresponding quasi-martingale. We can decompose (see [K-R]) $\xi_n = f_n + h_n$ where (f_n, Σ_n) is a martingale and $h_n \to 0$ in $L^1(P)_{L^1}$. Let $D : L^1(P) \to L^1$ be the operator corresponding to the martingale (f_n, Σ_n) i.e. $D\varphi = \lim_n \int \xi_n \varphi$, $\varphi \in L^1(P.)$

We want to prove that this operator is Dunford Pettis. It suffices to prove that the martingale f_n is Cauchy in the Pettis norm. (Recall for $f \in L^1_X$, the Pettis norm of f is $|||f||| = \sup_{\|x^*\|\leq 1} \{\int_\Omega |x^* f|\, dP$, $x^*\in X^*\}$).

It is clear that the martingale (f_n) is Cauchy in the Pettis norm iff the quasi martingale (ξ_n) is Cauchy in the Pettis norm.

Notice that $|||\xi_{n+1} - \xi_n||| = \sup_{\|g\|_\infty \leq 1} \int |\langle \xi_{n+1}(t) - \xi_n(t), g\rangle|\, dP =$

$$\sup_{\|g\|_\infty \leq 1} \int |\langle \sum_{|\beta|=n+1} x_\beta \chi_{U_\beta} - \sum_{|\alpha|=n} x_\alpha \chi_{U_\alpha}, g\rangle|\, dP.$$

For every α, $|\alpha| = n$ we have that $x_\alpha \chi_{U_\alpha} = x_\alpha \sum_{\beta\in S_\alpha} \chi_{U_\beta} = \sum_{\beta\in S_\alpha} x_\alpha \chi_{U_\beta}$ Also notice that $P(U_\beta) = \lambda_\beta P(U_\alpha)$ and $\sum_{\beta\in S_\alpha} \lambda_\beta = 1$.

It is clear that

$$\int \langle x_\alpha \chi_{U_\alpha} - \sum_{\beta\in S_\alpha} x_\beta \chi_{U_\beta}, g\rangle\, dP = \int \langle \sum_{\beta\in S_\alpha} x_\alpha \chi_{U_\alpha} - \sum_{\beta\in S_\alpha} x_\beta \chi_{U_\beta}, g\rangle\, dP$$

$$\int \langle \sum_{\beta\in S_\alpha} (x_\alpha - x_\beta)\chi_{U_\beta}, g\rangle\, dP = \langle \sum_{\beta\in S_\alpha} P(U_\beta)(x_\alpha - x_\beta), g\rangle .$$

Every term of the sum $\int |\langle \sum_{|\beta|=n+1} x_\beta \chi_{U_\beta} - \sum_{|\alpha|=n} x_\alpha \chi_{U_\alpha}, g \rangle| \, dP$ can be made smaller than $\| \sum_{\beta \in S_\alpha} \varepsilon_\beta \alpha_\beta (x_\alpha - x_\beta) \|$ for an appropriate choice of signs ε_β.

It follows that

$$|||\xi_{n+1} - \xi_n||| \leq \sum_{|\alpha|=n} \| \sum_{\beta \in S_\alpha} \varepsilon_\beta \alpha_\beta (x_\alpha - x_\beta) \| \leq (\sum_{|\alpha|=n} \frac{1}{2^{|\alpha|}} \cdot \frac{1}{\#\{\gamma : |\gamma| = n\}}) \leq \frac{1}{2^n}$$

This implies that the quasimartingale (ξ_n, Σ_n) is Cauchy in the Pettis norm and the operator D is Dunford-Pettis. The operator $T \circ D$ is not Bochner representable since it maps the "standard" bush in $L^1(P)$ into a δ-bush in X (see [D - U]). This is the end of the proof of theorem 1.

Corollary 1 : *Let* K *be a closed convex subset of a Banach space* X. *Suppose that if* $T : L^1 \to X$ *is an operator with* $T(F) \subset K$ *then the operator* $T \circ D$ *is representable for all Dunford-Pettis operators* $D : L^1 \to L^1$. *Then the* RNP *and the* KMP *are equivalent properties on the subsets of* K.

Proof : By Theorem 1 every operator $T : L^1 \to X$ such that $T(F) \subset K$ is strongly regular. By theorem V.3 of [G-G-M-S] every subset of K is strongly regular. Now any stronly regular set with the KMP has the RNP (see [S]).

As H. Rosenthal pointed out to the first author the proof of Theorem 2 in [R] actually gives the following

Theorem 2 : *Let* K *be a closed subset of a Banach space such that every closed convex bounded subset* K' *of* K *with the* PCP *(Point of Continuinity Property) has the* RNP. *Let* $\delta > 0$ *and* $(K_i)_{i \in I}$ *be a family of convex bounded* δ-*non-dentable subsets of* K *such that for every slice* S *of* K_i *there is a* $K_j \subset S$ *for some* $j \in I$. *Then there exists a* $\delta' > 0$ *and* $i_0 \in I$

such that if $K_{j_1}, K_{j_2}, \ldots, K_{j_n}$ *are subsets of* K_{i_0}, *then* $\mathrm{diam}\,(\sum_{s=1}^{n} a_s K_{j_s}) > \delta'$

for all non negative scalars $a_1, a_2, \ldots, a_n$ *such that* $\sum_{s=1}^{n} a_s = 1$.

Corollary 2 : *Let* K *be a closed convex subset of a Banach space* X. *Suppose that every closed convex bounded subset* K' *of* K *with the* PCP *has the* RNP. *Let* $T: L^1 \to X$ *be any non-representable operator with* $T(F) \subset X$. *Then there is a Dunford - Pettis operator* $D: L^1 \to L^1$ *such that the operator* $T \circ D$ *is not representable.*

Proof : Suppose $T: L^1 \to X$ is a non-representable operator with $T(F) \subset X$. There is a set $A \subset [0,1]$ of positive measure and a $\delta > 0$ such that for every subset B of A with $m(B) > 0$, $\mathrm{diam}(T(F_B)) > \delta$. It is well known (see [B2]) that every slice S of the set $\Gamma_B = T(F_B)$ contains a set of the form $\Gamma_{B'} = T(F_{B'})$ for some set $B' \subset B$ of positive measure. It is clear that the family $\{\Gamma_B : B \subset A,\ m(B) > 0\}$ satisfies the assumptions in the statement of Theorem 2. Therefore there is a $\delta' > 0$ and a set $B_0 \subset A$, $m(B_0) > 0$ such that if $B_1, B_2, \ldots, B_n$ are any subsets of B_0 of positive measure then $\mathrm{diam}(\sum_{i=1}^{n} a_i \Gamma_{B_i}) > \delta'$ for all nonegative real numbers $a_1, a_2, \ldots, a_n$ such that $\sum_{i=1}^{n} a_i = 1$. Let W be any weakly open subset of F_{B_0} . There are slices $S_1, S_2, \ldots, S_n$ of F_{B_0} such that $W \supset \sum_{i=1}^{n} a_i S_i$ for some $a_i \geq 0$ $i = 1, 2, \ldots, n$, $\sum_{i=1}^{n} a_i = 1$.

Every slice S_i contains a set of the form F_{B_i} , $B_i \subset B_0$, $m(B_i) > 0$. Hence $W \supset \sum_{i=1}^{n} a_i F_{B_i}$. Now $T(W) \supset T(\sum_{i=1}^{n} a_i F_{B_i}) \supset \sum_{i=1}^{n} a_i \Gamma_{B_i}$ and therefore

$\text{diam}(T(W)) > \delta'$. This proves that the operator T is not strongly regular. By Theorem 1 there is a Dunford Pettis operator $D : L^1 \to L^1$ such that the operator $T \circ D$ is not representable.

Remark 4 : Let X be any separable Banach lattice not containing c_0. In [G-M] it is proved that a closed convex bounded subset K of X has the RNP if and only if it has the point of continuity property. Let $\mathbf{T}$ be the torus with the Lebesgue measure. It follows from Corollary 2 that if $T : L^1(\mathbf{T}) \to X$ is any non representable operator then there is a Dunford - Pettis operator $D : L^1(\mathbf{T}) \to L^1(\mathbf{T})$ such that the operator $T \circ D$ is not representable. It is proved in [K-P-R-U] (see also [P]) that this operator D can be taken to be a convolution operator of the form D_μ, $D_\mu f = f * \mu$ for $f \in L^1(\mathbf{T})$, where μ is a probability measure on $\mathbf{T}$ and the Fourier coefficients of μ vanish at infinity. (This means the operator D_μ is Dunford - Pettis (see [D-U]))

The next example shows that in general we cannot expect that the operator D in the statement of Theorem 1 can be taken to be a convolution operator.

Example 1 : Let $T : L^1(\mathbf{T}) \to c_0$ be the operator $Tf = (\hat{f}(n))_{n \in \mathbb{N}}$, $f \in L^1(\mathbf{T})$ where $\hat{f}(n)$, $n \in \mathbb{N}$ are the Fourier coefficients of the function f. Let μ be any probability measure on $\mathbf{T}$ with $\hat{\mu}(\pm\infty) = 0$. Define the operator $D_\mu : L^1(\mathbf{T}) \to L^1(\mathbf{T})$ by $D_\mu f = f * \mu$ $f \in L^1(\mathbf{T})$. Notice that $T \circ D_\mu f = (\hat{f}(n) \cdot \hat{\mu}(n))_{n \in \mathbb{N}}$ for $f \in L^1(\mathbf{T})$.

Every sequence in the image under $T \circ D_\mu$ of the unit ball of $L^1(\mathbf{T})$ is dominated coordinatewise by the fixed sequence $(\hat{\mu}(n))_{n \in \mathbb{N}}$ in c_0.

This implies that the operator $T \circ D_{\bar{n}}$ is compact and therefore representable. On the other hand it is easy to see that the operator T is not Dunford - Pettis and hence it is not strongly regular.

Example 2 : We give an example of a subset K of c_0 with the following properties

1. K fails RNP

2. If $T : L^1 \to c_0$ is any operator with $T(F) \subset K$ and $D : L^1 \to L^1$ is any Dunford - Pettis operator, then the operator $T \circ D : L^1 \to c_0$ is representable.

By Theorem 1 every such operator T is strongly regular and therefore (see Theorem V.3 of [G-G-M-S]) the set K is strongly regular.

Let M[0,1] be the space of all Radon measures on [0,1].

Let $I_{n,k} = \left[\frac{k-1}{2^n}, \frac{k}{2^n} \right]$, $k = 1,2, \ldots, 2^n$, $n = 0,1,2, \ldots$ be the sequence of dyadic intervals. Define a map

$$\varphi : M[0,1] \to l^\infty \text{ by } \varphi(\mu) = (\mu(I_{0,1}), \mu(I_{1,1}), \mu(I_{1,2}), \ldots).$$

Let Q be the set of all diffuse measures of norm ≤ 1. It is clear that the restriction of φ to Q is an 1 - 1 , weak* - weak continuous map that takes values in c_0. Let $K = \varphi(Q)$. It is easy to see that the image of the "standard" dyadic tree in L^1 under φ is an $\frac{1}{2}$ - tree in K. Therefore K fails the RNP.

Assume now that $T : L^1 \to c_0$ is an operator with $T(F) \subset K$. Let $D : L^1 \to L^1$ be any Dunford - Pettis operator. We claim that the operator $T \circ D : L^1 \to c_0$ is representable : For $f \in L^1$ define $\tilde{T} f = \varphi^{-1} T f$. The map $\tilde{T} : L^1 \to M[0,1]$ is a bounded operator and maps the densities F of $L^1[0,1]$ into Q. The operator $\tilde{T} \circ D$ has separable range and we may assume that takes values in a space $L^1(v)$ for some diffuse measure v. Let M_0 be the set M[0,1] with the norm $\| \mu \|_0 = \sup \{ | \mu(I) |, \ I \text{ subinterval of } [0,1] \}$ and $W : M \to M_0$ be the identity map.

The operator $\tilde{T} \circ D$ is Dunford - Pettis and by Theorem 7 in [B1] (see also [Wei]) the operator $W \circ \tilde{T} \circ D : L^1 \to M_0$ is representable. Therefore the martingale $\xi_n(t) \cdot v$ corresponding to this operator is Cauchy in the M_0 norm for almost all t. (see [D-U]).

Let $\mu_n(t) = \xi_n(t) \cdot v$. For almost all t, the sequence $\mu_n(t)$ is bounded in M[0,1] and there is a $\mu(t) \in M[0,1]$ such that $\mu_n(t) \xrightarrow{\| \, \|_0} \mu(t)$.. Notice that the measures $\mu_n(t)$ are diffuse. By Remark 3.9 in [Wei] the measures $\mu(t)$ must be diffuse for almost all t. Notice that $\varphi(\mu(t)) \in c_0$ for almost all t and that $\varphi(\mu_n(t)) \xrightarrow{w} \varphi(\mu(t))$ since the map φ is w^*-w continuous. This implies that the martingale corresponding to the operator $T \circ D$ converges for almost all t and hence the operator $T \circ D$ is representable.

References

[B1] : J. Bourgain : Dunford - Pettis operators on L^1 and the RNP. Israel Journal of Math. Vol 37 (1980), 34-47.

[B2] : J. Bourgain : La propriété de Radon - Nikodym, Publications mathématiques de l' Université Pierre et Marie Curie no 36 (1979).

[Bo] : R. Bourgin : Geometric Aspects of Convex sets with the Radon-Nikodym Property, Lecture Notes 993, 1983.

[D-U] : J. Diestel and J.J. Uhl Jr. : Vector measures, Mathematical Surveys, Vol 15, Amer. Math. Soc. 1977.

[G-G-M] : N. Ghoussoub, G. Godefroy, B. Maurey : First class functions around a subset of a compact space and regular sets in Banach spaces (Preprint 1985).

[G-G-M-S] : N. Ghoussoub, G. Godefroy, B. Maurey, W Schachermayer : Some topological and geometric structures in Banach spaces, Memoirs of the A.M.S. Vol 70 Number 378, 1987.

[G-M] : N. Ghoussoub, B. Maurey : On the Radon - Nikodym property in function spaces Lecture Notes in Mathematics 1166 Banach spaces procceedings, Missouri 1984.

[K-R] : K. Kunen and H. Rosenthal : Martingale proofs of some geometrical results in Banach space theory, Pasific Journal of Mathematics Vol 100, 1982.

[K-P-R-U] : R. Kaufman, M Petrakis, L. Riddle, J.J. Uhl Jr. Nearly representable operators, Trans. Amer. Math. Soc. Vol. 312, Number 1, March 1989.

[P] : M. Petrakis : Nearly representable operators, Ph.D. Thesis, Univ. of Illinois (1987).

[P-S] : F. Lust Piquart - W. Schachermayer: Functions in $L^\infty(G)$ and associated convolution operators: Studia Mathematica (1989).

[R] : H. Rosenthal : On the structure of non-dentable closed bounded convex sets, Advances in Math. Vol. 70, July 1988.

[S] : W. Schachermayer : RNP and KMP are equiValent for strongly regular sets , Trans. Amer. Math. Soc. Vol. 303, Number 2, Oct 1987 p. 673.

[Wei] : L.Weis : On the representation of order continuous operators by random measures, Trans. Amer. Math. Soc. Vol 285, Number 2, October 1984.

[Wes] : A. Wessel : Some remarks on Dunford - Pettis operators, strong regularity and the RNP, Seminaire d' analyse Fonctionelle 1985/1986/1987 Paris VI - VII Publications Mathématiques de l' Université Paris VII, Paris.

The entropy of convex bodies with "few" extreme points

Keith Ball[1]
Department of Mathematics
Texas A&M University
College Station, TX 77843

and

Alain Pajor
U.E.R. de Mathématiques
Université de Paris VII
2 Place Jussieu
75251 PARIS CEDEX 05

---

(1) Supported in part by N.S.F. DMS-8807243

A (rather weak) consequence of a result of Talagrand [T], states that if $(x_i)_1^\infty$ is a sequence of points in Hilbert space with $\|x_i\| \leq \frac{1}{\sqrt{\log(1+i)}}$ for each i, then, for each $n \in \mathbb{N}$, the convex hull, conv $(x_i)_1^\infty$ can be covered by 2^n (Hilbertian) balls of radius about $\frac{1}{\sqrt{n}}$.

Dudley [D], considered a similar question in which he assumed a faster rate of decay on $(\|x_i\|)_1^\infty$. He showed that, if there is a positive α so that $\|x_i\| \leq i^{-\alpha}$ for each i, then for each n, conv $(x_i)_1^\infty$ can be covered by 2^n balls of radius about

$$\frac{(\log(1+n))^\beta}{n^{\frac{1}{2}+\alpha}}$$

for some constant β independent of $(x_i)_1^\infty$ and n.

The motivation for such an entropy estimate is that convex sets are frequently defined in terms of their extreme points. The appearance of the "extra" $n^{-\frac{1}{2}}$ in passing from estimates on the extreme points, to estimates on the convex hull, is what is most important in applications such as those described by Dudley. The purpose of this note is to observe that the logarithmic factor in Dudley's result can be removed. The result proved is the following.

Theorem 1. For each $\alpha > 0$, there is a constant $C = C(\alpha)$ so that if $(x_i)_1^\infty$ is a sequence in ℓ_2 satisfying $\|x_i\| \leq i^{-\alpha}$ for each i, then for each $n \in \mathbb{N}$, conv $(x_i)_1^\infty$ can be covered by 2^n balls in ℓ_2 of radius $Cn^{-\frac{1}{2}-\alpha}$. For each α, this estimate is best possible for all n, apart from the value of $C(\alpha)$.

The argument given below uses standard methods for estimating entropy numbers developed by Carl [C1]. The removal of the "usual" logarithmic factor depends principally on the care with which these methods are applied.

Dudley actually proved a more general result than the one stated earlier, hypothesising only that, for each i, the sequence $(x_i)_1^\infty$ can be covered by at most i balls of radius $i^{-\alpha}$. This generalization can also be dealt with by the methods used here, since the important point is that the operator considered should be well-approximated by operators which factor through low-dimensional ℓ_1 spaces: for the sake of clarity only the proof of Theorem 1 will be given.

To see that the estimate $Cn^{-\frac{1}{2}-\alpha}$ is best possible for each fixed α and every n, observe that if $e_1, \ldots, e_n$ is the standard basis of $\mathbf{R}^n$ then the volume of conv $\{n^{-\alpha}e_1, \ldots, n^{-\alpha}e_n, 0\}$ is at least $(n^{-1-\alpha})^n$ while the volume of an Euclidean unit ball in $\mathbf{R}^n$ is at most $(\frac{2\pi e}{n})^{\frac{n}{2}}$.

The estimates obtained here for $C(\alpha)$ satisfy (for some constant B)

$$C(\alpha) \leq \frac{B}{\sqrt{\alpha}} \qquad \text{as} \qquad \alpha \to 0$$
$$C(\alpha) \leq \alpha^{B\alpha} \qquad \text{as} \qquad \alpha \to \infty.$$

Both of these estimates are best possible, apart from the value of B, if $C(\alpha)$ is required to be independent of n.

It will be convenient to couch the argument in the language of entropy numbers of operators between normed spaces. If T: $X \to Y$ is such an operator, the nth (dyadic) entropy number of T (if it exists) is $e_n(T) = \inf\{\varepsilon > 0\colon\ T(B_x)$ can be covered by 2^{n-1} balls in Y of radius $\varepsilon\}$, where B_x is the unit ball of X. The conventional choice of 2^{n-1} balls in the definition ensures that $e_1(T) = \|T\|$ and is convenient for several other reasons.

For a sequence $(x_i)_1^\infty$ in ℓ_2, define T: $\ell_1 \to \ell_2$ by $T(e_i) = x_i, i \in \mathbf{N}$, where e_i is the ith standard basis vector of ℓ_1. Then $T(B_{\ell_1})$ includes the absolutely convex hull of $(x_i)_1^\infty$. It is the entropy numbers of this operator T that will actually be estimated.

The basic properties of entropy numbers are contained in Lemmas 2, 3, and 4.

Lemma 2. If S, T: $X \to Y$ are operators between normed spaces, $n \in N$ and $1 \leq k \leq n$ then

$$e_n(S+T) \leq e_k(S) + e_{n-k+1}(T). \qquad \square$$

Lemma 3. If T: $X \to Y$ has rank n and $k \geq n$ then

$$e_k(T) \leq 8.2^{-\frac{k}{n}} e_n(T). \qquad \square$$

Lemma 4. Let $\alpha \geq 0, n \in \mathbf{N}$ and for an operator U: $X \to Y$ between normed spaces set

$$\Phi(U) = \max_{1 \leq k \leq n} k^\alpha e_k(U).$$

Then if $p = \frac{1}{1+\alpha}$ and S, T: $X \to Y$,

$$\Phi(S+T)^p \le \Phi(S)^p + \Phi(T)^p. \qquad \square$$

Remark. Lemma 2 is just the triangle inequality and Lemma 4 follows from Lemma 2 since if

$$\lambda = \frac{\Phi(S)^p}{\Phi(S)^p + \Phi(T)^p} \quad \text{and} \quad k = \lceil \lambda n \rceil$$

then $k \ge \lambda n, n-k+1 \ge (1-\lambda)n$ and

$$\begin{aligned}(n^\alpha e_n(S+T))^p &\le (n^\alpha e_k(S) + n^\alpha e_{n-k+1}(T))^p \\ &\le (\lambda^{-\alpha}\Phi(S) + (1-\lambda)^{-\alpha}\Phi(T))^p \\ &= \Phi(S)^p + \Phi(T)^p.\end{aligned}$$

Lemma 4 substitutes for the triangle inequality: for each $n > 1, e_n$ is not a norm on the space of operators on which it is finite. Lemma 3 depends upon the fact that the ball of radius 1 in an n-dimensional normed space can be covered by 2^{k-n} balls of radius $8.2^{-k/n}$.

The crucial lemma needed for the proof of Theorem 1 is a result of Carl [C2] which was proved using an averaging argument of Maurey.

Lemma 5. There is an absolute constant C so that if $T: \ell_1^n \to \ell_2$ and $k \le n$ then

$$e_k(T) \le \frac{C\sqrt{1+\log\frac{n}{k}}}{\sqrt{k}}\|T\|. \qquad \square$$

The appearance of the factor $\log\frac{n}{k}$, rather than $\log n$ is critical for the proof of Theorem 1 since the expression

$$\max_{1\le k\le n}\left(\frac{k}{n}\right)^\alpha \sqrt{\log\frac{n}{k}}$$

will be estimated from above by $\frac{1}{\sqrt{\alpha}}$.

Maurey's argument only uses the type 2 property of Hilbert space and there is a corresponding, weaker, estimate for type p spaces. This transfers, through the proof below, to an estimate for the entropy of convex sets in type p spaces.

The first step in the proof of Theorem 1 deals with operators of small rank: the result is essentially known but a proof is included for completeness.

Lemma 6. There is an absolute constant K so that if $\alpha > 0, n \in \mathbb{N}$ and T: $\ell_1^n \to \ell_2$ satisfies $\|Te_i\| \leq i^{-\alpha}, 1 \leq i \leq n$ then

$$e_n(T) \leq (K + \alpha^2)^{1+\alpha} n^{-\frac{1}{2}-\alpha}.$$

Proof. Let K be at least 2, to be chosen. The proof proceeds by induction on n. For $n \leq K + \alpha^2$ there is nothing to prove. So assume $n > K + \alpha^2$ and set

$$m = \left[\frac{n}{K + \alpha^2}\right] \leq \frac{n}{2}.$$

Write $T = U_1 + U_2$ where U_1 agrees with T on the first m unit vectors of ℓ_1^n and U_2 agrees with T on the remainder. Then $\|U_2\| \leq (m+1)^{-\alpha} \leq (\frac{n}{K+\alpha^2})^{-\alpha}$: since U_2 maps ℓ_1^n into ℓ_2, Lemma 5 shows that, for some absolute constant C', and for $r = \lceil \frac{n}{2} \rceil$,

$$\begin{aligned} e_r(U_2) &\leq \frac{C'}{\sqrt{n}}\left(\frac{n}{K + \alpha^2}\right)^{-\alpha} \\ &= C'(K + \alpha^2)^{\alpha} n^{-\frac{1}{2}-\alpha} \\ &\leq \frac{1}{2}(K + \alpha^2)^{1+\alpha} n^{-\frac{1}{2}-\alpha} \end{aligned} \tag{1}$$

if K is sufficiently large.

On the other hand, the restriction of U_1 to the first m coordinates of ℓ_1^n satisfies the hypothesis of the lemma with m in place of n, so by the inductive hypothesis

$$e_m(U_1) \leq (K + \alpha^2)^{1+\alpha} m^{-\frac{1}{2}-\alpha}.$$

By Lemma 3,

$$\begin{aligned} e_r(U_1) &\leq 8.2^{-\frac{r}{m}}(K + \alpha^2)^{1+\alpha} m^{-\frac{1}{2}-\alpha} \\ &\leq 8.2^{-\frac{n}{2m}}\left(\frac{n}{m}\right)^{\frac{1}{2}+\alpha}(K + \alpha^2)^{1+\alpha} n^{-\frac{1}{2}-\alpha}. \end{aligned}$$

For K sufficiently large, the fact that $\frac{n}{m} \geq K + \alpha^2$ implies that $8.2^{-\frac{n}{2m}}(\frac{n}{m})^{\frac{1}{2}+\alpha} \leq \frac{1}{2}$ and hence

$$e_r(U_1) \leq \frac{1}{2}(K+\alpha^2)^{1+\alpha} n^{-\frac{1}{2}-\alpha}. \tag{2}$$

Finally, from (1) and (2) and Lemma 2,

$$e_n(T) \leq (K+\alpha^2)^{1+\alpha} n^{-\frac{1}{2}-\alpha}$$

completing the inductive step. □

Proof of Theorem 1. For a sequence $(x_i)_1^\infty$ in ℓ_2 with $\|x_i\| \leq i^{-\alpha}, i \in \mathbb{N}$, define $T\colon \ell_1 \to \ell_2$ by $Te_i = x_i$. It will be shown that, for each n

$$e_n(T) \leq C(\alpha) n^{-\frac{1}{n}-\alpha}$$

so that the absolutely convex hull of $(x_i)_1^\infty$ can be covered by 2^n balls of radius $C(\alpha)n^{-\frac{1}{n}-\alpha}$. It will be convenient to estimate $e_{2n}(T)$ rather than e_n.

Let q be an integer greater than 2 (to be specified later). Partition the unit vector basis of ℓ_1 into successive blocks, the ith block having length ni^q. Let $U_i\colon \ell_1 \to \ell_2$ be the operator which agrees with T on the unit vectors of the ith block and is zero on the remainder. Then U_i factors through $\ell_1^{ni^q}$ (isometrically) and for $i > 1$,

$$\|U_i\| \leq (n(i-1)^q)^{-\alpha}.$$

Also $T = \sum_1^\infty U_i$.

By Lemma 2,

$$e_{2n}(T) \leq e_n(U_1) + e_n\left(\sum_2^\infty U_i\right).$$

Lemma 6 shows that $e_n(U_1) \leq (K+\alpha^2)^{1+\alpha} n^{-\frac{1}{2}-\alpha}$. So it suffices to estimate the second term. It will be shown that for some absolute constant B,

$$\sup_{k \leq n} k^{\alpha+\frac{1}{2}} e_k\left(\sum_2^\infty U_i\right) \leq B\left(1+\frac{1}{\sqrt{\alpha}}\right) \tag{3}$$

so the estimate behaves well as $\alpha \to \infty$ but exhibits the aforementioned growth as $\alpha \to 0$.

By Lemma 4 with $p = \frac{1}{1+\alpha+\frac{1}{2}} = \frac{2}{3+2\alpha}$,

$$\left(\sup_{k\le n} k^{\alpha+\frac{1}{2}} e_k\left(\sum_2^\infty U_i\right)\right)^p \le \sum_{i=2}^\infty \left(\sup_{k\le n} k^{\alpha+\frac{1}{2}} e_k(U_i)\right)^p. \tag{4}$$

Now since U_i is non-zero on only ni^q basis vectors of ℓ_1 and has norm at most $(n(i-1)^q)^{-\alpha}$, Lemma 5 shows that for $k \le n$, and some absolute constant C'

$$e_k(U_i) \le \frac{C'\sqrt{\log \frac{ni^q}{k}}}{\sqrt{k}}(n(i-1)^q)^{-\alpha}.$$

So

$$\begin{aligned} k^{\alpha+\frac{1}{2}} e_k(U_i) &\le C'\left(\frac{k}{n}\right)^\alpha \sqrt{\log\frac{ni^q}{k}}\frac{1}{(i-1)^{\alpha q}} \\ &\le C'\left(\frac{k}{n}\right)^\alpha \left(\sqrt{\log\frac{n}{k}} + \sqrt{q\log i}\right)\frac{1}{(i-1)^{\alpha q}}. \end{aligned}$$

Now for any $k \le n$,

$$\left(\frac{k}{n}\right)^\alpha \sqrt{\log\frac{k}{n}} \le \frac{1}{\sqrt{2e\alpha}} \le \frac{1}{\sqrt{\alpha}}.$$

Hence

$$\sup_{k\le n} k^{\alpha+\frac{1}{2}} e_k(U_i) \le C'\left(\frac{1}{\sqrt{\alpha}} + \sqrt{q\log i}\right)\frac{1}{(i-1)^{\alpha q}}.$$

Now choose $q = \frac{2}{\alpha p} = 2 + \frac{3}{\alpha}$. Then for some constant C'', the majorant in the preceding inequality is at most

$$C''\left(1 + \frac{1}{\sqrt{\alpha}}\right)\frac{\sqrt{\log i}}{(i-1)^{\frac{2}{p}}}.$$

Since the sums

$$\sum_2^\infty \left(\frac{\sqrt{\log i}}{(i-1)^{\frac{2}{p}}}\right)^p$$

are uniformly bounded for $0 < p < 1$, inequality (4) yields the desired estimate (3). □

The same approach can be used to estimate entropy numbers for slower rates of decay of $(\|x_i\|)_1^\infty$. For example, if $\alpha > 0$ and

$$\|x_i\| \leq (\log(1+i))^{-\frac{1}{2}-\alpha}$$

then, with the obvious notation,

$$e_n \leq \frac{C(\alpha)}{\sqrt{n}(\log(1+n))^\alpha}.$$

It is natural to ask whether a general estimate can be obtained which includes all such rates of decay (and preferably Lemma 5 as well): such an estimate would be something like

$$e_n \leq \left(\sum_{i=1}^{\infty} e^{-\frac{n}{i}}\left(1+\frac{i}{n}\right)\frac{\|x_i\|^2}{i^2}\right)^{\frac{1}{2}}.$$

There seems to be a problem with such estimates if $\|x_i\|$ decays extremely rapidly.

References

[C1] B. Carl, Entropy numbers, s-numbers and eigenvalue problems, J. Funct. Anal. 41 (1981), 290-306.

[C2] B. Carl, Inequalities of Bernstein-Jackson-type and the degree of compactness of operators in Banach spaces, Ann. Inst. Fourier 35 (1985), 79-118.

[D] R.M. Dudley, Universal Donsker classes and metric entropy, Ann. Prob. 15 (1987), 1306-1326.

[T] M. Talagrand, Regularity of Gaussian processes, Acta. Math. 159 (1987), 99-149.

SPACES OF VECTOR VALUED ANALYTIC FUNCTIONS AND APPLICATIONS.

by

Oscar Blasco*

ABSTRACT: Spaces of vector valued analytic functions verifying the following certain growth conditions are considered.

$$(1) \quad \int_0^1 \int_{-\pi}^{\pi} (1-r)^{\alpha} \, \| F'(re^{it}) \| \, dt \, dr < \infty \quad , \alpha > -1.$$

$$(2) \quad \| F''(z) \| = O((1-|z|)^{\alpha-2}) \ , \quad \alpha < 2$$

We deal with the representability of operators from the first space into a general Banach space in terms of functions of the second space. We apply this to the study of multipliers and certain Besov-Lipschizt classes.

INTRODUCTION

In this paper we are concerned with the study of certain spaces of analytic functions from the unit disc into a Banach space. We shall consider functions whose derivative have certain growth at the boundary. We are going to look at two kinds of dual conditions:

$$(0.1) \qquad \| F'(z) \| = O((1-|z|)^{\alpha-2}) \ , \quad \alpha < 2$$

$$(0.2) \qquad \int_0^1 \int_{-\pi}^{\pi} (1-r)^{\alpha} \, \| F'(re^{it}) \| \, dt \, dr < \infty \quad , \alpha > -1.$$

Spaces of this type were considered in the scalar valued case by several authors (see [H-L], [Z2], [D-R-S], [S], [A-C-P], [F1]). Some analogous conditions in the vector-valued case have been already used (see [K], [Bu]).

* Partially supported by the grant C.A.I.Y.T. PB-85-0338 and M.S.R.I. at Berkeley. This paper was iniciated in the fall of 1987, during my stay in the Mathematical Science Research Institute at Berkeley.

Taking the second derivative in (0.1) allows us to unify proofs and look at three very interesting spaces under the same scope: Bloch functions, the Zygmund class and Lipschizt classes of order $0 < \alpha < 1$.

The paper has two main objectives. To extend to the vector valued case some of the classical results, getting even unified proofs, what is done in the second section, and to apply the vector valued analysis to get easily known results as corollaries, what is the content of Section 3.

The first section is devoted to introduce the spaces $A_\alpha(X)$ and $J_\alpha(X)$, give examples and prove some elementary properties. In section 2 we show that the space of bounded operators from $J_\alpha(X)$ into another Banach space Y can be identified with $A_\alpha(L(X,Y))$. This result leads to some corollaries about duality and representation of operators. In the last section we obtain most of the results on multipliers in [D-S] as consequence of representability of operators and results of duality for certain Besov-Lipschizt classes -which were proved in [F2] - are obtained from previous vector-valued results.

Throughout the paper $(X, \|\ \|)$ denotes a complex Banach space, $M_p(F,r)$ means $\int_{-\pi}^{\pi} \|F(re^{it})\| \frac{dt}{2\pi}$, and C will be a constant not necessarily the same at each ocurrence.

SPACES OF VECTOR VALUED ANALYTIC FUNCTIONS.

DEFINITION 1.1. Let $\alpha < 2$ and $F: D \to X$ analytic . We shall say that F belongs to $A_\alpha(X)$ if

$$\text{(1.1)} \qquad \| F''(z) \| = O((1-|z|)^{\alpha-2})$$

This space becomes a Banach space endowed with the norm

$$\text{(1.2)} \quad \| F \|_\alpha = \| F(0) \| + \| F'(0) \| + \sup \{ (1-|z|)^{2-\alpha} \| F''(z) \| \ : |z| < 1 \}$$

The restriction $\alpha < 2$ comes from the fact that (1.2) for $\alpha > 2$ would imply $F''(z) = 0$ for $|z| < 1$. Hence the only functions in the space would be $F(z) = a + bz$, with a, b in X. The case $\alpha=2$ is not considered since corresponds to F'' in $H^\infty(X)$.

Let us recall two basic tools we have at our disposal when working with analytic functions

$$(*) \quad F'(re^{i\theta}) = \int_{-\pi}^{\pi} \frac{F(\rho e^{it})}{(\rho e^{it} - re^{i\theta})^2} \rho e^{it} \frac{dt}{2\pi} \quad , \rho = \frac{1+r}{2}$$

$$(**) \quad F(re^{i\theta}) = \int_0^r F'(se^{i\theta})\, e^{i\theta} ds \; + F(0)$$

Using this two facts one easily shows the following lemma

LEMMA 1.1.-

(1) If $\| F(z) \| = O((1-|z|)^{-\alpha})$ *then* $\| F'(z) \| = O((1-|z|)^{-(\alpha+1)})$.

(2) For $\beta > 0$, *if* $\| F'(z) \| = O((1-|z|)^{-\beta})$ *then*

(i) $\| F(z) \| = O((1-|z|)^{-(\beta-1)})$ *for* $\beta > 1$,

(ii) $\| F(z) \| = O(-\log(1-|z|))$ *for* $\beta = 1$,

(iii) $\| F(z) \| = O(1)$ *for* $\beta < 1$.

Using lemma 1.1 one can see the following special cases:

Case $\alpha = 0$ corresponds to the space of vector valued Bloch functions (see [A-C-P])

$$A_0(X) = B(X) = \{ F:D \to X \text{ analytic} : \| F'(z) \| = O((1-|z|)^{-1}) \}$$

Case $\alpha = 1$ corresponds to the vector valued Zygmund class (see [Z2], [D,page 74])

$$A_1(X) = Z(X) = \{ F:D \to X \text{ analytic} : \| F''(z) \| = O((1-|z|)^{-1}) \}$$

Case $0 < \alpha < 1$ corresponds to vector valued Lipschizt classes (see [H-L], [D,page 74])

$$A_\alpha(X) = \Lambda_\alpha(X) = \{ F:D \to X \text{ analytic} : \| F'(z) \| = O((1-|z|)^{-(1-\alpha)}) \}$$

Most of results of this paper are valid for all values of α and unify some proofs for the previous spaces.

Remark 1.1 Using the simple fact that if $\omega_n = O(n^\beta)$, $\beta \geq 0$, then

$\sum \omega_n r^n = O((1-r)^{-(\beta+1)})$, one can easily give examples of functions in

these classes taking values in spaces of sequences , say $X = c_o, l^p, \ldots$

Remark 1.2- An interesting example is given for $X = H^1(D)$, the classical Hardy space of analytic functions f on the disc . For $\alpha < 1$ denote by $C_{z,\alpha}(\xi) = (1-\xi z)^{\alpha-1}$ and define the function from the disc into $H^1(D)$ given by $F(z) = C_{z,\alpha}$. It is easy to see that F belongs to $A_\alpha(H^1)$ using the following estimate:

$$\int_{-\pi}^{\pi} \frac{dt}{|1 - e^{i\theta}|^{3-\alpha}} = O((1-|z|)^{\alpha-2}) \quad , \quad \text{(see [D,page 65]).}$$

Note that, in particular, the Cauchy kernel $C_{z,0}$ defines a H^1-valued Bloch function.

Remark1.3. Let X be any Banach space, $\alpha \leq 0$ and x_k a sequence in X . Consider

$$(1.1) \qquad F(z) = \sum_{k=0}^{\infty} x_k z^{2^k}$$

Then F belongs to $A_\alpha(X)$ if and only if $\sup_k \{ \| x_k \| 2^{k\alpha} \} < \infty$.

Indeed for any function $G(z) = \sum_{n=0}^{\infty} a_n z^n$ we have

$$n(n-1)\, a_n = \int_{-\pi}^{\pi} G''(re^{i\theta})\, r^{-(n-2)}\, e^{-i(n-2)\theta} \frac{d\theta}{2\pi}$$

Therefore if G belongs to $A_\alpha(X)$ then

$$n(n-1)\, \|a_n\| \leq C\,(1-r)^{\alpha-2}\, r^{-(n-2)}$$

and taking $r = 1 - 1/n$ we get $\| a_n \| \leq C\ n^{-\alpha}$, what obviously implies the result when we restrict to F. On the other hand

$$F''(z) = \sum_{k=2}^{\infty} 2^k(2^k-1)x_k z^{2^k-2}$$

Hence

$$\| F''(z) \| \leq \sum_{k=2}^{\infty} 2^{-\alpha k} 2^k(2^k-1)|z|^{2^k-2}$$

$$\leq C \sum_{n=1}^{\infty} n^{-\alpha}(n-1)\ |z|^{n-2} = O(\ (1-r)^{\alpha-2}\ ).$$

In the study of duality problems and representation of operators associated to the space $A_\alpha(X)$ it shall appear in a natural way the following space of vector valued functions.

DEFINITION 1.2.- Let $-1 < \alpha$ and $F: D \to X$ analytic. We shall say that F belongs to $J_\alpha(X)$ if

(1.2) $$\int_0^1 \int_{-\pi}^{\pi} (1-r)^\alpha \ \| F'(re^{it}) \| \ dt\, dr \ < \infty$$

The restriction $-1 < \alpha$ comes from the fact that we want to have the analytic polynomials in the space.

Observe that we are simply assuming that the derivatives of the functions belongs to certain weigthed Bergman spaces (see [S]), whose predecesors are the B_p spaces considered in [D-R-S].

Since our space depends on the derivative then in order to get a norm we consider

(1.3) $${}_\alpha\| F \| = \| F(0) \| + \int_0^1 \int_{-\pi}^{\pi} (1-r)^\alpha \ \| F'(re^{it}) \| \frac{dt}{2\pi}\, dr$$

Using (*) and (**) one easily obtains the next result.

LEMMA 1.2.- *Let* $\alpha > 0$. *The following are equivalent*

$$(1) \quad \int_0^1 \int_{-\pi}^{\pi} (1-r)^{\alpha} \| F'(re^{it}) \| \, dt \, dr < \infty$$

$$(2) \quad \int_0^1 \int_{-\pi}^{\pi} (1-r)^{\alpha-1} \| F(re^{it}) \| \, dt \, dr < \infty$$

From this lemma we see that for $\alpha > 0$ we have that $J_\alpha = B^p$ where $p=1/(1+\alpha)$ (see [D-R-S]).

Let us start off by mentioning some properties of $J_\alpha(X)$. They follow very closely the ideas in [D-R-S] and [S], and therefore we omit the proof.

PROPOSITION 1.1.- *Let* $-1 < \alpha$. *and let* F *be in* $J_\alpha(X)$ *Then*

(1) $\| F(z) \| \leq C \, (1-|z|)^{-(\alpha+1)}$

(2) *If* F *belongs to* $J_\alpha(X)$ *and* F_r *denotes the function* $F_r(z) = F(rz)$ *for any* $0 < r < 1$ *then* F_r *belongs to* $J_\alpha(X)$ *and* $F_r \to F$ *in* $J_\alpha(X)$ *as* $r \to 1$.

(3) *The analytic polynomials are dense in* $J_\alpha(X)$.

Next result, on the contrary to this previous one, says that the convergence of F_r to F is not always guaranteed in $A_\alpha(X)$. The proof is easy and we omit it (see [ACP] for the case $\alpha=0$).

PROPOSITION 1.2.- *Let* $\alpha < 2$ *and* F *belong to* $A_\alpha(X)$. *The following are equivalent*

(1) $\| F''(z) \| = o((1-|z|)^{\alpha-2})$

(2) F_r *converges to* F *in* $A_\alpha(X)$ *as* $r \to 1$.

(3) F *can be approached by analytic polynomials in* $A_\alpha(X)$.

REPRESENTATION OF VECTOR VALUED ANALYTIC FUNCTIONS AS OPERATORS.

Our next objective is to identify the functions in $A_\alpha(X)$ with operators from J_α. The main theorem of this section allows us to get the duality results in [D-R-S] and [A-C-P] from a unified point of view. We shall denote by L(X,Y) the space of bounded linear operators from X to Y.

Let us recall first a pair of useful and standard estimates

LEMMA 2.1. *Let $\alpha > -1$ and $v > 1 + \alpha$. Then*

$$(2.1) \quad \int_{-\pi}^{\pi} \frac{dt}{|1 - we^{it}|^\alpha} = O((1-|w|)^{\alpha-1}) \quad \text{(see [D,page 65])}$$

$$(2.2) \quad \int_0^1 (1-r)^\alpha (1-rs)^{-v} dr = O((1-s)^{1+\alpha-v}) \quad \text{(see [S-W],page 291)}$$

THEOREM 2.1. *Let $-1 < \alpha < 2$ and let X, Y be complex Banach spaces. Then*

$$L(J_\alpha(X),Y) = A_\alpha(L(X,Y)) \quad \text{(with equivalent norms).}$$

Proof.-Let us take a bounded operator T from $J_\alpha(X)$ into Y. Given x in X and n in N denote by $u_{x,n}(z) = x.z^n$. Define Φ_n: X→ Y by

$$(2.3) \qquad \Phi_n(x) = T(u_{x,n})$$

Since it is clear that ${}_\alpha\| u_{x,n} \| \leq C \|x\| (n+1)^{-\alpha}$ then Φ_n belongs to L(X,Y) and $\| \Phi_n \| \leq C(n+1)^{-\alpha}$. This allows us to define the following L(X,Y) valued analytic function

$$(2.4) \qquad F(z) = \sum_{n=0}^{\infty} \Phi_n z^n$$

Let us check that F belongs to $A_\alpha(L(X,Y))$.

$$\| F''(z) \|_{L(X,Y)} = \sup_{\|x\|=1} \| \sum_{n=2}^{\infty} n(n-1)\Phi_n(x) z^{n-2} \|_Y .$$

Using now the continuity of T and the fact that $\sum_{n=2}^{\infty} n(n-1)u_{x,n}z^{n-2}$ is absolutely convergent in $J_\alpha(X)$ we can write

$$\| F''(z) \|_{L(X,Y)} = \sup_{\|x\|=1} \| T(\sum_{n=2}^{\infty} n(n-1)u_{x,n}z^{n-2})\|_Y$$

$$= \sup_{\|x\|=1} \| T(x.C_z'') \| \leq \| T \| \,.\, {}_\alpha\| C_z'' \|$$

where $C_z(w) = \dfrac{1}{1-zw}$.

Then everything reduces to estimate the norm in J_α of the scalar valued function C_z''. To do so we use (2.1) and (2.2) and we get

$$\int_{-\pi}^{\pi}\int_0^1 (1-r)^\alpha |C_z'''(re^{it})| drdt$$

$$\leq C \int_0^1 \Big(\int_{-\pi}^{\pi} \frac{dt}{|1-zre^{it}|^4} \Big)(1-r)^\alpha \, dr$$

$$\leq C \int_0^1 (1-r)^\alpha (1-r|z|)^{-3} dr = O((1-|z|)^{2-\alpha})$$

Conversely let us take F in $A_\alpha(L(X,Y))$ and G in $J_\alpha(X)$ given by $F(z) = \sum_{n=0}^{\infty} \Psi_n z^n$ and $G(z) = \sum_{n=0}^{\infty} x_n z^n$. Easy estimates on the Taylor coefficients of F and G allow us to show that the following series converges

$$(2.5) \qquad y_r = \sum_{n=1}^{\infty} \Psi_n(x_n) r^{n-1} + \Psi_0(x_0) \quad , 0 < r < 1.$$

Our next procedure will be based on the following elementary equality

$$n(n+1)(n+2) \int_0^1 (1-s^2)^2 s^{2n-1} \, ds = 1 \quad \text{for all } n \geq 1.$$

Hence

$$\sum_{n=1}^{\infty} \Psi_n(x_n)r^{n-1} = \int_0^1 \sum_{n=1}^{\infty} n(n+1)(n+2)\Psi_n(x_n)(rs)^{n-1} s^n(1-s^2)^2 \, ds =$$

$$\int_0^1 (1-s^2)^2 \left(\int_{-\pi}^{\pi} F''_1(se^{-it})\{G'(rse^{it})\}e^{-it} \, dt)ds\right)$$

where $F_1(z) = z^2(F(z) - F(0))$.

Thus $F_1''(z) = z^2F''(z) + 4zF'(z) + F(z) - F(0)$ and then

$$\| F_1''(z) \|_{L(X,Y)} \leq C (\| F''(z) \|_{L(X,Y)} + \| F'(z) \|_{L(X,Y)})$$

Part (1) in Corollary 1.1. shows that F_1 also belong to $A_\alpha(L(X,Y))$ and

$\| F_1 \|_\alpha \leq C \| F \|_\alpha$

Finally observe that

$\| y_r - y_{r'} \| \leq$

$$C \int_0^1 \left(\int_{-\pi}^{\pi} (1-s)^2 \|F''_1(se^{-it})\|_{L(X,Y)} \|G'(rse^{it})-G'(r'se^{it}) \|_X dt\right)ds$$

$$\leq C \| F \|_\alpha \int_0^1 \left(\int_{-\pi}^{\pi} (1-s)^\alpha \|G'(rse^{it})-G'(r'se^{it})\|_X dt\right)ds$$

$$\leq C \| F \|_\alpha \cdot {}_\alpha\| G_r - G_{r'} \|$$

This last inequality implies, using Prop.1.1. (2), that there exists the limit of y_r.

For fixed F this limit obviously defines a linear operator from $J_\alpha(X)$ into Y. The boundedness of it follows from the next estimate

$$\| y_r \| \leq C \int_0^1 \left(\int_{-\pi}^{\pi} (1-s)^2 \|F''_1(se^{-it})\|_{L(X,Y)}\|G'(rse^{it})\|_X dt\right)ds$$

$$\leq C \| F \|_\alpha \int_0^1 \left(\int_{-\pi}^{\pi} (1-s)^\alpha \|G'(rse^{it})\|_X dt \right) ds$$

$$\leq C \| F \|_\alpha \cdot {}_\alpha\| G_r \| \leq C \| F \|_\alpha \cdot {}_\alpha\| G \|.$$

and the fact that $|\Psi_0(x_0)| \leq \| F(0) \| \| G(0) \| \leq \| F \|_\alpha \cdot {}_\alpha\| G \|$.

COROLLARY 2.1.- *Let $-1 < \alpha < 2$. Then*

(2.6) $$L(J_\alpha, X) = A_\alpha(X).$$

(2.7) $$(J_\alpha(X))^* = A_\alpha(X^*).$$

COROLLARY 2.2.- *Let $-1 < \alpha < 2$. Then* $J_\alpha(X) = J_\alpha \hat{\otimes} X$

(where $J_\alpha \hat{\otimes} X$ means the tensor product with the projective norm).

Proof.- It follows easily from the density of $J_\alpha \hat{\otimes} X$ in $J_\alpha(X)$ together with the fact that they both have the same dual $L(J_\alpha, X^*)$.

APPLICATIONS

Given a sequence $\lambda = (\lambda_n)$ of complex numbers we can associate two kinds of operators acting on analytic polynomials and a formal power series

(3.1) $$T_\lambda\left(\sum_{n=1}^{k} a_n z^n\right) = \sum_{n=1}^{k} \lambda_n a_n z^n$$

(3.2) $$R_\lambda\left(\sum_{n=1}^{k} a_n z^n\right) = (\lambda_n a_n)_{1 \leq n \leq k}$$

(3.3) $$f_\lambda(z) = \sum_{n=1}^{k} \lambda_n z^n$$

Let us recall that a sequence is said to be a multiplier from the space J_α into either a space of holomorphic functions or a space of sequences if either T_λ or R_λ extends to a bounded operator from J_α into the other space. Multipliers from B_p $(0 < p < 1)$ into B_q, H^q and l^q were studied in [D-S]. We shall get most of the results proved there, using our techniques but only for the range $1/3 < p < 1$, because of the restriction $\alpha < 2$.

PROPOSITION 3.1.- *Let $-1 < \alpha < 2$, $\beta > 0$, $p \geq 1$ and (λ_n) a sequence. Then*

(i) $T_\lambda \in L(J_\alpha, H^p)$ if and only if $M_p(f_\lambda'', r) = O(\ (1-r)^{\alpha-2})$

(ii) $T_\lambda \in L(J_\alpha, J_\beta)$ if and only if $M_1(f_\lambda''', r) = O(\ (1-r)^{\alpha-\beta-3})$

Proof.- Using Theorem 2.1, we know that associated to T_λ we have the function

$$(3.4) \qquad F_\lambda(z) = \sum_{n=0}^{\infty} \lambda_n u_n z^n \quad \text{where} \quad u_n(z) = z^n.$$

(i) follows by observing that $\| F''_\lambda(re^{it})\|_{H^p} = M_p(f_\lambda'', r)$ and imposing that F_λ belongs to $A_\alpha(H^p)$.

(ii) Notice that

$$\| F''_\lambda(re^{i\theta})\|_{J_\beta} = \int_0^1 \int_{-\pi}^{\pi} (1-s)^\beta \, |f_\lambda'''(rse^{i(\theta+t)})| \frac{dt}{2\pi} ds$$

$$= \int_0^1 (1-s)^\beta \, M_1(f_\lambda''', rs)\, ds$$

Therefore from $M_1(f_\lambda''', r) = O(\ (1-r)^{\alpha-\beta-3})$ and (3.2) we have

$$\| F''_\lambda(re^{i\theta})\|_{J_\beta} \leq C \int_0^1 (1-s)^\beta \, (1-rs)^{\alpha-\beta-3} ds \leq C\ (1-r)^{2-\alpha}$$

On the other hand if we assume that F_λ belongs to $A_\alpha(J_\beta)$ then

$$M_1(f_\lambda''', r^2)\ (1-r)^{\beta+1} \leq (1+\beta) \int_r^1 (1-s)^\beta \, M_1(f_\lambda''', rs)\, ds$$

$$\leq (1+\beta)\ \| F''_\lambda(re^{i\theta})\|_{J_\beta} \leq C\ (1-r)^{2-\alpha}$$

Then it follows that $M_1(f_\lambda''', r) = O(\ (1-r)^{\alpha-\beta-3})$.

To deal with R_λ let us first state the following elementary lemma whose proof is a simple exercise.

LEMMA3.1.- *Let $\beta > 0$ and $\mu_\kappa \geq 0$. Then*

$$\sum_{n=0}^{\infty} \mu_n r^n = O((1-r)^{-\beta}) \text{ if and only if } \sum_{k=0}^{n} \mu_k = O(n^\beta).$$

PROPOSITION 3.2.- *Let $-1 < \alpha < 2$, $p \geq 1$ and λ_n a sequence. Then*

$$R_\lambda \in L(J_\alpha, l^p) \text{ if and only if } \sum_{k=0}^{n} k^{2p} |\lambda_k|^p = O(n^{(2-\alpha)p})$$

Proof.- The analytic function associated to R_λ is

(3.5) $G_\lambda(z) = \sum_{n=0}^{\infty} \lambda_n e_n z^n$ where e_n is the canonic basis in l^p.

Hence $\| G''_\lambda(z) \|_{l^p} = \left(\sum_{n=0}^{\infty} n^p (n-1)^p |\lambda_n|^p |z|^{(n-2)p} \right)^{1/p}$

Notice now that $R_\lambda \in L(J_\alpha, l^p)$ is equivalent to $G_\lambda \in A_\alpha(l^p)$, that is

$$\sum_{n=0}^{\infty} n^p (n-1)^p |\lambda_n|^p |z|^{(n-2)p} = O((1-|z|)^{(\alpha-2)p})$$

and the result follows from the previous lemma.

In the literature several authors have been studying spaces similar to those considered until now but defined using the L^p norm instead of the L^1 and L^∞ norm. (See [F1],[T],[H-L]).

Let us define for $|\alpha| < 1$ and $1 \leq p \leq \infty$,

$A_\alpha{}^p = \{ F: D \to C$ analytic : $M_p(F',r) = O((1-r)^{\alpha-1}) \}$

$J_\alpha{}^p = \{ F: D \to C$ analytic : $\int_0^1 (1-s)^\alpha M_p(F',r)\, dr < \infty \}$

Our next objective is to show that some results for these spaces follow from the vector valued case. To do so let us first observe that if $F:D\to \mathbb{C}$ is analytic we can consider the following vector-valued analytic function

(3.6) $$\mathbf{F}(z) = F_z \quad \text{where} \quad F_z(\theta) = F(ze^{i\theta}).$$

In other words if $F(z) = \sum_{n=1}^{\infty} \lambda_n z^n$ then

$$\mathbf{F}(z) = \sum_{n=1}^{\infty} \lambda_n e_n z^n \quad \text{where} \quad e_n(\theta) = e^{in\theta}.$$

Notice that $\mathbf{F}'(z)(\theta) = e^{i\theta} F'(ze^{i\theta})$, what allows to write for any $|w| = r$

$$M_p(F',r) = \| \mathbf{F}'(w) \|_{L^p} = \int_{-\pi}^{\pi} \| \mathbf{F}'(re^{it}) \|_{L^p} \frac{dt}{2\pi} .$$

In other words

$$F \in J_\alpha^p \text{ if and only if } \mathbf{F} \in J_\alpha(L^p)$$

$$F \in A_\alpha^p \text{ if and only if } \mathbf{F} \in A_\alpha(L^p)$$

We can now obtain the following duality result which was essentially proved in [F2] using fractional integration.

THEOREM 3.1.- *Let $1 \leq p < \infty$, $1/p + 1/q = 1$ and $-1 < \alpha < 1$.*

$$(J_\alpha^p)^* = A_\alpha^q$$

Proof.- Let us consider $G(z) = \sum_{n=0}^{\infty} a_n z^n$ in A_α^q

and $F(z) = \sum_{n=0}^{\infty} b_n z^n$ a function in J^p_α. Using (2.7) for $X = L^p$

$$\left|\sum_{n=0}^{\infty} a_n b_n\right| = \left|\sum_{n=0}^{\infty} \langle a_n e_n, b_n e_n\rangle\right| \leq C.\| \mathbf{G} \|_{\alpha\,\alpha} \| \mathbf{F} \| = C.\| G \|_{A_\alpha^q} \| F \|_{JP_\alpha}.$$

Conversely let ϕ belong to $(JP_\alpha)^*$. Take $\lambda_n = \phi(u_n)$ where $u_n(z) = z^n$ and define

$$G(z) = \sum_{n=0}^{\infty} \lambda_n z^n$$

To verify that G belongs to A^q_α, we shall see that G belongs to $A_\alpha(L^q)$. Now according to (2.6) we must show that the map

$$T: \sum_{n=0}^{\infty} \beta_n z^n \rightarrow \sum_{n=0}^{\infty} \lambda_n \beta_n e_n$$

defines a bounded linear operator from J_α into L^q. Now using duality we find a function f in the unit ball of L^p such that

$$\left\| \sum_{n=0}^{\infty} \lambda_n \beta_n e_n \right\|_q = \left| \sum_{n=0}^{\infty} \lambda_n \beta_n \hat{f}(-n) \right| = \left| \phi\left(\sum_{n=0}^{\infty} \hat{f}(-n) \beta_n u_n \right) \right|$$

Hence

$$(3.7) \qquad \left\| \sum_{n=0}^{\infty} \lambda_n \beta_n e_n \right\|_q \leq \| \phi \| \left\| \sum_{n=0}^{\infty} \hat{f}(-n) \beta_n u_n \right\|_{J_\alpha^p}$$

Notice that if $g(z) = \sum_{n=0}^{\infty} \hat{f}(-n)\beta_n z^n$, then $g(z) = h_r * f(\theta)$ where $h(z) = \sum_{n=0}^{\infty} \beta_n z^n$ and $z = re^{i\theta}$. Hence we also have $g'(z) = h'_r * f(\theta)$.

Therefore $M_p(g',r) \leq M_1(h',r) \| f \|_p$ and consequently

$$\| g \|_{JP_\alpha} \leq \int_0^1 (1-s)^\alpha M_p(g',s)ds \leq \int_0^1 (1-s)^\alpha M_1(h',s)ds = \| h \|_{J_\alpha}$$

Combining now this last inequality and (3.7) the proof is completed.

Remark 3.1. The same proof as in Theorem 3.1 can be done in the vector valued case. Note that we only have applied that L^q is embedded in $(L^p)^*$, what is still true in the vector valued case (see [D-U]). Hence we can see that $A_\alpha{}^p(X^*)$ is always a dual no matter the geometry of X^*, what shows the difference with $H^p(X^*)$.

REFERENCES.

[A-C-P] J.M ANDERSON, J CLUNIE and CH. POMMERENKE On Bloch functions and normal functions. J. Reine Angew. Math. 270 (1974), 12-37.

[Bu] A. V. BUKHVALOV, The duals to spaces of analytic vector valued functions and the duality of functors generated by these spaces. Zap. Nauch. Semi. LOMI 92 (1979), 30-50.

[D-U] J.DIESTEL and J.J. UHL. "Vector Measures " Math. surveys, no 15, Amer. Math. Soc. , Providence, R.I. 1977.

[D] P. L. DUREN, "Theory of H^p spaces". Academic Press, New York , 1970.

[D-S] P.L. DUREN and A.L. SHIELDS. Coefficient multipliers on H^p and B^p spaces. Pacific J. Math. 32 (1970), 69-78.

[D-R-S] P.L. DUREN, B.W. ROMBERG and A.L. SHIELDS. Linear functionals on H^p spaces $0 < p < 1$ .J. Reine Angew. Math. 238 (1969), 32-60.

[F1] T.M. FLETT. On the rate of growth of mean values of holomorphic and harmonic functions. Proc. London Math. Soc. 20 (1970),749-768.

[F2] T.M. FLETT. Lipschizt spaces of functions on the circle and the disc.; J. Math. Analysis and applic.39 (1972) 125-158.

[G] J.B. GARNETT. "Bounded analytic functions." Academic Press. New York 1981

[H-L] G.H. HARDY and J.E. LITTLEWOOD. Some properties of fractional integrals, II. Math. Z. 34 (1932) 403-439.

[K] N. KALTON. Analytic functions in non locally convex spaces and applications. Studia Math. 83 (1986) 275-303.

[S] J.H. SHAPIRO. Mackey topologies, reproducing kernels and diagonal maps on Hardy and Bergman spaces. Duke Math. J. 43(1976), 187-202.

[S-W] A.L. SHIELDS and D.L. WILLIAMS. Bounded projection, duality and multipliers in spaces of analytic functions. Trans. Amer. Math. Soc. 162 (1971), 287-302.

[T] M. TAIBLESON. On the theory of Lipschizt spaces and distributions on euclidean n-space I,II,III. J. Math. Mech. 13(1964), 407-479, 14(1965),821-839, 15 (1966), 973-981.

[Z1] A. ZYGMUND, "Trigonometric Series." Cambrigde Univ. Press, London and New York. 1959.

[Z2] A. ZYGMUND, Smooth functions. Duke Math. J. 12(1945), 47-76.

Oscar Blasco

Departamento de Matematicas

Universidad de Zaragoza

Zaragoza-50009 (SPAIN)

AMS Classification (1980): 46E40, 42A45

Key Words: Vector-Valued Analytic Function, boundary values problems, Lipschizt-Besov classes, Multipliers.

Notes on approximation properties in separable Banach spaces

P.G. Casazza and N.J. Kalton*
Department of Mathematics, University of Missouri
Columbia, Mo. 65211, U.S.A.

(*): The research of the first author was supported by NSF-grant DMS 8702329 and the research of the second author was supported by NSF-grant DMS 8901636

1. Introduction, definitions and discussion of results.

Although the example given by Enflo in 1973 [5] settled the approximation problem and the basis problem for Banach spaces, a number of closely related problems have continued to arouse interest. If X is a separable Banach space, there are a number of natural properties intermediate between X having the approximation property and having a basis.

Let us first make some definitions. Suppose X is a separable Banach space. Then X has the *approximation property (AP)* if there is a net of finite-rank operators T_α so that $T_\alpha x \to x$ for $x \in X$, uniformly on compact sets. X is said to have the *bounded approximation property (BAP)* if this net can be replaced by a sequence T_n; alternatively X has (BAP) if there is a sequence of finite-rank operators, T_n, such that $\sup \|T_n\| < \infty$ and $T_n x \to x$ for $x \in X$. A sequence T_n with these properties will be called an *approximating sequence.* If X has an approximating sequence T_n with $\lim_{n\to\infty} \|T_n\| = 1$ then X has the *metric approximation property (MAP).*

An important principle [15] that we will use frequently is that if T_n is any approximating sequence for X then there is an approximating sequence S_n satisfying $S_m S_n = S_n$ whenever $m > n$ and such that for some subsequence T_{k_n} of T_n then $\lim_{n\to\infty} \|T_{k_n} - S_n\| = 0$. (See Lemma 2.4 of [15]).

A slight weakening of the basis property is to require that X has a *finite-dimensional decomposition (FDD)* i.e. that X has an approximating sequence T_n satisfying $T_m T_n = T_{\min(m,n)}$ for $m, n \in \mathbf{N}$. Szarek [24] has given an example to show that a space with an (FDD) need not have a basis. Between (FDD) and (BAP) we can isolate two other natural properties. We say X has the *π−property* if X has an approximating sequence of projections, and the *commuting bounded approximation property (CBAP)* if it has a commuting approximating sequence. We may add to both these properties the corresponding metric properties (π_1) and (CMAP) where we also have $\lim_{n\to\infty} \|T_n\| = 1$. In general if X has a commuting approximating sequence T_n with $\liminf \|T_n\| = \lambda$ we say X has λ-CBAP.

Johnson [12] showed that a space with the (π_1)-property has an (FDD). However it is not known whether every π−space has an (FDD).

The (CBAP) property was first isolated by Rosenthal and Johnson [14] in the early seventies and has most recently been studied by the first author [2]. Let us note that X has λ−CBAP if and only if it has an approximating sequence T_n such that

$T_mT_n = T_{\min(m,n)}$ for $m \neq n$, and $\limsup \|T_n\| \leq \lambda$. (This is doubtless well-known; it is proved in Proposition 2.1 below.) This suggests that it is quite close to the (FDD) property. However very recently Read [22] gave an example of a Banach space with (CBAP) but failing (FDD). Conversely Casazza [2] showed that a space with both π and (CBAP) has an (FDD), so Read's space is not a $\pi-$space.

It is in general not known if (BAP) implies (CBAP). However certain hypotheses on X do give this implication: it holds if X is reflexive or is a separable dual space [13]. Coincidentally, the same hypotheses give that (AP) implies (MAP) (Grothendieck [11]). It has also been shown by Johnson [14] that any space with (CBAP) can be renormed to have (CMAP). These results suggest a close relationship between the properties (CBAP) and (MAP). Our main result is that X has (CBAP) if and only if it can be equivalently normed to have (MAP), so that (CBAP) is the isomorphic version of (MAP). Further if X has (MAP) it has (CMAP). The proofs of these results (Theorem 2.4 and Corollary 2.5) are quite simple modifications of techniques from the study of approximate identities in Banach algebras (due to Sinclair [23]; see also [4]). We also give another condition, the reverse metric approximation property, which implies (CBAP).

Pelczynski [21] and Johnson, Rosenthal and Zippin [15] showed that any space with (BAP) is isomorphic to a complemented subspace of a space with a basis. Johnson [14] has shown that if X has (BAP) then there is a reflexive space Y so that $X \oplus Y$ is a $\pi-$space. In fact Y can be taken to be the space $C_p\,(1 < p < \infty)$ defined in Section 3. He conjectures that in fact $X \oplus C_p$ has an (FDD) and hence a basis. However, we show that $X \oplus C_p$ has a basis if and only if X has (CBAP). This shows that several possible conjectures are equivalent.

We then give some results on renormings of spaces with (CBAP) and conclude by studying spaces X which have an approximating sequence T_n with $\lim \|I - 2T_n\| = 1$. This condition is closely related to unconditional forms of the approximation property. We say that X has the *unconditional approximation property (UnAP)* if there is an approximating sequence T_n such that if $A_n = T_n - T_{n-1}$ $(T_0 = 0)$ then

$$\sup \|\sum_{i=1}^{N} \eta_i A_i\| < \infty$$

where the supremum is taken over all N and all $\eta_i = \pm 1$, $i = 1, 2, \ldots, N$. We introduce the metric version of (UnAP) and relate our work to recent results of Cho, Johnson, Godefroy, P. Saab and Li ([3],[8],[9] and [18]).

2. Equivalent formulations of (CBAP).

We will write $[A, B] = AB - BA$ and $\prod_{j=a}^{b} T_j = T_aT_{a+1}\ldots T_b$.

PROPOSITION 2.1. *Suppose X is a separable Banach space and T_n is an approximating sequence for X with $T_mT_n = T_n$ for $m > n$. Let $\lambda = \liminf \|T_n\|$. If $\sum \|[T_n, T_{n+1}]\| < \infty$, then X has $\lambda-$CBAP (and, further has an approximating sequence R_n for which $R_mR_n = R_{\min(m,n)}$ for $m \neq n$ and $\limsup \|R_n\| \leq \lambda$.)*

PROOF: We first show that we can suppose that $T_n(X) = T_n^2(X)$. Let P_n be any bounded projection of X onto $T_n(X)$ and choose a sequence $0 < \alpha_n < 1$, so that

$\sum \alpha_n \|P_n\| < \infty$ and $-\alpha_n/(1-\alpha_n)$ is not an eigenvalue of T_n. Then we may replace T_n by $(1-\alpha_n)T_n + \alpha_n P_n$ and the hypotheses of the Proposition will still hold with the additional constraint that $T_n(X) = T_n^2(X)$.

Now let $\epsilon_n = \|[T_n, T_{n+1}]\|$. For $n \in \mathbf{N}$ and $k \geq 1$ we define $A(n,k) = \prod_{j=n}^{n+k-1} T_j$. Let $A(n,0) = I$. Then for $k \geq 1$,

$$A(n,k+1) = A(n,k) + A(n,k-1)[T_{n+k-1}, T_{n+k}].$$

Thus if $M_n(k) = \max_{1\leq l\leq k} \|A(n,l)\|$ then $M_n(1) = \|T_n\|$ and $M_n(k+1) \leq M_n(k)(1+\epsilon_{n+k-1})$. Hence

$$\|A(n,k)\| \leq \|T_n\| \prod_{j=n}^{\infty} (1+\epsilon_j).$$

It now follows that $A(n,k)$ is norm-convergent and we can define $S_n = \prod_{j=n}^{\infty} T_j = \lim_{k\to\infty} A(n,k)$. Clearly the operators S_n are finite-rank with $S_n(X) = S_n^2(X) = T_n(X)$. Further if $m > n$ we have $S_m S_n = S_n$ and S_n is an approximating sequence with $\|S_n\| \leq \beta_n \|T_n\|$ where $\lim_{n\to\infty} \beta_n = 1$.

A simple calculation also shows that if $m > n$ then

$$S_n S_m = \left(\prod_{j=n}^{m-1} T_j \right) T_m^2 S_{m+1}$$

so that

$$[S_m, S_n] = S_n(I - S_m) = A(n, m-n-1)[T_{m-1}, T_m] S_m.$$

Thus for a suitable constant K, independent of m, n we have

$$\|[S_m, S_n]\| \leq K\epsilon_{m-1}$$

whenever $m > n$. In particular we have $\lim_{m\to\infty} \|[S_m, S_n]\| = 0$ for each fixed n.

We now pass to a subsequence V_n of S_n so that $\limsup \|V_n\| \leq \lambda$. We will also have $V_m V_n = V_n$ for $m > n$ and $\lim_{m\to\infty} \|[V_m, V_n]\| = 0$ for each n.

Finally, by induction we will choose an increasing sequence of positive integers (n_k) and operators R_k so that:

(1) $R_k^2(X) = R_k(X) = V_{n_k}(X)$.
(2) R_k is a polynomial in $V_{n_1}, \ldots, V_{n_k}$.
(3) For $1 \leq l < k$, $R_k R_l = R_l R_k = R_l$.
(4) $\|R_k\| \leq \|S_{n_k}\| + 2^{-k}$.

To start let $n_1 = 1$ and $R_1 = V_1$. Now suppose $n_1, \ldots, n_k$ and $R_1, \ldots, R_k$ have been determined to satisfy (1-4). Since $R_k(X) = R_k^2(X)$ we can find an operator W_k which is a polynomial in R_k so that $W_k R_k x = x$ for $x \in R_k(X)$. Now, $\lim_{m\to\infty} \|R_k(I - V_m)\| = 0$ since R_k is a polynomial in $V_{n_1}, \ldots, V_{n_k}$. Thus we may pick n_{k+1} so that

$$\|R_k(I - V_{n_{k+1}})\| < 2^{-(k+1)} \|W_k\|^{-1}.$$

We then define

$$R_{k+1} = V_{n_{k+1}} + W_k R_k (I - V_{n_{k+1}}).$$

Clearly conditions (2) and (4) above hold. For condition (3) note that $I - R_{k+1} = (I - W_k R_k)(I - V_{n_{k+1}})$ from which it follows that $R_k(I - R_{k+1}) = (I - R_{k+1})R_k = 0$ or $R_k R_{k+1} = R_{k+1} R_k = R_k$. Now if $1 \le l < k$ then $R_l R_{k+1} = R_l R_k R_{k+1} = R_l R_k = R_l$ and similarly $R_{k+1} R_l = R_l$. Thus (3) is verified.

For (1) we clearly have $R_{k+1}(X) \subset V_{n_{k+1}}(X)$. It suffices to show that R_{k+1} is injective on $V_{n_{k+1}}(X)$. Indeed suppose $x \in V_{n_{k+1}}(X)$ and $R_{k+1}x = 0$. Then since $W_k R_k (I - V_{n_{k+1}})x \in R_k(X) = V_{n_k}(X)$ we have $V_{n_{k+1}} x \in V_{n_k}(X)$. Thus $V_{n_{k+1}}^2 x = V_{n_{k+1}} x$ and by the fact that $V_{n_{k+1}}$ is injective on $V_{n_{k+1}}(X)$ we have $x = V_{n_{k+1}} x$ and hence $x = 0$ as required.

It is now immediate that X has $\lambda-$CBAP.∎

COROLLARY 2.2. *Suppose X has an approximating sequence T_n for which*

$$\lim_{m,n\to\infty} \|[T_m, T_n]\| = 0$$

and $\liminf_{n\to\infty} \|T_n\| = \lambda$. *Then X has $\lambda-$CBAP.*

REMARK: By $\lim_{m,n\to\infty} a_{mn} = 0$ we mean that given $\epsilon > 0$ there exists N so that if $m, n \ge N$ then $|a_{mn}| < \epsilon$.

PROOF: We may find a subsequence T_{n_k} and finite rank operators S_k so that $\lim_k \|T_{n_k}\| = \lambda$, $\|S_k - T_{n_k}\| \to 0$, and $S_l S_k = S_k$ for $l > k$. Then $\lim_{m,n\to\infty} \|[S_m, S_n]\| = 0$ and so by passing to a further subsequence we can apply the Proposition.

COROLLARY 2.3. *Suppose X has an approximating sequence T_n for which*

$$\lim_{m\to\infty} \|[T_m, T_n]\| = 0$$

for each fixed n and $\liminf \|T_n\| = \lambda$. *Then X has $\lambda-$CBAP.*

This Corollary is immediate from the Proposition. The same result without the precise estimate on the constant was shown in [2].

We now come to our main result. The argument in the next theorem is a simple modification of a theorem of Sinclair [23] on approximate identities in Banach algebras. Sinclair shows that if A is a Banach algebra with a bounded two-sided sequential approximate identity then it has a commuting approximate identity with the same bound. This result can be applied directly to the algebra $\mathcal{K}(X)$ of compact operators on X when X^* is separable and has (AP), and hence (MAP). Under these circumstances $\mathcal{K}(X)$ has a norm-one two-sided approximate identity, and we can recover Theorem 2 of [2]. In general, however, some modification of Sinclair's approach is necessary.

THEOREM 2.4. *Suppose X is a separable Banach space with (MAP). Then X has (CMAP).*

PROOF: We shall suppose that X has an approximating sequence T_n with $T_m T_n = T_n$ for $m > n$ and $\|T_n\| \le 1 + \epsilon_n$ where $\sum \epsilon_n = \beta < \infty$. For $t > 0$ define the operators

$$V_n(t) = e^{-nt} \exp(t \sum_{k=1}^{n} T_k) = e^{-nt} \sum_{j=0}^{\infty} \frac{t^j}{j!}(T_1 + \cdots + T_n)^j.$$

Then

$$\|V_n(t)\| \le e^{-nt}\exp(t\sum_{k=0}^{n}\|T_k\|) \le e^{\beta t}.$$

Let $E_n = T_n(X)$. Then each E_n is an invariant subspace for every T_m and hence also for every $V_m(t)$. Rewriting $V_m(t)$ as $\exp(t\sum_{k=1}^{m}(T_k - I))$ it is clear that if $x \in E_n$ and $m > n$ then $V_m(t)x = V_n(t)x$. It follows therefore from the bound on the norms of $V_n(t)$ that we can define $S(t)$ by $S(t)x = \lim_{n\to\infty} V_n(t)x$ for all $x \in X$. Clearly $\|S(t)\| \le e^{\beta t}$. Furthermore $S(t)$ has the semigroup property $S(t_1 + t_2) = S(t_1)S(t_2)$ since each $V_n(t)$ is a semigroup and the property is preserved by strong limits.

We further claim that each $S(t)$ is compact for $t > 0$. Indeed suppose $l \in \mathbf{N}$ and that $x \in E_n$ where $n > l$. Then $d(S(t)x, E_l) = d(V_n(t)x, E_l)$. It is then easy to see, by expansion, that the operator $\exp t(T_1 + \cdots + T_n) - \exp t(T_{l+1} + \cdots + T_n)$ has range contained in E_l. Thus

$$\begin{aligned} d(S(t)x, E_l) &= e^{-nt}d(\exp t(T_{l+1} + \cdots + T_n)x, E_l) \\ &\le e^{-nt}\|\exp t(T_{l+1} + \cdots + T_n)\|\|x\| \\ &\le e^{-nt}\exp t(\|T_{l+1}\| + \cdots + \|T_n\|)\|x\| \\ &\le e^{\beta t}e^{-lt}\|x\|. \end{aligned}$$

Hence for all $x \in X$,

$$d(S(t)x, E_l) \le e^{\alpha t}e^{-lt}\|x\|$$

and hence $S(t)$ is compact.

Note that as $t \to 0$ $\|S(t)\| \to 1$. Also if $x \in E_n$ we have $S(t)x = V_n(t)x \to x$. Hence for all $x \in X$, we have $\lim_{t\to 0} S(t)x = x$.

Since X has (MAP) there exist finite-rank operators R_n so that $\|R_n - S(1/n)\| \to 0$. Then R_n is an approximating sequence, $\lim\|R_n\| = 1$ and $\lim_{m,n\to\infty}\|[R_m, R_n]\| = 0$ since the operators $S(1/n)$ commute. Hence by Corollary 2.2, X has (CMAP).∎

COROLLARY 2.5. *Let X be a separable Banach space. Then X has (CBAP) if and only if X can be equivalently normed to have (MAP).*

Corollary 2.5 follows directly from Theorem 2.4 and the result of Johnson [14] that every space with (CBAP) can be renormed to have (MAP).

THEOREM 2.6. *Suppose X is a separable Banach space with an approximating sequence T_n satisfying $T_mT_n = T_n$ for $m > n$ and such that*

$$\sup_{i_1<i_2<\cdots<i_k}\|T_{i_1}T_{i_2}\dots T_{i_k}\| = \lambda < \infty.$$

Then X has $\lambda-$CBAP.

PROOF: The proof is essentially that of Theorem 2.4. We first estimate $\|T_1+\cdots+T_n\|^k$. On expanding this consists of n^k terms of the form $T_1^{l_1}T_2^{l_2}\dots T_n^{l_n}$. It is clear that if we define the weight, $w(l_1,\dots,l_n) = \sum_{l_j>1}(l_j - 1)$ then by grouping terms,

$$\|\prod_{j=1}^{n}T_j^{l_j}\| \le \lambda^{(1+w(l_1,\dots,l_n))}.$$

Now $(\prod T_j^{l_j})(T_1 + \cdots + T_n)$ consists of n terms of which $n-1$ have at most the same weight $w = w(l_1, \cdots, l_n)$ and one has weight $w+1$. Thus if if we define W_k by

$$W_k = \sum \lambda^{w(l_1,\ldots,l_n)}$$

where the sum is over all terms $\prod T_j^{l_j}$ in $(\sum T_j)^k$ then $W_{k+1} \leq (n-1+\lambda)W_k$ for every k. As $W_1 \leq n-1+\lambda$ we obtain $W_k \leq (n-1+\lambda)^k$ for $k \geq 1$. Thus

$$\|(T_1 + \cdots + T_n)^k\| \leq \lambda(n-1+\lambda)^k$$

and so

$$\|\exp t(T_1 + \cdots + T_n)\| \leq \lambda e^{t(n-1+\lambda)}.$$

Thus repeating the proof of Theorem 2.4 we obtain

$$\|V_n(t)\| \leq \lambda e^{t(\lambda-1)}.$$

The proofs now goes through unchanged since $\|S(t)\| \leq \lambda e^{t(\lambda-1)}$ and $\lim_{t\to 0} \|S(t)\| = \lambda$. Hence X has λ−CBAP.∎

We will use Theorem 2.6 to give another characterization of (CBAP). We say that X has the *reverse monotone approximation property* (RMAP) if there is an approximating sequence T_n with $\lim_{n\to\infty} \|I - T_n\| = 1$. We first prove a simple equivalence for (RMAP).

PROPOSITION 2.7. *(i) X has (MAP) if and only if there exists $\alpha > 0$ and an approximating sequence (T_n) with $\lim \|I + \alpha T_n\| = 1 + \alpha$.*

(ii) X has (RMAP) if and only if there exists $\alpha > 0$ and an approximating sequence (T_n) with $\lim \|I - \alpha T_n\| = 1$.

PROOF: The proof of (i) is again somewhat similar to the proof of Theorem 2.3. By passing to a subsequence one may suppose that $\|I + \alpha T_n\| \leq 1 + \alpha(1 + \epsilon_n)$ where $\sum \epsilon_n = \beta < \infty$. Then, defining $V_n(t)$ as in Theorem 2.3 one obtains, by estimating $\|\exp(t \sum_{k=1}^n (I + \alpha T_k))\|$, that $\|V_n(\alpha t)\| \leq e^{\alpha\beta t}$ and the proof goes through as before.

For (ii) it suffices to consider the case $\alpha < 1$ by a simple convexity argument. We may further suppose $T_m T_n = T_n$ for $m > n$. Then pick a sequence of integers l_n so that $\lim l_n = \infty$ and $\lim \|(I - \alpha T_n)^{l_n}\| = 1$. Then set $S_n = I - (I - \alpha T_n)^{l_n}$. Clearly (S_n) is an approximating sequence with $\lim \|I - S_n\| = 1$.∎

THEOREM 2.8. *Let X be a separable Banach space with (RMAP). Then X has (CBAP).*

To prove Theorem 2.8 we require the following lemma:

LEMMA 2.9. *Suppose X has (RMAP). Then X has an approximating sequence T_n with $\lim \|I - T_n\| = 1$ and $\limsup \|(T_n - T_n^2)^2\| \leq \frac{15}{16}$.*

PROOF: We assume S_n is an approximating sequence with $S_m S_n$ for $m > n$ and $\|I - S_n\| \leq 1 + \epsilon_n$ where $\epsilon_n \downarrow 0$. Put $T_n = \frac{1}{2n} \sum_{j=n+1}^{3n} S_j$. Then the properties of T_n specified in the lemma are clear except possibly for the last.

Consider

$$(I - T_n)T_n = \frac{1}{4n^2} \sum_{i=n+1}^{3n} \sum_{j=n+1}^{3n} (I - T_i)T_j.$$

$$(T_n - T_n^2)^2 = \frac{1}{16n^4} \sum_{i=n+1}^{3n} \sum_{j=n+1}^{3n} \sum_{k=n+1}^{3n} \sum_{l=n+1}^{3n} (I - T_i)T_j(I - T_k)T_l.$$

Now $(I - T_i)T_j(I - T_k)T_l$ vanishes if either $i > j$ or $k > l$; this eliminates all but $n^2(2n+1)^2$ terms. Consider those remaining terms where $k < l \leq 2n < i < j$. In this case $(I - T_i)T_jT_kT_l = (I - T_i)T_kT_l = 0$ and hence the term becomes $(I - T_i)T_jT_l = 0$. There are $\frac{1}{4}n^2(n-1)^2$ such terms. Thus there remain at most $n^2(2n+1)^2 - \frac{1}{4}n^2(n-1)^2$ terms of norm at most $(2 + \epsilon_n)^2(1 + \epsilon_n)^2$. Hence

$$\|(T_n - T_n^2)^2\| \leq \frac{1}{16}((2 + \frac{1}{n})^2 - \frac{1}{4}(1 - \frac{1}{n})^2)(2 + \epsilon_n)^2(1 + \epsilon_n)^2.$$

Thus

$$\limsup_{n \to \infty} \|(T_n - T_n^2)^2\| \leq \frac{15}{16}. \blacksquare$$

PROOF OF THEOREM 2.8: We may suppose X has an approximating sequence T_n with $\|I - T_n\| \leq 1 + \epsilon_n$ where $\prod(1 + \epsilon_n) \leq 2$, $T_mT_n = T_n$ for $m > n$ and $\|(T_n - T_n^2)^2\| < 19/20$ for all n. This is possible by the lemma. We then define a new approximating sequence (S_n) by

$$I - S_n = \left(\prod_{k=1}^{n-1} (I - T_k)^2\right)(I - T_n).$$

Thus $\|I - S_n\| \leq 4$ and so $\|S_n\| \leq 5$.

Now if $A_n = I - S_n$ we have, provided $l_n \geq 1$,

$$\prod_{k=1}^{n} A_k^{l_k} = \left(\prod_{k=1}^{n-1} (I - T_k)^{2+l_k}\right)(I - T_n)^{l_n}.$$

Thus if $p_1, \ldots, p_{n-1}$ are any polynomials,

$$\left(\prod_{k=1}^{n-1} p_k(A_k)\right) A_n = \left(\prod_{k=1}^{n-1} [p_k(I - T_k)](I - T_k)^2\right)(I - T_n).$$

In particular if $i_1 < i_2 < \cdots < i_m = n$,

$$\left(\prod_{k=1}^{m-1} S_{i_k}^2\right)(I - S_{i_m}) = \left(\prod_{k=1}^{n-1} (I - T_k)^2 T_k^{2\beta_k}\right)(I - T_n)^2,$$

where $\beta_k = 1$ if $k \in \{i_1, \ldots, i_{m-1}\}$ and $\beta_k = 0$ otherwise. Hence

$$\|\left(\prod_{k=1}^{m-1} S_{i_k}^2\right)(I - S_{i_m})\| \leq 4(\frac{19}{20})^{m-1}.$$

This implies an estimate, since $\|I + S_{i_m}\| \leq 6$,

$$\|\prod_{k=1}^{m} S_{i_k}^2 - \prod_{k=1}^{m-1} S_{i_k}^2\| \leq 24(\frac{19}{20})^{m-1}$$

from which we obtain

$$\|\prod_{k=1}^{m} S_{i_k}^2\| \leq 1 + 24\sum_{k=0}^{\infty}(\frac{19}{20})^k = 481.$$

Notice that $S_n(X) \subset T_n(X)$; hence if $m > n$, $(I - S_m)S_n = 0$ so that $S_mS_n = S_n$ Hence $S_m^2S_n^2 = S_n^2$ and so we may apply Theorem 2.6 to to the approximating sequence S_n^2 to deduce that X has (CBAP).∎

3. Complementation and renormings.

PROPOSITION 3.1. *Suppose X is a separable Banach space and Y is a separable reflexive Banach space so that $X \oplus Y$ has (CBAP). Then X has (CBAP).*

PROOF: Suppose S_n is an approximating sequence for $X \oplus Y$ such that $S_mS_n = S_{\min(m,n)}$ for $m \neq n$. Let P be the projection onto X. Consider the operators $PS_n : Y \to X$. Then for every $y \in Y$ we have $\lim \|PS_ny\| = 0$. Hence since Y is reflexive we have $\lim(PS_n)^*x^* = 0$ weakly for every $x^* \in X^*$. Thus ([17]) PS_n converges weakly to zero in $\mathcal{L}(Y, X)$. Hence if $Q = I - P$, PS_nQ converges weakly to zero in $\mathcal{L}(X \oplus Y)$. Now we may pass to a sequence of convex combinations R_n which is still an approximating sequence for $X \oplus Y$ and so that $\lim \|PR_nQ\| = 0$. Define $T_n : X \to X$ by $T_nx = PR_nx$. Then T_n is an approximating sequence for X and

$$[T_m, T_n] = PR_nQR_mP - PR_mQR_nP \mid_X .$$

Hence $\lim_{m,n\to\infty} \|[T_m, T_n]\| = 0$ and the result follows by Corollary 2.2.∎

Our next result is a slight modification of an argument in [2].

PROPOSITION 3.2. *Let X be a separable Banach space with an approximating sequence T_n such $T_mT_n = T_{\min(m,n)}$ for $m \neq n$. Let $E_n = (T_n - T_n^2)(X)$. Then for $1 < p < \infty$, $X \oplus \ell_p(E_n)$ has an FDD. Furthermore if we denote by S_n the associated partial sum projections, we have*

$$\lim_{n\to\infty} \|\frac{1}{n}\sum_{k=1}^{n}(T_n - S_n \mid_X)\| = 0.$$

PROOF: We use an argument which dates back to Johnson [13] and is exploited in [2]. Define projections S_n on $X \oplus \ell_p(E_n)$ by

$$S_n(x, y_1, y_2, \ldots) = (T_nx + y_n, y_1, \ldots, y_{n-1}, (T_n - T_n^2)x + (I - T_n)y_n, 0, 0, \ldots).$$

Then $S_m S_n = S_{\min(m,n)}$ and S_n is an approximating sequence so that $X \oplus \ell_p(E_n)$ has an (FDD). For $x \in X$

$$T_n x - S_n x = (0, 0, \ldots, 0, (T_n - T_n)^2 x, 0, 0, \ldots)$$

where the only non-zero entry is in the position of E_n. Since $p > 1$ the last part follows easily.■

Let C_p denote the space $\ell_p(F_n)$ where F_n is a sequence of finite-dimensional Banach spaces dense in the Banach-Mazur sense in the collection of all finite-dimensional Banach spaces (we may assume each space is repeated infinitely often). This space has been studied extensively by Johnson and Zippin (see [16]); it is noted by Johnson [13] that $X \oplus C_p$ has an (FDD) if and only if $X \oplus C_p$ has a basis. The next Corollary combines the two preceding results.

COROLLARY 3.3. *Let X be a separable Banach space and suppose $1 < p < \infty$. Then X has (CBAP) if and only if $X \oplus C_p$ has a basis.*

We remark that Lusky [20] has shown that if C_∞ denotes $c_0(F_n)$ then X has (BAP) if and only if $X \oplus C_\infty$ has a basis. Let us note a brief proof of Lusky's theorem. Let T_n be an approximating sequence for X with $T_m T_n = T_n$ for $m > n$ and let $E_n = T_n(X)$. We define $S_n : X \oplus c_0(E_n) \to X \oplus c_0(E_n)$ by

$$S_n(x, y_1, y_2, \ldots) = (T_n x + y_n, z_1, z_2, \ldots)$$

where

$$z_k = T_k(I - T_n)x - T_k y_n + y_k$$

for $1 \leq k \leq n$ and $z_k = 0$ for $k \geq n+1$. Then $S_m S_n = S_{\min(m,n)}$ and so $X \oplus c_0(E_n)$ has an (FDD). Thus $X \oplus C_\infty$ has an (FDD) and hence also a basis.

We now apply the complementation results to give a renorming theorem. We require first the following lemma.

LEMMA 3.4. *Suppose X has an (FDD) with partial sum operators S_n, and suppose $0 < \alpha < 2$. Then X can be equivalently renormed so that $\|S_n\| = \|I - \alpha S_n\| = 1$ for all n.*

PROOF: It suffices to show that the semigroup of operators generated by S_n, $I - \alpha S_n$, $n \in \mathbb{N}$ is bounded. To do this it suffices to consider a product

$$T = \prod_{k=1}^{n} (I - \alpha S_{i_k})$$

where $i_1 \leq i_2 \leq \cdots \leq i_n$. We can rewrite $I - \alpha S_m = I - S_m + \beta S_m$ where $\beta = 1 - \alpha$. Then

$$T = (I - S_{i_n}) + \sum_{k=1}^{n-1} \beta^{n-k} (S_{i_k} - S_{i_{k-1}})$$

where S_{i_0} is defined to be zero. Thus

$$\|T\| \leq M + 1 + 2M|\beta|(1 - |\beta|)^{-1}$$

where $M = \sup_n \|S_n\|$.∎

THEOREM 3.5. *Suppose X has (CBAP) and $1 \leq \alpha < 2$. Then X can renormed so it has a commuting approximating sequence (T_n) with*

$$\lim_{n\to\infty} \|T_n\| = \lim_{n\to\infty} \|I - \alpha T_n\| = 1$$

and

$$\limsup_{n\to\infty} \|T_n - T_n^2\| \leq \frac{1}{4}.$$

In particular X can be renormed to have both (MAP) and (RMAP).

PROOF: We use Proposition 3.2. We can find an (FDD) of a space $X \oplus Y$ with partial sum projections S_n and a commuting approximating sequence T_n for X so that

$$\lim_{n\to\infty} \|\frac{1}{n} \sum_{k=1} S_n |x - T_n\| = 0.$$

Here we have replaced the original sequence T_n by its sequence of Cesaro means. By Lemma 3.4 we can renorm $X \oplus Y$ so that $\|S_n\| = \|I - \alpha S_n\| = 1$ for all n. Then under the same renorming restricted to X we easily get that the first equation holds for the sequence T_n. Further more if $R_n = \frac{1}{n}\sum_{k=1}^n S_k$,

$$\begin{aligned} R_n(I - R_n) &= \frac{1}{n^2} \sum_{i=1}^{n} \sum_{j=1}^{n} S_i(I - S_j) \\ &= \frac{1}{n^2} \sum_{j<i} (S_i - S_j) \\ &= \frac{1}{n^2} \sum_{i=1}^{n} (2i - n) S_i \\ &= \frac{1}{n^2} \sum_{2i<n} (n - 2i) S_{n-i}(I - S_i) \end{aligned}$$

so that

$$\|R_n(I - R_n)\| \leq \frac{1}{n^2} \sum_{2i<n} (n - 2i) \to \frac{1}{4}.$$

Thus $\limsup \|T_n - T_n^2\| \leq \frac{1}{4}$.∎

We now demonstrate the limits of this renorming by using a simple modification of an argument of Beauzamy [1] and Esterle [6].

PROPOSITION 3.6. *Suppose X is a Banach space and that T is a bounded operator on X. Suppose $\|T - T^2\| = \theta < \frac{1}{4}$. Then there is a projection P on X such that $\{x : Tx = x\} \subset P(X) \subset T(X)$ and*

$$\|P\| \leq \frac{1}{2}\left(1 + \frac{1 + 2\|T\|}{(1 - 4\theta)^{1/2}}\right).$$

PROOF: Define

$$S=\sum_{m=0}^{\infty}\binom{2m}{m}(T-T^2)^m$$

and

$$P=\frac{1}{2}(I-(I-2T)S).$$

Since $(1-4z)^{-\frac{1}{2}}$ has a power series expansion $\sum_{m=0}^{\infty}\binom{2m}{m}z^m$ valid for $|z|<\frac{1}{4}$, it is clear that $\|S\|\leq(1-4\theta)^{-\frac{1}{2}}$ and by a power series manipulation that $(I-2T)^2S^2=I$. Hence P is a projection on X and the estimate on $\|P\|$ follows. Note also that

$$P=3T^2-2T^3+\frac{1}{2}(I-2T)\sum_{m=2}^{\infty}\binom{2m}{m}(T-T^2)^m.$$

The remaining properties of P follow easily.∎

A result of Casazza [2] asserts that a separable Banach space has an (FDD) if and only if it has (CBAP) and the π-property. In view of Proposition 3.6 we obtain:

THEOREM 3.7. *Let X be a separable Banach space. Suppose X has an approximating sequence T_n for which* $\limsup_{n\to\infty}\|T_n-T_n^2\|<\frac{1}{4}$. *Then X has the π-property, and if X has (CBAP) then X has an (FDD).*

We remark that Read [22] gives an example of a reflexive Banach space with (CBAP) but having no (FDD). Thus this corollary shows that Read's space cannot be renormed to have an approximating sequence T_n for which $\limsup\|T_n-T_n^2\|<\frac{1}{4}$.

Motivated by Theorem 3.5 we introduce the *unconditional metric approximation property* (UMAP). We shall say that X has (UMAP) provided it has an approximating sequence T_n for which $\lim_{n\to\infty}\|I-2T_n\|=1$. The justification for this terminology lies in the following:

THEOREM 3.8. *A separable Banach space X has (UMAP) if and only if for every $\epsilon>0$ there exists an approximating sequence (T_n) so that if $A_n=T_n-T_{n-1}$ for $n\in\mathbb{N}$ (with $T_0=0$) then for every $N\in\mathbb{N}$ and $\eta_i=\pm1$, $i=1,2,\ldots,N$ then*

$$\|\sum_{i=1}^{N}\eta_iA_i\|\leq1+\epsilon.$$

PROOF: First suppose X has (UMAP), and $\epsilon>0$. Then X has an approximating sequence T_n for which $T_mT_n=T_n$ for $m>n$ and $\|I-2T_n\|=1+\delta_n$ where $\prod(1+\delta_n)<1+\epsilon$. Defining $A_n=T_n-T_{n-1}$ as above with $T_0=0$ we have for $N\in\mathbb{N}$ and $\eta_i=\pm1$,

$$\eta_N\prod_{i=1}^{N-1}(I-(1-\eta_{N-i}\eta_{N-i-1})T_{N-i})=\eta_N\prod_{i=1}^{N-1}(I-T_{N-i}+\eta_{N-i}\eta_{N-i-1}T_{N-i})$$

$$=\eta_N(I-T_{N-1})+\sum_{i=1}^{N-1}\eta_iA_i.$$

Thus

$$\begin{aligned}\|\sum_{i=1}^{N}\eta_iA_i\| &= \|T_N\prod_{i=1}^{N-1}(I-(1-\eta_{N-i}\eta_{N-i-1})T_{N-i})\| \\ &\leq (1+\frac{1}{2}\delta_N)\prod_{i=1}^{N-1}(1+\delta_i) \\ &\leq 1+\epsilon.\end{aligned}$$

For the converse direction suppose (T_n) is an approximating sequence for which for every N and $\eta_i = \pm 1$ we have

$$\|\sum_{i=1}^{N}\eta_iA_i\| \leq 1+\epsilon.$$

Then for any n and $m > n$

$$T_m - 2T_n = \sum_{i=n+1}^{m} A_i - \sum_{i=1}^{n} A_i$$

so that $\|I - 2T_n\| \leq \liminf_{m\to\infty}\|T_m - 2T_n\| \leq 1+\epsilon$. It follows easily that X has an approximating sequence S_n for which $\|I - 2S_n\| \to 1$.∎

Let us say that X has $UCMAP$ if for every $\epsilon > 0$ there is a commuting approximating sequence T_n for which if $A_n = T_n - T_{n-1}$ and $\eta_i = \pm 1$, $i = 1, 2, \ldots, N$ then $\|\sum_{i=1}^{N}\eta_iA_i\| \leq 1+\epsilon$.

We also introduce the notion of *u-ideal*. We say that if X is a subspace of Y then X is a u-ideal in Y if there is a projection Π of Y^* onto $X^\perp$ satisfying $\|I - 2\Pi\| = 1$. Equivalently there is a complementary subspace M for $X^\perp$ such that if $\phi \in M$ and $\psi \in X^\perp$ then $\|\phi + \psi\| = \|\phi - \psi\|$. For example, if X is an M-ideal in Y then X is a u-ideal. The following result is suggested by results of Feder [7], Godefroy-P. Saab [9] and Godefroy-Li [8] and Li [18]. We are grateful to Gilles Godefroy for correcting an error in the original proof of Theorem 3.9.

THEOREM 3.9. *Let X be a separable reflexive Banach space with the approximation property. Then the following conditions are equivalent:*
(i) X has (UMAP).
(ii) X has (UCMAP).
(iii) X is isometric to a subspace of a Banach space with (UMAP).
(iv) $\mathcal{K}(X)$ is a u-ideal in $L(X)$.

PROOF: Plainly we have $(ii) \Rightarrow (i) \Rightarrow (iii)$. Let us prove $(iii) \Rightarrow (ii)$. Suppose X is (isometric to) a subspace of Y where Y has (UMAP). Since X is reflexive it has (MAP) and hence (CMAP). Let T_n be an approximating sequence for Y such that $\lim\|I - 2T_n\| = 1$ and let S_n be a commuting approximating sequence for X. Then $T_n - S_n \in \mathcal{K}(X, Y)$ and $T_n - S_n \to 0$ in the strong operator topology. Hence as X is reflexive, $T_n^* - S_n^* \to 0$ in the weak-operator topology and thus [17] $T_n - S_n \to 0$ weakly. It follows that we can find $V_n \in \text{co}(T_n, T_{n+1}, \ldots)$ and $R_n \in \text{co}(S_n, S_{n+1}, \ldots)$

so that $\|R_n - V_n\| \to 0$. Then R_n is a commuting approximating sequence for which $\|I - 2R_n\| \to 1$ and the argument of Theorem 3.8 shows that X has (UCMAP).

Now let us show $(i) \Rightarrow (iv)$. Let T_n be an approximating sequence for X such that $\lim \|I - 2T_n\| = 1$. For $\phi \in \mathcal{L}(X)^*$ we define $\Pi(\phi) \in \mathcal{L}(X)^*$ by $\Pi(\phi)(A) = \lim_{\mathcal{U}} \phi(A - T_n A)$ where $\mathcal{U}$ is a non-principal ultrafilter on $\mathbf{N}$. Clearly $(\phi - 2\Pi(\phi))(A) = \lim_{n\in\mathcal{U}} \phi((2T_n - I)A)$ so that $\|I - 2\Pi\| = 1$. It is also easy to check that Π is a projection onto $\mathcal{K}(X)^{\perp}$.

Conversely, to show $(iv) \Rightarrow (i)$, suppose $\Pi : \mathcal{L}(X)^* \to \mathcal{K}(X)^{\perp}$ is a projection satisfying $\|I - 2\Pi\| = 1$. Since X is reflexive with (AP) we can identify $\mathcal{L}(X)$ with $\mathcal{K}(X)^{**}$; thus if $j : \mathcal{K}(X) \to \mathcal{L}(X)$ denotes the natural inclusion, then j^* induces a projection of $\mathcal{L}(X)^*$ onto its subspace M of weak* continuous linear functionals. Now, according to a result of Godefroy and Saphar [10], Corollary 5.4, $\mathcal{K}(X)$ has the unique extension property (U.E.P.) so that j^* is the unique projection of norm one on $\mathcal{L}(X)^*$ with kernel $\mathcal{K}(X)^{\perp}$ (see the remark on p. 681 of [10]). It follows that $j^* = I - \Pi$.

Now since X has (AP) it also has (BAP) and so there is an approximating sequence (T_n). It will suffice to show that for any n, $\inf \|I - 2A\| = 1$ where A runs through all convex combinations of $(T_n, T_{n+1}, \ldots)$. If this fails, then by the Hahn-Banach theorem there exists $\phi \in \mathcal{L}(X)^*$ with $\|\phi\| = 1$, and $\phi(I - 2T_n) \geq 1 + \delta$ where $\delta > 0$. Then since $T_n \to I$ for the weak*-topology, we have $j^*\phi(I - 2T_n) \to -j^*\phi(I)$. However $\Pi(\phi)(I - 2T_n) = \Pi(\phi)(I)$. Hence $\lim \phi(I - 2T_n) = (\Pi - j^*)(\phi)(I) = (2\Pi(\phi) - \phi)(I)$. This is a contradiction since $\|\phi - 2\Pi(\phi)\| = \|\phi\| = 1$.∎

4. Concluding remarks.

We first make some comments on the example of Read [22]. Read shows that there is a subspace X of C_2 so that X has (BAP) but no (FDD). This answers a question raised by Johnson and Zippin [16], (cf. also [19], p. 86) as to whether every subspace of C_2 with the approximation property is of the form $\ell_2(E_n)$ with E_n finite-dimensional. Clearly X is reflexive and has (CBAP) and even (UCMAP) by Theorem 3.9. It follows by Corollary 3.3 that $X \oplus C_2$ has a basis. Hence by the results of Johnson-Zippin [16] ([19] p.85), $X \oplus C_2$ is of the form $\ell_2(E_n)$ and hence $X \oplus C_2$ is isomorphic to C_2. Hence X is isomorphic to a complemented subspace of C_2. Thus we have an example of a complemented subspace of a space with a (UFDD) which fails to have an (FDD).

The major remaining open problem here seems to be whether (BAP) implies (CBAP). Let us note that this is equivalent to the problem of whether a π−space has an (FDD). Indeed if X has (BAP) but not (CBAP) then for any $1 < p < \infty$ $X \oplus C_p$ is a π−space (cf. Johnson [13]) but must fail (CBAP) by Theorem 3.1. For the converse, Casazza [2] shows that a π−space with (CBAP) has an (FDD). Yet another form of this problem [13] is whether for any X with (BAP) $X \oplus C_p$ has a basis (when $1 < p < \infty$.) A related problem is whether (UnAP) implies (CBAP).

We also may raise the question of whether (UnAP) and (CBAP) together will imply the existence of a commuting approximating family (T_n) for which the series $\sum(T_n - T_{n-1})$ is weakly unconditionally Cauchy. In a similar vein, does (UMAP) imply (UCMAP) in general? This is proved for reflexive spaces in Theorem 3.9; obviously (UMAP) implies (CBAP).

References.

1. B. Beauzamy, *Introduction to operator theory and invariant subspaces,* North Holland, 1989.
2. P. G. Casazza, The commuting B.A.P. for Banach spaces, *Analysis at Urbana II,* (E. Berkson, N.T. Peck and J.J. Uhl, editors) London Math. Soc. Lecture Notes, 138 (1989) 108-127.
3. C.-M. Cho and W.B. Johnson, A characterization of subspaces X of ℓ_p for which $K(X)$ is an M-ideal in $L(X)$, Proc. Amer. Math. Soc. 93 (1985) 466-470.
4. R.S. Doran and J. Wichmann, *Approximate identities and factorization in Banach modules,* Springer Lecture Notes 768, Berlin 1979
5. P. Enflo, A counterexample to the approximation property in Banach spaces, Acta Math. 130 (1973) 309-317.
6. J. Esterle, Quasi-multipliers, representations of H^∞, and the closed ideal problem for commutative Banach algebras, Springer Lecture Notes No. 975, (1983) 66-162
7. M. Feder, On subspaces of spaces with an unconditional basis and spaces of operators, Illinois J. Math. 24 (1980) 196-205.
8. G. Godefroy and D. Li, Banach spaces which are M-ideals in their biduals have property (u), Ann. Inst. Fourier (Grenoble) 39 (1989) 361-371.
9. G. Godefroy and P. Saab, Weakly unconditionally converging series in M-ideals, Math. Scand. 64 (1989) 307-318.
10. G. Godefroy and P.D. Saphar, Duality in spaces of operators and smooth norms on Banach spaces, Illinois J. Math. 32 (1988) 672-695.
11. A. Grothendieck, *Produits tensoriels topologiques et espaces nucleaires,* Memoir Amer. Math. Soc. No. 16, 1955.
12. W.B. Johnson, Finite-dimensional Schauder decompositions in π_λ and dual π_λ spaces, Illinois J. Math. 14 (1970) 642-647.
13. W.B. Johnson, Factoring compact operators, Israel J. Math. 9 (1971) 337-345.
14. W.B. Johnson, A complementably universal conjugate Banach space and its relation to the approximation property, Israel J. Math. 13 (1972) 301-310.
15. W.B. Johnson, H.P. Rosenthal and M. Zippin, On bases, finite-dimensional decompositions and weaker structures in Banach spaces, Israel J. Math. 9 (1971) 488-506.
16. W.B. Johnson and M. Zippin, On subspaces and quotients of $(\sum G_n)_{\ell_p}$ and $(\sum G_n)_{c_0}$, Israel J. Math. 13 (1972) 311-316.
17. N.J. Kalton, Spaces of compact operators, Math. Ann. 208 (1974) 267-278.
18. D. Li, Quantitative unconditionality of Banach spaces E for which $\mathcal{K}(E)$ is an M-ideal in $\mathcal{L}(E)$, Studia Math 96 (1990) 39-50.
19. J. Lindenstrauss and L. Tzafriri, *Classical Banach spaces I: sequence spaces* Springer, Berlin-Heidelberg-New York 1977.
20. W. Lusky, A note on Banach spaces containing c_0 or C_∞, J. Functional Analysis 62 (1985) 1-7.
21. A. Pelczynski, Any separable Banach space with the bounded approximation property is a complemented subspace of a Banach space with a basis, Studia Math. 40 (1971) 239-242.
22. C.J. Read, Different forms of the approximation property, to appear

23. A.M. Sinclair, Bounded approximate identities, factorization and a convolution algebra, J. Functional Analysis, 29 (1978) 308-318.
24. S.J. Szarek, A Banach space without a basis which has the bounded approximation property, Acta Math. 159 (1987) 81-98.

Moduli of Complex Convexity

WILLIAM J. DAVIS

Department of Mathematics
The Ohio State University
Columbus, OH 43210, USA

Abstract. In our original paper on complex uniform convexity, the author, D. J. H. Garling and N. Tomczak-Jaegermann introduced several possible moduli of complex convexity. The only moduli studied in that paper were the $H_p(\epsilon)$'s. Remarks were made concerning two other moduli, the four point modulus and the tangential modulus. No proofs were given for the assertions about these moduli which appeared in that paper. Here we remedy that situation by giving those proofs.

In **[D-G-J]**, a study of complex notions of uniform convexity was begun. In particular, a (complex) quasi-normed linear space E is said to be $p-uniformly-PLconvex$ $(0 \leq p \leq \infty)$ if the moduli

$$H_p(\epsilon) = \inf\{(\frac{1}{2\pi}\int_0^{2\pi} ||x + e^{i\theta}y||^p d\theta)^{\frac{1}{p}} - 1 : ||x|| = 1, ||y|| = \epsilon\}$$

for $p < \infty$, (respectively,

$$H_\infty(\epsilon) = \inf\{\sup_\theta ||x + e^{i\theta}y|| - 1 : ||x|| = 1, ||y|| = \epsilon\}$$

for $p = \infty$) are positive for positive ϵ. It was then shown, for example, that for p in the range $0 < p < \infty$, these moduli are equivalent, and so the space E is said to be $uniformlyPL-convex$ if it is $p-uniformly-PL-convex$ for any p in $(0, \infty)$, and $uniformlyH_\infty - convex$ in the last case. The study began with the examination of the results of J. Globevnik **[Gl]** in which, rather than examining moduli defined on complex circles of the form $\{x + e^{i\theta}y : \theta \in \mathbf{T}\}$, the moduli were defined using only the four points $\{x + i^k y : k = 0, 1, 2, 3\}$. We shall show first that the moduli defined in this way are equivalent to the H_p moduli above.

In the case of *real* Banach spaces, and *real* moduli of convexity, it is natural to consider the moduli defined by line segments tangent to the ball of the space as well as the usual definitions using mid points of segments whose endpoints lie on the sphere. In the complex case, it is impossible in general to ask that all of the points $x + e^{i\theta}y$ lie on the surface of the sphere. However, it is possible to examine moduli of complex convexity defined by pairs x, y, as above, with the added restriction that the vector y is tangent to the sphere at the point x. The interesting thing about these moduli is that they are all equivalent in the range $0 < p \leq \infty$ and that they are all equivalent not to H_p with p finite, but with H_∞.

One of the problems with the concepts of complex uniform convexity defined in [**D-G-J**] has been the difficulty of establishing ties, with that definition, with other concepts such as the analytic Radon-Nikodym property, with complex martingale convergence theorems and with renorming theorems in the spirit of the Enflo-Pisier [**E,P**] theorems in the real case. Recent work on such questions has led to exciting advances, mainly in the direction of complex martingales such as the Hardy martingales of Garling [**G**] (see also the work of Xu [**X**]). There are, for example, in [**X**], more severe moduli defined which might also be compared with those of [**D-G-J**] as given above. We shall see easily that they are also comparable in the same way with their four-point and tangential counterparts.In looking at them we may find the essential difference between uniform PL-convexity and Hardy convexity. Throughout this note, we shall assume that the spaces involved are infinite dimensional, or at least of dimension large enough to render our definitions meaningful.

Definitions: Here we recall the relevant definitions of the moduli we shall compare. First, for the four-point moduli, we define

$$C_p(\epsilon) = \inf\{(\frac{1}{4}\sum_{k=0}^{3} ||x + i^k y||^p)^{\frac{1}{p}} - 1 : ||x|| = 1, ||y|| = \epsilon\}$$

for $0 < p < \infty$, and with the obvious change for $p = \infty$. For the tangential moduli, we define $T_p(\epsilon)$ exactly as we defined $H_p(\epsilon)$, except that we restrict y's that are used to being tangent to the ball of E at x.

Before we define the analogous moduli for the Hardy versions of complex convexity, let us recall some definitions. If we let $\mathbf{T}$ denote the unit circle in the complex plane with its Haar measure, σ, and if we denote by $\mathbf{T}^{\mathbf{N}}$ the product of countably many copies of the circle, we can define *Analytic Martingales* :

$$f_n(\Theta) = \sum_{k=0}^{n} \phi_n(\Theta) e^{i\theta_n},$$

where $\phi_n(\Theta) = \phi_n(\theta_1, \theta_2, \ldots, \theta_{n-1})$. Notice that these are nothing more than martingale transforms of the Steinhaus sequence $\{e^{i\theta_n}\}$. Due to inherent difficulties of proving convergence of such martingales, and of relating the boundedness of them to the complex geometric structure, Garling [**G**] defined a broader class of such martingales, the *Hardy Martingales* as

$$h_n(\Theta) = \sum_{k=0}^{n} \Psi_n(\Theta),$$

where now

$$\Psi_n(\Theta) = \sum_{k \in \mathbf{N}} \psi_k(\theta_1, \ldots, \theta_{n-1}) e^{ik\theta_n}.$$

The corresponding complex moduli are then defined as

$$h_1(\epsilon) = \inf\{(\int_{\mathbf{T}} ||x + P(\theta)|| d\sigma) - 1 : ||x|| = 1, \int_{\mathbf{T}} ||P(\theta)|| d\sigma \geq \epsilon\},$$

where $P(\theta) = \sum_{n\geq 1} x_n e^{in\theta}$ is a trigonometric polynomial with values in E. Then, as space E is called *uniformly H_p convex* if these moduli are positive for positive ϵ. The advantage of this stronger definition is that blocking techniques may be used with the Hardy martingales to establish the links with the geometric theory which were missing in the previous instances.

As with the H_p moduli, we define the analogous four-point and tangential moduli, now called $c_p(\epsilon)$ and $t_p(\epsilon)$ in exact correspondence with these new *Hardy* moduli. At this time, we do not see the equivalence with their capitalized counterparts. Part of the difference lies in the fact that in the equivalence of, for example, C_p with H_p, only convexity is used. In attempting to establish equivalence for c_p and h_p, it appears that subharmonicity must be used, and similar arguments must fail due to the fact that purely atomic measures cannot be Jensen measures.

Proofs of the equivalence. In this section, we shall prove the equivalences announced above for the moduli T_p and C_p. We shall also indicate the changes in proof necessary to handle their lower-case counterparts.

THEOREM 1. *Let E be a Banach space. For $0 < p \leq \infty$, the moduli $T_p(\epsilon)$ are equivalent. Further, they are all equivalent to the modulus H_∞.*

PROOF: Let $x, y \in E, ||x|| = 1, ||y|| = \epsilon$. Since $\frac{d\theta}{2\pi = d\sigma}$ is a probability measure, it is clear that $(\int_{\mathsf{T}} ||x + e^{i\theta}y||^p d\sigma)^{\frac{1}{p}} \geq (\int_{\mathsf{T}} ||x + e^{i\theta}y||^q d\sigma)^{\frac{1}{q}}$ if $p \geq q$. Thus we need only demonstrate the equivalence for the L_∞ and L_0 metrics. We will show that $\sigma\{\theta : ||x + e^{i\theta}y|| \geq 1 + \Delta\} \geq \frac{1}{3}$, where $1 + \Delta = \sup_\theta ||x + e^{i\theta}y/2||$. Assume that Δ is assumed at $\theta = 0$, and consider all the line segments through $x + y/2$ which have one endpoint on the circle $\{x + e^{i\theta}y/2\}$ and the other on the larger circle, $\{x + e^{i\phi}y\}$. For each such segment, we have $||x + e^{i\theta}y/2|| \leq 1 + \Delta$, so by convexity of the norm, we also have $||x + e^{i\phi}y|| \geq 1 + \Delta$. As we let θ run through all possible values, the set of ϕ's we run through on the larger circle is $(-\frac{\pi}{3}, \frac{\pi}{3})$. Since y is tangent to the sphere at x, we have automatically that $||x + e^{i\phi}y|| \geq 1$ for all ϕ. This completes the proof of the first part, since it now follows easily that for all $\epsilon > 0$,

$$T_\infty(\epsilon) \geq T_p(\epsilon) \geq AT_\infty(\frac{\epsilon}{2})$$

for all $p \in (0, \infty)$.

Now we must check the equivalence of the T_p's with H_∞. First, we must recall that $H_\infty(\epsilon) = o(\epsilon)$ in infinite dimensional spaces. Select $x, y \in E$ with $||x|| = 1, ||y|| = \epsilon$ such that $||x + y|| = sup_\theta ||x + e^{i\theta}y|| \leq 1 + 2H_\infty(\epsilon)$. Let $f \in E^*, ||f|| = 1 = f(x)$. Then we have that $||x + y|| \geq 1 + |f(y)|$, so that $|f(y)| \leq 2H_\infty(\epsilon) \leq \eta\epsilon$, where η's smallness will become apparent. In the linear span of x and y in E, let $x + u(\theta)$ denote the intersections of the lines through $x + e^{i\theta}y$ with the (complex) line $[f = 1]$. If η is small enough, we have $\frac{\epsilon}{2} \leq ||u(\theta)|| \leq 2\epsilon$. We see, then, that for all θ, $||x + u(\theta)|| \leq ||x + e^{i\theta}y||(1 + 2H_\infty(\epsilon)) \leq (1 + 2H_\infty(\epsilon))^2$. It follows that $T_\infty(\frac{\epsilon}{2}) \leq 5H_\infty(\epsilon)$. □

Now we move to the equivalences of the moduli H_p and C_p.

THEOREM 2. *For $1 \leq p \leq \infty$, we have $H_p(\epsilon) \geq C_p(\epsilon) \geq H_p(\frac{\epsilon}{\sqrt{2}})$.*

PROOF: The case of $p = \infty$ is rather easy: Let $||x|| = 1, ||y|| = \epsilon$ and consider the circles $\{x + e^{i\theta}y\}, \{x + \frac{1}{\sqrt{2}}e^{i\theta}y\}$ and the square $co\{x + i^k y : k = 0, 1, 2, 3\}$. Clearly we have $\sup_\theta ||x + e^{i\theta}y|| \geq \max_k ||x + i^k y||$, and simple convexity gives us that $\max_k ||x + i^k y|| \geq \sup_\theta ||x + \frac{1}{\sqrt{2}}e^{i\theta}y||$. Letting x and y range over all suitable choices proves the equivalence of H_∞ and C_∞.

The case for $1 \leq p < \infty$ is true for the same reasons, but demands some further explanation. It is easiest to see for the case $p = 1$, and is not essentially different for the other values of p, so we present that case only. Again, let x and y be suitably chosen. Then,

$$\int_{\mathbf{T}} ||x + e^{i\theta}y||d\sigma \geq \int_{\mathbf{T}} [\frac{1}{4}\sum_{k=0}^{3} ||x + i^k e^{i\theta}y||]d\sigma \geq \min_\theta \frac{1}{4}\sum_{k=0}^{3} ||x + i^k e^{i\theta}y||.$$

It follows that $H_1(\epsilon) \geq C_1(\epsilon)$. For the second part, notice that each point on the circle $\{x+\frac{1}{\sqrt{2}}e^{i\theta}y\}$ can be written as a convex combination of the four points $\{x+i^k y : 0 \leq k \leq 3\}$ as follows:

$$x + \frac{1}{\sqrt{2}}e^{i\theta}y = \sum_{k=0}^{3} \lambda_k(\theta)(x + i^k y).$$

Thus,

$$\int_{\mathbf{T}} ||x + e^{i\theta}y||d\sigma \leq \sum_{k=0}^{3} \int_{\mathbf{T}} \lambda_k(\theta)||x + i^k y||d\sigma.$$

By symmetry, and the fact that $\sum \lambda_k(\theta) = 1$, we get

$$\int_{\mathbf{T}} \lambda_k(\theta)d\sigma = \frac{1}{4}, \quad 0 \leq k \leq 3$$

which proves that $H_1(\frac{\epsilon}{\sqrt{2}}) \leq C_1(\epsilon)$.

□

Comments on the Hardy moduli: Let us look at the Hardy modulus, $h_1(\epsilon)$. In its definition, the infimum is taken over trigonometric polynomials of the form $p(\theta) = \sum_{n\geq 1} x_n e^{in\theta}$ with $||p||_{L_1(\mathbf{T})} \geq \epsilon$. A key element of the proof of the equivalence of C_p with H_p was the fact that convexity allows us to estimate $||x + re^{i\theta}y||_{L_1^E(\mathbf{T})}$ above by $||x + e^{i\theta}y||_{L_1(\mu)}$ for any of the four point supported probability measures one gets by rotating the natural one. Since no four point measure can be a Jensen measure, it is clear that one cannot extend that equivalence proof as it stands to the case of the Hardy moduli.

The question about the equivalence of the various definitions of complex uniform convexity remains open. It may very well be that the key to establishing that they are different lies in the above sort of observation. This line of investigation is worthy of pursuit.

REFERENCES

[D-G-J] W. J. Davis, D. J. H. Garling and N. Tomczak-Jaegermann, *The complex convexity of quasi-normed spaces*, J. Funct. Anal. **55** (1984), 110 - 150.

[E] Per Enflo, *Banach spaces which can be given an equivalent uniformly convex norm*, Israel J. Math. **13** (1972), 281 -288.

[G] D. J. H. Garling, *On martingales with values in a complex Banach space*, Proc. Camb. Phil. Soc. (1988).

[Gl] J. Globevnik, *On complex strict and uniform convexity*, Proc. AMS **47** (1975), 175 - 178.

[P] G. Pisier, *Martingales with values in uniformly convex spaces*, Israel J. Math. **20** (1975), 326 -350.

[X] Xu Quanhua, *Inégalités pour les martingales de Hardy et renomage des espaces quasi-normés*, C. R. Acad. Paris **306** (1988), 601 -604.

Grothendieck type inequalities
and
weak Hilbert spaces

Martin Defant
Marius Junge

Summary: Grothendieck's inequality characterizes operators which factor through a Hilbert space. We prove modified Grothendieck type inequalities for hilbertian operators in the sense of Pietsch. In particular, every operator from l_1 into a weak Hilbert space is (r,1)-summing for every $1 < r < \infty$. Classical results of Kwapień and Pelczyński can be deduced from this concept.

Introduction: In the theory of summing operators one of the most important theorems is the following inequality due to Grothendieck (1956):

Let $n \in \mathbb{N}$ and $A = (a_{ij})_{i,j=1}^n$ be a matrix such that for all sequeunces of scalars $(s_i)_{i=1}^n$ and $(t_j)_{j=1}^n$

$$\left| \sum_{i,j=1}^n a_{ij} \, s_i \, t_j \right| \le \sup_{i=1,\dots,n} |s_i| \sup_{j=1,\dots,n} |t_j| \, .$$

Then for all Hilbert spaces (H, $(\circ . \circ)$) and seqeunces $(x_i)_{i=1}^n \subset H$ and $(y_j)_{j=1}^n \subset H$

$$\left| \sum_{i,j=1}^n a_{ij} \, (x_i \, . \, y_j) \right| \le K_G \sup_{i=1,\dots,n} \|x_i\| \sup_{j=1,\dots,n} \|y_j\| \, ,$$

where the smallest possible constant K_G is called Grothendieck constant.

In the language of linear bounded operators this reads as follows:

For an operator T acting between to Banach spaces X an Y the following conditions are equivalent.

1) T factors through a Hilbert space.
2) There is a constant $c \ge 0$ such that for all $n \in \mathbb{N}$ and each matrix A: $l_\infty^n \to l_1^n$
   $\| A \otimes T: l_\infty^n(X) \to l_1^n(Y) \| \le c \, \| A: l_\infty^n \to l_1^n \|$.
3) For every operator $S: l_1 \to X$ the composition TS is 1-summing.

Here $A\otimes T$ denotes the tensor produkt of A and T . For more precise definitions see the prelimimaries below.

The aim of our paper is to apply Grothendieck's inequality to a larger class of operators than those factorizing through a Hilbert space. Hence the above conditions 2) and 3) have to be modified appropriately.

In chapter 1 we investigate the following conditions which we call Grothendieck type inequalities (theorem1.4.).

For this let $1 \le r \le 2 \le p \le \infty$ *with* $1/r = 1/p + 1/2$. *Then for an operator* $T: X \to Y$ *the following conditions are equivalent.*

1) *For every operator* $u: Z \to X$ *whose dual operator* u^* *is 2-summing the composition* Tu *is* $(p,2)$*-summing.*
2) *There is a constant* $c \ge 0$ *such that for all* $n \in \mathbb{N}$ *and each matrix* $A: l_\infty^n \to l_1^n$
   $\| A\otimes T: l_\infty^n(X) \to l_r^n(Y) \| \le c \, \| A: l_\infty^n \to l_1^n \|$.
3) *For every operator* $S: l_1 \to X$ *the compostion* TS *is* $(r,1)$*-summing.*

Note that by a result of Kwapień [KW2] condition 1) for p=1 and r=2 means that T factors through a Hilbert space (however, Kwapień's result is basic for this paper).

In chapter 2 we study the relation of this Grothendieck type inequalities to the so-called hilbertian operators defined by Pietsch [PI2]. They are a generalization of operators factorizing through a Hilbert space and weak Hilbert operators defined by Pisier [PS2]. This concept turns out to be useful to apply the above inequalities to this larger class of operators.

For $1 \le r,s \le \infty$ an operator $T: X \to Y$ is said to be (r,s)-*hilbertian* ($T \in \mathfrak{H}_{r,s}(X,Y)$) if there is a constant $c \ge 0$ such that for all operators $u: l_2 \to X$ whose dual operator u^* is 2-summing and for all 2-summing operator $v: Y \to l_2$

$$\Big(\sum_{k \in \mathbb{N}} (k^{1/r-1/s} a_k(vTu))^s \Big)^{1/s} \le c \; \pi_2(u^*) \; \pi_2(v) ,$$

where a_k denotes the k-th approximation number. In this case $h_{r,s}(T)$ is the infimum taken over all constants c satisfying the inequality above.
($(\mathfrak{H}_r , h_r) := (\mathfrak{H}_{r,r} , h_{r,r})$, the r-hilbertian operators)

We prove that for every r-hilbertian operator $T: X \to Y$ *and for every operator* $S: l_1 \to X$ *the composition is* TS $(r,1)$*-summing.*

Since for every $1 \le r \le 2 \le p \le \infty$ with $1/r = 1/p + 1/2$ the spaces l_p and $l_{p'}$ $(1/p + 1/p' = 1)$ are r-hilbertian [PI2] we obtain the well-known result of Kwapień [KW1] that every operator $S: l_1 \to l_p$ or $S: l_1 \to l_{p'}$ is (r,1)-summing.

The class of $(1,\infty)$-hilbertian operators is just the class of weak Hilbert operators. *We show that for every weak Hilbert operator $T: X \to Y$, every operator $S: l_1 \to X$ and $1 < r \le 2$ the composition TS is $(r,1)$-summing and*

$$(*) \qquad \pi_{r,1}(TS) \le K_G \frac{r}{r-1} \, h_{1,\infty}(T) \, \|S\| .$$

In particular, the sum operator

$$\Sigma : l_1 \to l_\infty , \ (\alpha_n)_{n \in \mathbb{N}} \mapsto \Big(\sum_{k=1}^{n} \alpha_k \Big)_{n \in \mathbb{N}}$$

is (p,q)-summing for all $1 \le q < p < \infty$, which was proved by Kwapień and Pelczyński [KWP].

In this context it is natural to ask whether (*) characterizes weak Hilbert spaces (i.e. id_X is a weak Hilbert operator). This we discuss in chapter 3 and we show that the interpolation space $A_{1/2,2}$ of Pisier and Xu [PSX] satisfies (*) but is not a weak Hilbert space.

Preliminaries: We use standard Banach space notations . Standard references on operator ideals is the monograph of Pietsch [PI1]. The ideal of linear bounded, finite rank operators are denoted by $\mathfrak{L}, \mathfrak{F}$, respectively.

Let (A,α) be a quasi-Banach ideal. The component $A^*(X,Y)$ of the conjugate ideal (A^*,α^*) is the class of all operators $T \in \mathfrak{L}(X,Y)$ such that

$$\alpha^*(T) := \sup \{ |tr(TS)| \mid S \in \mathfrak{F}(Y,X), \ \alpha(S) \le 1 \} .$$

The component $A^d(X,Y)$ of the dual ideal (A^d,α^d) is the class of all operators $T \in \mathfrak{L}(X,Y)$ such that $T^* \in A(Y^*,X^*)$. We set

$$\alpha^d(T) := \alpha(T) .$$

The Lorentz sequence space $(l_{p,q}, \|\circ\|_{p,q})$, $0 < p,q \le \infty$ are defined in the usual way. In particular,

$$\alpha \in l_{1,\infty} \quad \text{iff} \quad \sup_{k\in\mathbb{N}} k\, \alpha_k^* < \infty ,$$

$$\alpha \in l_{\infty,1} \quad \text{iff} \quad \sum_{k\in\mathbb{N}} \frac{\alpha_k^*}{k} < \infty ,$$

where $\alpha^* = (\alpha_k^*)_{k\in\mathbb{N}}$ denotes the non increasing rearrangement of $\alpha \in l_\infty$.
Let $(e_k)_k$ be the sequence of unit vectors in $l_{p,q}$ ($l_{p,q}^n$).

For a Banach space X the vector-valued Lorentz sequence space ($l_{p,q}(X)$, $\|\circ\|_{p,q}$) consists of all sequences $x = (x_k)_{k\in\mathbb{N}} \subset X$ such that

$$\|x\|_{p,q} := \| (\|x_k\|)_{k\in\mathbb{N}} \|_{p,q} < \infty .$$

For all $n\in\mathbb{N}$ the spaces $l_{p,q}^n$, $l_{p,q}^n(X)$ are the subspaces of $l_{p,q}$, $l_{p,q}(X)$, respectively, of sequences of length n.

Let $n\in\mathbb{N}$, $A = (a_{ij})_{i,j=1}^n \in \mathbb{K}^{n\times n}$ a matrix and $T \in \mathfrak{L}(X,Y)$. Then the operator

$$A\otimes T : l_p^n(X) \to l_q^n(Y)$$

is for $x = (x_k)_{k=1}^n = \sum_{k=1}^n e_k \otimes x_k \in l_p^n(X)$ defined by

$$A\otimes T\ (x) := \sum_{k=1}^n A(e_k) \otimes Tx_k = \Big(\sum_{k=1}^n a_{ik}\, Tx_k \Big)_{i=1}^n \in l_q^n(Y) .$$

For $n\in\mathbb{N}$ the n-th approximation number of an operator $T \in \mathfrak{L}(X,Y)$ is given by

$$a_n(T) := \inf \{ \|T-S\| \mid S \in \mathfrak{F}(X,Y) ,\ \mathrm{rank}(S) < n \}.$$

For a Hilbert space H and $0 < p,q \le \infty$ the Schatten-von Neumann class $\mathfrak{S}_{p,q}(H,H)$ consists of all operators $T \in \mathfrak{L}(H,H)$ such that

$$\sigma_{p,q}(T) := \| (a_k(T))_{k\in\mathbb{N}} \|_{p,q} < \infty .$$

Note that for $1 \le p < \infty$ and $1 < p' \le \infty$ with $1/p + 1/p' = 1$

(1) $\quad \mathfrak{S}_p^*(H,H) = \mathfrak{S}_{p'}(H,H)$ with equal norms.

For $1 \le q \le p < \infty$ an operator $T \in \mathfrak{L}(X,Y)$ is said to be (p,q)-summing ($T \in \Pi_{p,q}(X,Y)$) if there is a constant $c \ge 0$ such that for all $n\in\mathbb{N}$, $(x_k)_{k=1}^n \subset X$

$$\Big(\sum_{k=1}^n \|Tx_k\|^p \Big)^{1/p} \le c \sup_{x^*\in B_{X^*}} \Big(\sum_{k=1}^n |<x_k , x^*>|^q \Big)^{1/q} .$$

We denote by $\pi_{p,q}(T) := \inf c$, where the infimum is taken over all c such that the above ineqality holds.

The following natural inclusion were proved by Kwapień [KW1].
Let $1 \le p \le r < \infty$, $1 \le q \le s < \infty$ with $1/q - 1/p \le 1/s - 1/r$, then

(2) $\Pi_{p,q} \subset \Pi_{r,s}$ and $\pi_{r,s} \le \pi_{p,q}$.

We set $(\Pi_r , \pi_r) := (\Pi_{r,r} , \pi_{r,r})$, the r-summing operators. Recall that in this case $\Pi_2^* = \Pi_2$ with equal norms.

The key result of 2-summing operators is the following result of Pietsch [PI1].
An operator $T \in \mathfrak{L}(X,Y)$ is 2-summing iff there are a probability measure μ on B_{X^*} and a constant $c \ge 0$ such that for all $x \in X$

(3) $$\|Tx\| \le c \left(\int_{B_{X^*}} |\langle x , x^* \rangle|^2 \, d\mu(x^*) \right)^{1/2} .$$

In this case $\pi_2(T) = \inf c$, where the infimum is taken over all c satisfying the above inequality (and is even obtained as a minimum). One may suppose μ to have its support in the $\sigma(X , X^*)$- closure of the extreme points of B_{X^*}.

A consequence is the factorization theorem of Pietsch.
An operator $T \in \mathfrak{L}(X,Y)$ is 2-summing iff there is a compact space K, a probability measure μ on K, an isometry $I \in \mathfrak{L}(X , C(K))$ and an operator $T \in \mathfrak{L}(L_2(K, \mu), Y)$ with $\|S\| = \pi_2(T)$ such that

(4) $T = SJI$,

where J denotes the canonical mapping from C(K) into $L_2(K, \mu)$.

Note that for all $n \in \mathbb{N}$ and for each operator $v \in \mathfrak{L}(X , \ell_2^n)$

(5) $$\pi_2(v) \le \left(\sum_{k=1}^n \|v^*(e_k)\|^2 \right)^{1/2} .$$

A result of Tomczak-Jaegermann [TOJ] states that for $n \in \mathbb{N}$ an operator $T \in \mathfrak{L}(X,Y)$ with $\mathrm{rank}(T) \le n$ only n vectors are needed to compute the 2-summing norm. More precisely,

(6) $\pi_2(T) \le 2 \, \pi_2^{(n)}(T)$,

where $$\pi_2^{(n)}(T) := \sup \left\{ \left(\sum_{k=1}^n \|Tx_k\|^2 \right)^{1/2} \Big| \sup_{x^* \in B_{X^*}} \left(\sum_{k=1}^n |\langle x_k , x^* \rangle|^2 \right)^{1/2} \right\} .$$

Let $(g_k)_{k \in \mathbb{N}}$ be a sequence of independent, normalized gaussian variables on a measure space (Ω, μ) . Then an operator $T \in \mathfrak{L}(X,Y)$ is said to be γ-summing ($T \in \Pi_\gamma(X,Y)$) if there is a constant $c \ge 0$ such that for all $n \in \mathbb{N}$, $(x_k)_{k=1}^n \subset X$

$$\Big\| \sum_{k=1}^n g_k \, Tx_k \Big\|_{L_2(\mu,X)} \le c \sup_{x^* \in B_{X^*}} \left(\sum_{k=1}^n |\langle x_k , x^* \rangle|^2 \right)^{1/2} .$$

We denote by $\pi_\gamma(T) := \inf c$, where the infimum is taken over all c such that the above inequality holds.

Note that all for $n \in \mathbb{N}$ and an operator $u \in \mathfrak{L}(\ l_2^n\ , X\ )$ the γ-summing norm can be computed as follows [LPI]

$$\pi_\gamma(u) = \| \sum_{k=1}^{n} g_k\ u(e_k) \|_{L_2(\mu,X)} .$$

By Pisier [PS2] a Banach space X is K-convex if there is a constant $c \geq 0$ such that for all $n \in \mathbb{N}$, $v \in \mathfrak{L}(X, l_2^n)$

(7) $\qquad \pi_\gamma(v^*) \leq c\ \pi_\gamma^*(v)$.

In this case $K(X) = \inf c$, where the infimum is taken over all c satisfying the above inequality.

Finally, we give a reformulation of the Grothendieck inequality which will be needed later.

For every $n \in \mathbb{N}$, $\varepsilon > 0$ and all operators $T \in \mathfrak{L}(l_\infty^n , l_1^n)$ there are opertors $R \in \mathfrak{L}(l_2^n , l_1^n)$, $S \in \mathfrak{L}(l_\infty^n , l_2^n)$ such that

(8) $\qquad T = RS \quad$ and $\quad \pi_2(S)\ \pi_2(R^*) \leq (1+\varepsilon)\ K_G\ \|T\|$,

where K_G denotes the Grothendieck constant.

1. Grothendieck type inequalities

The following lemma seems to be implicitely known. However, since it is basic for what follows we give a proof.

1.1. Lemma: *Let* $n \in \mathbb{N}$ *and* $T \in \mathfrak{L}(\ l_\infty^n\ , X)$. *Then there are a sequence* $\alpha \in l_2^n$ *and an operator* $A \in \mathfrak{L}(l_2^n , X)$ *such that*

$$T = AD_\alpha \quad \textit{and} \quad \|\alpha\|_2\ \|A\| = \pi_2(T) ,$$

where $D_\alpha = \sum_{k=1}^{n} e_k \otimes \alpha_k\ e_k$ *denotes the diagonal operator from* l_∞^n *into* l_2^n .

Proof: By the factorization theorem of Pietsch (3) there is a probability measure μ on $K = \{ e_k \mid k = 1,..,n \}$ (the extreme points of $B_{(l_\infty^n)^*}$) such that for all $x \in l_\infty^n$

$$\|Tx\| \le \pi_2(T) \left(\int_K |<x, x^*>|^2 \, d\mu(x^*) \right)^{1/2} .$$

Let us define

$$\alpha_k := (\mu(\{e_k\}))^{1/2} \quad \text{for all} \quad k = 1,..,n ,$$

$$D_\alpha := \sum_{k=1}^n e_k \otimes \alpha_k e_k \quad \in \mathfrak{L}(l_\infty^n , l_2^n) \quad \text{and}$$

$$A := \sum_{\substack{k=1 \\ \alpha_k > 0}}^n e_k \otimes \frac{1}{\alpha_k} Te_k \quad \in \mathfrak{L}(l_2^n , X) .$$

Note that $\|\alpha\|_2 = \sum_{k=1}^n \mu(\{e_k\}) = \mu(K) = 1$. Furthermore we have for all $(\beta_k)_{k=1}^n \subset K$

$$\| A (\sum_{k=1}^n \beta_k e_k) \| = \| T \left(\sum_{\substack{k=1 \\ \alpha_k > 0}}^n \frac{\beta_k}{\alpha_k} e_k \right) \| \le \pi_2(T) \left(\sum_{\substack{k=1 \\ \alpha_k > 0}}^n |\frac{\beta_k}{\alpha_k}|^2 \alpha_k^2 \right)^{1/2}$$

$$\le \pi_2(T) \|\beta\|_2 .$$

Hence $\|D_\alpha\| = \|\alpha\|_2 = 1$ and $\|A\| \le \pi_2(T)$ and therefore $\|\alpha\|_2 \|A\| \le \pi_2(T)$.

Since obviously $T = AD_\alpha$ and (by (5)) $\pi_2(T) \le \|A\| \pi_2(D_\alpha) \le \|A\| \|\alpha\|_2$ we have proved the assertion. □

From lemma 1.1. we deduce easily the following factorization for operators acting from l_∞^n into l_1^n .

1.2. Corollary: *Let* $n \in \mathbb{N}$, $T \in \mathfrak{L}(l_\infty^n , l_1^n)$ *and* $\varepsilon > 0$. *Then there are sequences* $\alpha, \beta \in l_2^n$ *and an operator* $A \in \mathfrak{L}(l_2^n , l_2^n)$ *such that*

$$T = D_\beta A D_\alpha \quad \text{and} \quad \|\alpha\|_2 \|A\| \|\beta\|_2 \le (1+\varepsilon) K_G \|T\| ,$$

where $D_\beta = \sum_{k=1}^n e_k \otimes \beta_k e_k$, $D_\alpha = \sum_{k=1}^n e_k \otimes \alpha_k e_k$ *denote the diagonal operators from* l_2^n *into* l_1^n , l_∞^n *into* l_2^n , *respectively, and* K_G *denotes the Grothendieck constant.*

Proof: By Grothendieck's inequality (8) we find operators $v \in \mathfrak{L}(X , l_2^n)$ and $u \in \mathfrak{L}(l_2^n , l_1^n)$ such that

$$T = uv \quad \text{and} \quad \pi_2(u^*) \, \pi_2(v) \le (1+\varepsilon) K_G \|T\|$$

Hence by lemma 1.1. there are sequences $\alpha, \beta \in l_2^n$ and operators $A_1 , A_2 \in \mathfrak{L}(l_2^n , l_2^n)$

such that $v = A_1 D_\alpha$, $u = D_\beta A_2$ and

$$\|\alpha\|_2 \, \|A_1\| = \pi_2(v) \quad , \quad \|\beta\|_2 \, \|A_2\| = \pi_2(u^*) \quad .$$

Defining $A := A_1 A_2 \in \mathfrak{L}(\ell_2^n, \ell_2^n)$ yields $T = D_\beta A D_\alpha$ and

$$\|\beta\|_2 \, \|A\| \, \|\alpha\|_2 \le (1+\varepsilon) \, K_G \, \|T\| \quad .$$

□

The next lemma contains only a simple observation. But since this is the key of this paper ,we want to point it out.

1.3. Lemma: *Let* $1 \le p \le \infty$, $T \in \mathfrak{L}(X, Y)$, $n \in \mathbb{N}$, $A = (a_{ij})_{i,j=1}^n \in \mathfrak{L}(\ell_2^n, \ell_2^n)$, $(x_i)_{i=1}^n \subset X$ *and* $(y_j^*)_{j=1}^n \subset Y^*$, *then we have*

$$1) \quad \Big(\sum_{j=1}^n \big\| \sum_{i=1}^n a_{ij} \, Tx_i \big\|^p \Big)^{1/p} = \Big\| A^* \otimes T \Big(\sum_{i=1}^n e_i \otimes x_i \Big) \Big\|_{\ell_p^n(Y)}$$

(with the usual modification for $p = \infty$ *)* .

$$2) \quad \Big\langle A^* \otimes T \Big(\sum_{i=1}^n e_i \otimes x_i \Big) \, , \, \sum_{j=1}^n e_j \otimes y_j^* \Big\rangle = \mathrm{tr}(vTuA) \quad ,$$

where $u = \sum_{i=1}^n e_i \otimes x_i \in \mathfrak{L}(\ell_2^n, \ell_2^n)$ *and* $v = \sum_{j=1}^n y_j^* \otimes e_j$.

Proof:

$$1) \quad A^* \otimes T \Big(\sum_{i=1}^n e_i \otimes x_i \Big) = \sum_{i=1}^n A^*(e_i) \otimes Tx_i = \sum_{i=1}^n \Big(\sum_{j=1}^n a_{ij} \, e_j \Big) \otimes Tx_i$$

$$= \sum_{j=1}^n e_j \otimes \sum_{i=1}^n a_{ij} \, Tx_i \quad .$$

Hence assertion 1) is proved.

$$2) \quad \Big\langle A^* \otimes T \Big(\sum_{i=1}^n e_i \otimes x_i \Big) \, , \, \sum_{j=1}^n e_j \otimes y_j^* \Big\rangle$$

$$= \Big\langle \sum_{j=1}^n e_j \otimes \sum_{i=1}^n a_{ij} \, Tx_i \, , \, \sum_{j=1}^n e_j \otimes y_j^* \Big\rangle$$

$$= \sum_{j=1}^n \sum_{i=1}^n a_{ij} \, \langle Tx_i \, , \, y_j^* \rangle = \sum_{j=1}^n \sum_{i=1}^n a_{ij} \, \langle Tu(e_i) \, , \, v^*(e_j) \rangle$$

$$= \sum_{j=1}^n \Big\langle Tu \Big(\sum_{i=1}^n a_{ij} \, e_i \Big) \, , \, v^*(e_j) \Big\rangle = \sum_{j=1}^n \langle TuA(e_j) \, , \, v^*(e_j) \rangle$$

$$= \mathrm{tr}(vTuA) \quad .$$

But this proves assertion 2) .

□

Now we are prepared to prove the Grothendieck type inequalities:

1.4. Theorem: *Let* $T \in \mathfrak{L}(X,Y)$ *and* $1 \le r \le 2 \le p \le \infty$ *such that* $1/r = 1/p + 1/2$. *Then the following are equivalent*

1) *There is a constant* $c_1(T) \ge 0$ *such that for all* $n \in \mathbb{N}$ *and operators* $A \in \mathfrak{L}(l_2^n, l_2^n)$
   $$\| A \otimes T: l_2^n(X) \to l_p^n(Y) \| \le c_1(T) \, \| A: l_2^n \to l_2^n \| .$$
2) *There is a constant* $c_2(T) \ge 0$ *such that for all* $n \in \mathbb{N}$ *and operators* $A \in \mathfrak{L}(l_\infty^n, l_1^n)$
   $$\| A \otimes T: l_\infty^n(X) \to l_r^n(Y) \| \le c_2(T) \, \| A: l_\infty^n \to l_1^n \| .$$
3) *There is a constant* $c_3(T) \ge 0$ *such that for all operators* $S \in \mathfrak{L}(l_1, X)$
   $$\pi_{r,1}(TS) \le c_3(T) \, \|S\| .$$
4) *There is a constant* $c_4(T) \ge 0$ *such that for all operators* $S \in \mathfrak{L}(l_1, X)$
   $$\pi_{p,2}(TS) \le c_4(T) \, \|S\| .$$
5) *There is a constant* $c_5(T) \ge 0$ *such that for all Banach spaces* Z *and for all operators* $u \in \Pi_2^d(Z, X)$
   $$\pi_{p,2}(Tu) \le c_5(T) \, \pi_2(u^*) .$$

Furthermore we have for the best constants
$$c_1(T) \le c_5(T) \le c_4(T) \le c_3(T) \le c_2(T) \le K_G \, c_1(T) .$$

Proof: **1) ⇒ 2)** Let $n \in \mathbb{N}$, $A \in \mathfrak{L}(l_\infty^n, l_1^n)$ and $\varepsilon > 0$. Then by corollary 1.2. there are $\alpha, \beta \in l_2^n$ and $B \in \mathfrak{L}(l_2^n, l_2^n)$ such that $A = D_\beta B D_\alpha$ and
$$\|\beta\|_2 \, \|B\| \, \|\alpha\|_2 \le (1+\varepsilon) \, K_G \, \|A\| .$$
This implies
$$\begin{aligned} &\| A \otimes T: l_\infty^n(X) \to l_r^n(Y) \| \\ \le\ &\| D_\alpha \otimes id_X: l_\infty^n(X) \to l_2^n(X) \| \\ &\| B \otimes T: l_2^n(X) \to l_p^n(Y) \| \\ &\| D_\beta \otimes id_Y: l_p^n(Y) \to l_r^n(Y) \| \\ \le\ &\|\alpha\|_2 \, c_1(T) \, \| B: l_2^n \to l_2^n \| \, \|\beta\|_2 \\ \le\ &c_1(T) \, (1+\varepsilon) \, K_G \, \| A: l_2^n \to l_2^n \| . \end{aligned}$$
Hence assertion 2) is proved.

2) ⇒ 3) W.l.o.g. it is enough to prove assertion 3) for $n \in \mathbb{N}$ and $S \in \mathfrak{L}(l_1^n, X)$. For this let $(x_k)_{k=1}^n \subset l_1^n$ and
$$A := \sum_{k=1}^n e_k \otimes x_k \in \mathfrak{L}(l_\infty^n, l_1^n) .$$

Then lemma 1.3. yields

$$\Big(\sum_{j=1}^{n} \| TSx_j \|^r \Big)^{1/r} = \Big(\sum_{j=1}^{n} \| TSA(e_j) \|^r \Big)^{1/r}$$

$$= \Big(\sum_{j=1}^{n} \| \sum_{i=1}^{n} a_{ij}\ TS(e_i) \|^r \Big)^{1/r} = \| A^* \otimes T \Big(\sum_{i=1}^{n} e_i \otimes x_i\Big) \|_{\ell_r^n(Y)}$$

$$\le c_2(T)\ \| A^*: \ell_\infty^n \to \ell_1^n \| \sup_{k=1,..,n} \| S(e_k) \| = c_2(T)\ \|A\|\ \|S\|$$

$$= c_2(T)\ \|S\| \sup_{x^* \in B_{(\ell_\infty^n)^*}} \Big(\sum_{k=1}^{n} |\langle x_k, x^* \rangle|^2 \Big)^{1/2} .$$

Hence the definition of (r,2)-summing operators completes the proof of assertion 3) .

3) ⇒ 4) The implication 3) ⇒ 4) is obvious since by Kwapień (2) always $\pi_{p,2} \le \pi_{r,1}$.

4) ⇒ 5) Let Z be a Banach space and $u \in \Pi_2^d(Z , X)$. By the factorization theorem (4) of Pietsch we find a compact set K, a probabilty measure μ on K and operators $i \in \mathfrak{L}(X^* , C(K))$, $S \in \mathfrak{L}(L_2(K, \mu) , Z^*)$ such that $u^* = S\, J\, i$, where $J \in \mathfrak{L}(C(K) , L_2(K, \mu))$ is the natural mapping . Furthermore we have

$$\|S\| = \pi_2(u^*) .$$

Since $C(K)^*$ is a $\mathfrak{L}_1$-Space it is simple to deduce that assumption 4) yields

$$\pi_{p,2} (T^{**} i^*) \le c_4(T) \| i^* \| .$$

(In fact, condition 4) describes an injective and maximal Banach ideal.) Hence we get

$$\pi_{p,2}(Tu) = \pi_{p,2} (T^{**} u^{**}) = \pi_{p,2} (T^{**} i^* J^* S^*)$$

$$\le \pi_{p,2} (T^{**} i^*) \| J^* \| \| S^* \| \le c_4(T) \| i^* \| \| J^* \| \| S^* \|$$

$$= c_4(T)\ \pi_2(T) .$$

5) ⇒ 1) Let $n \in \mathbb{N}$, $A \in \mathfrak{L}(\ell_2^n , \ell_2^n)$ and $(x_k)_{k=1}^n \subset X$. Defining $u := \sum_{k=1}^{n} e_k \otimes x_k \in \mathfrak{L}(\ell_2^n , X)$ assertion 1) can be deduced from lemma 1.3. and (5)

$$\| A \otimes T \Big(\sum_{i=1}^{n} e_i \otimes x_i\Big) \|_{\ell_p^n(Y)} = \Big(\sum_{j=1}^{n} \| \sum_{i=1}^{n} a_{ji}\ Tx_i \|^p \Big)^{1/p}$$

$$= \Big(\sum_{j=1}^{n} \| Tu \Big(\sum_{i=1}^{n} a_{ji}\ e_i\Big) \|^p \Big)^{1/p} = \Big(\sum_{j=1}^{n} \| TuA^*(e_j) \|^p \Big)^{1/p}$$

$$\le \pi_{p,2} (TuA^*) \sup_{x^* \in B_{\ell_2^n}} \Big(\sum_{k=1}^{n} |\langle e_k , x^* \rangle|^2 \Big)^{1/2}$$

$$\le c_5(T)\ \pi_2((uA^*)^*) \le c_5(T)\ \pi_2(u^*)\ \| A: \ell_2^n \to \ell_2^n \|$$

$$\le c_5(T)\ \| A: \ell_2^n \to \ell_2^n \| \Big(\sum_{k=1}^{n} \|x_k\|^2 \Big)^{1/2} .$$

□

2. Application to hilbertian operators

Recall the definition of hilbertian operators. We mentioned that $\mathfrak{H}_1$ is the class of operators which factor through a Hilbert space and $\mathfrak{H}_{1,\infty}$ is just the class of weak Hilbert operators. Although weak Hilbert spaces are rather difficult to construct there are canonical examples of hilbertian operators.

2.1. Examples: **i)** *Pietsch* [PI2] *proved that for* $1 \le r \le 2 \le p \le \infty$ *with* $1/r = 1/p + 1/2$ *every* L_p *or* $L_{p'}$ *space is r-hilbertian. We can even show that*

$$h_r(L_p) = h_r(L_{p'}) = 1$$

ii) *Using a simple extreme point argument (an observation of Geiss) it is easy to check that the Grothendieck numbers of the sum opertor*

$$\sum : l_1 \to l_\infty \ , \ (\alpha_n)_{n\in\mathbb{N}} \mapsto \ (\sum_{k=1}^{n} \alpha_k \ )_{n\in\mathbb{N}}$$

are bounded. But by a result stated in [DJ] *this implies that* $\sum$ *is a weak Hilbert operator, which was first remarked in* [PS1].

The crucial link between the hilbertian operators and the Grothendieck type inequalities is given by the following

2.2. Proposition: *Let* $1 \le r \le 2 \le p \le \infty$ *with* $1/r = 1/p + 1/2$. *For every r-hilbertian operator* $T \in \mathfrak{H}_r(X,Y)$, $n \in \mathbb{N}$ *and* $A \in \mathfrak{L}(l_2^n, l_2^n)$

$$\| A\otimes T: l_2^n(X) \to l_p^n(Y) \| \le h_r(T) \ \| A: l_2^n \to l_2^n \| .$$

Proof: Let $n\in\mathbb{N}$, $x = (x_k)_{k=1}^n \subset B_{l_2^n(X)}$ and $y^* = (y_j^*)_{j=1}^n \subset B_{l_{p'}^n(Y^*)}$. In order to apply lemma 1.3. we set

$$u := \sum_{k=1}^{n} e_k \otimes x_k \in \mathfrak{L}(l_2^n, X) \quad \text{and} \quad v := \sum_{k=1}^{n} y_k^* \otimes e_k \in \mathfrak{L}(Y, l_2^n) .$$

Inequality (5) implies

$$\pi_2(u^*) = \pi_2 \left(\sum_{k=1}^{n} e_k \otimes x_k \right) \le \left(\sum_{k=1}^{n} \|x_k\|^2 \right)^{1/2} \le 1 .$$

For $(0^{-1} 0 := 0)$

$$v_o := \sum_{j=1}^{n} \| y_j^* \|^{-1} \ y_j^* \otimes e_j \in \mathfrak{L}(Y, l_\infty^n) ,$$

$$D_\rho := \sum_{j=1}^{n} e_j \otimes \| y_j^* \|^{p'/2} e_j \in \mathfrak{L}(l_\infty^n, l_2^n) ,$$

$$D_\tau := \sum_{j=1}^{n} e_j \otimes \| y_j^* \|^{p'/r'} e_j \in \mathfrak{L}(\ell_2^n, \ell_2^n)$$

we have with $1/_{p'} = 1/_{r'} + 1/_2$ and again (5)

$$v = D_\tau D_\rho v_o \quad \text{and}$$

$$\pi_2(D_\rho v_o) \le \pi_2(D_\rho) \|v_o\| \le \Big(\sum_{j=1}^{n} (\| y_j^* \|^{p'/2})^2 \Big)^{1/2} \| v_o^* \|$$

$$= (\| y^* \|_{\ell_{p'}^n(Y^*)})^{p'/2} \sup_{j=1,..,n} \| y_j^* \|^{-1} y_j^* \le 1 .$$

This yields with lemma 1.4. , (1) and the definition of $\mathfrak{H}_r$

$$|\langle A \otimes T \Big(\sum_{k=1}^{n} e_k \otimes x_k \Big) , \sum_{j=1}^{n} e_j \otimes y_j^* \rangle| = | \operatorname{tr}(vTuA^*) |$$

$$= | \operatorname{tr}(D_\tau D_\rho v_o TuA^*) | \le \sigma_{r'} (D_\tau) \; \sigma_r (D_\rho v_o TuA^*)$$

$$\le (\| y^* \|_{\ell_{p'}^n(Y^*)})^{p'/r'} \; h_r(T) \; \pi_2(D_\rho v_o) \; \pi_2(u^*) \; \|A\|$$

$$\le h_r(T) \; \|A\| .$$

□

As an immediate cosequence of proposition 2.2. and theorem 1.4. we get the following

2.3. Theorem: *Let* $1 \le r \le 2$. *For every r-hilbertian operator* $T \in \mathfrak{H}_r(X,Y)$ *and every operator* $S \in \mathfrak{L}(\ell_1, X)$ *the composition* TS *is (r,1)-summing with*

$$\pi_{r,1}(TS) \le K_G \; h_r(T) \; \|S\| .$$

2.4. Corollary: *For every weak Hilbert operator* $T \in \mathfrak{H}_{1,\infty}(X,Y)$, *every operator* $S \in \mathfrak{L}(\ell_1, X)$ *and every* $1 < r \le 2$ *the composition* TS *is* (r,1)*-summing with*

$$\pi_{r,1}(TS) \le K_G \frac{r}{r-1} \; h_{1,\infty}(T) \; \|S\| .$$

Proof: First of all we note that for a positive non increasing sequence $\alpha \in \ell_{1,\infty}$ and every $1 < r < \infty$

$$\|\alpha\|_r \le \Big(\sum_{k \in \mathbb{N}} \alpha_k^{\,r} \Big)^{1/r} \le \Big(\sum_{k \in \mathbb{N}} k^{-r} \Big)^{1/r} \|\alpha\|_{1,\infty} \le \frac{r}{r-1} \|\alpha\|_{1,\infty} .$$

But this implies that every $(1,\infty)$-hilbertian operator $T \in \mathfrak{H}_{1,\infty}(X,Y)$ is r-hilbertian with $h_r(T) \le \frac{r}{r-1} h_{1,\infty}(T)$. Therefore the assertion follows from theorem 2.3. . □

Using the examples 2.1. more classical results can be deduced from theorem 2.3. and its corollary 2.4.

2.5. Applications: i) [KW1] *Let* $1 \le r \le 2 \le p \le \infty$ *and* $1/r = 1/p + 1/2$, *then every operator* $S \in \mathfrak{L}(l_1, l_p)$ *or* $S \in \mathfrak{L}(l_1, l_{p'})$ *is (r,1)-summing and*

$$\pi_{r,1}(S) \le K_G \, \|S\| .$$

ii) [KWP] *By corollary 2.4. the sum operator is (r,1)-summing and*

$$\pi_{r,1}(\Sigma) \le 10 \frac{r}{r-1} \qquad \textit{for all } 1 < r \le 2 .$$

Moreover, by (2) this implies that for all $1 \le q < p < \infty$ *the sum operator is (p,q)-summing and*

$$\pi_{p,q}(\Sigma) \le 10 \frac{pq}{p-q}$$

3. Supplementary results

In this chapter we want to show that the condition of corollary 2.4. does not characterize weak Hilbert spaces or operators, whereas the classical result characterizes Hilbert spaces. More precisely, we shall show that an interpolation space between the space of bounded variation and the space of bounded sequences satisfy the assertion of corollary 2.4. , but is not a weak Hilbert space.

For the following we need the exact describtion of the (sequence) dual of $l_{\infty,1}$ ([GAR])

$$l^{+}_{\infty,1} := \{ \alpha \in l_\infty \mid \|\alpha\|^{+}_{\infty,1} := \sup_{n \in \mathbb{N}} \Big(\sum_{k=1}^{n} \frac{1}{k} \Big)^{-1} \sum_{k=1}^{n} \alpha^*_k < \infty \} .$$

The Banach space ($l^{+}_{\infty,1}$, $\|\circ\|^{+}_{\infty,1}$) is the convexification of the Lorentz sequence space ($l_{1,\infty}$, $\|\circ\|_{1,\infty}$) . For the canonical identity $\iota: l^{+}_{\infty,1} \to l_r$, ($1 < r \le \infty$) we have

$$\| \iota: l^{+}_{\infty,1} \to l_r \| \le \| \iota: l_{r'} \to l_{\infty,1} \| \le \frac{r}{r-1} .$$

For an abitrary Banach space X we set

$$l^{+}_{\infty,1}(X) := \{ x = (x_k)_{k \in \mathbb{N}} \subset X \mid \|x\|^{+}_{\infty,1} := \| (\|x_k\|)_{k \in \mathbb{N}} \|^{+}_{\infty,1} < \infty \} .$$

For $n \in \mathbb{N}$ the spaces $(l^{+}_{\infty,1})^n$ and $(l^{+}_{\infty,1})^n (X)$ are defined in the usual way.

The following proposition is a modified version of theorem 1.4.

3.1. Proposition: *For an operator* $T \in \mathfrak{L}(X,Y)$ *the following conditions are equivalent.*

1) *There is a constant* $c_1(T) \geq 0$ *such that for all* $n,m \in \mathbb{N}$ *and operators* $A \in \mathfrak{L}(l_2^m, l_2^n)$

$$\| A \otimes T: l_2^m(X) \to l_2^n(Y) \| \leq c_1(T)\, (1+\ln n)\, \| A: l_2^m \to l_2^n \| .$$

2) *There is a constant* $c_2(T) \geq 0$ *such that for all* $m \in \mathbb{N}$ *and operators* $A \in \mathfrak{L}(l_\infty^m, l_1^m)$

$$\| A \otimes T: l_\infty^m(X) \to (l_{\infty,1}^{+})^m(Y) \| \leq c_2(T)\, \| A: l_\infty^m \to l_1^m \| .$$

3) *There is a constant* $c_3(T) \geq 0$ *such that for all operators* $S \in \mathfrak{L}(l_1, X)$ *and all* $1 < r \leq 2$

$$\pi_{r,1}(TS) \leq c_3(T)\, \frac{r}{r-1}\, \|S\| .$$

4) *There is a constant* $c_4(T) \geq 0$ *such that for all* $n \in \mathbb{N}$ *and operator* $u \in \mathfrak{L}(l_2^n, X)$

$$\pi_2(Tu) \leq c_4(T)\, (1+\ln n)\, \pi_2(u^*) .$$

Furthermore we have for the best constants

$$c_1(T) \leq c_4(T) \leq 6e\, c_3(T) \leq 6e\, c_2(T) \leq 12e\, K_G\, c_1(T) .$$

Proof: The implication 1) ⇒ 2) can be proved analogously to the implication 1) ⇒ 2) of theorem 1.4. by the use of a slight modification of lemma 1.3. and the inequality

$$1+\ln n \leq 2 \sum_{k=1}^{n} \frac{1}{k} .$$

For the implication 2) ⇒ 3) note that for every $1 < r \leq 2$, $m \in \mathbb{N}$ and $A \in \mathfrak{L}(l_\infty^n, l_1^n)$

$$\| A \otimes T: l_\infty^m(X) \to l_r^m(Y) \|$$
$$\leq \| A \otimes T: l_\infty^m(X) \to (l_{\infty,1}^{+})^m(Y) \| \; \| id_Y \otimes \iota: (l_{\infty,1}^{+})^m(Y) \to l_r^m(Y) \|$$
$$\leq c_3(T)\, \|A\|\, \frac{r}{r-1} .$$

Therefore 2) ⇒ 3) follows from theorem 1.4. 2) ⇒ 3) .

3) ⇒ 4) Let $n \in \mathbb{N}$, $u \in \mathfrak{L}(l_2^n, X)$. By theorem 1.4. 3) ⇒ 5) we have for every $2 < p < \infty$

$$\pi_{p,2}(Tu) \leq c_3(T)\, \frac{2p}{p-2}\, \pi_2(u^*) .$$

Choosing $2 < p < \infty$ with $\frac{2p}{p-2} = 3+\ln n$ we obtain by (6)

$$\pi_2(Tu) \leq 2\, \pi_2^{(n)}(Tu) \leq 2\, n^{1/2-1/p}\, \pi_{p,2}(Tu)$$
$$\leq 2\, c_3(T)\, n^{\frac{p-2}{2p}}\, \frac{2p}{p-2}\, \pi_2(u^*)$$
$$\leq 6e\, c_3(T)\, (1+\ln n)\, \pi_2(u^*) .$$

4) ⇒ 1) Let $m \in \mathbb{N}$, $(x_k)_{k=1}^m \subset B_{l_2^m(X)}$ and $A \in \mathfrak{L}(l_2^m, l_2^n)$. Then we have for

$$u := \sum_{k=1}^m e_k \otimes x_k \in \mathfrak{L}(l_2^m, X) ,$$

by lemma 1.3. and (5)

$$\| A \otimes T \left(\sum_{k=1}^m e_k \otimes x_k \right) \|_{l_2^n(Y)} = \left(\sum_{j=1}^n \| TuA^*(e_j) \|^2 \right)^{1/2}$$

$$\leq \pi_2(TuA^*) \leq c_4(T) (1+\ln n) \pi_2((uA^*)^*) \leq c_4(T) (1+\ln n) \|A\| . \quad \square$$

3.2. Remark: *By a standard trace duality argument one can easily check that condition 3.1.4) means that for all* $u \in \Pi_2^d(l_2, X)$ *and* $v \in \Pi_2(X, l_2)$

$$\sup_{n \in \mathbb{N}} \left(\sum_{k=1}^n \frac{1}{k} \right)^{-1} \sum_{k=1}^n a_k(vTu) \leq c_4(T) \pi_2(u^*) \pi_2(v) .$$

Therefore proposition 3.1. can be considered as a Grothendieck type inequality which characterizes this sort of ($l_{\infty,1}^+$ -) *hilbertian operators.*

In the following we shall closely follow the approach of Pisier and Xu [PSX] to the interpolation spaces between v_p spaces. They denoted by v_1 the space of bounded variation

$$v_1 := \{ (\alpha_k)_{k \in \mathbb{N}} \subset l_\infty \mid |\alpha_o| + \sum_{k=1}^\infty |\alpha_k - \alpha_{k-1}| < \infty \}$$

Via the sum operator v_1 is isometric to l_1 . For our purpose it is enough to consider only the interpolation space $A_2 := (v_1, l_\infty)_{1/2, 2}$. Here the real interpolation method is used. We want to point out the following facts (see [PSX]) :

(3.3.) A_2 is a K-convex Banach space which is isomorphic to its dual.

(3.4.) There is a constant $c > 0$ such that for all $n \in \mathbb{N}$ and $(x_k)_{k=1}^n \subset A_2$

$$(c(1+\ln n))^{-1/2} \| \sum_{k=1}^n g_k x_k \|_{L_2(\mu, A_2)}$$

$$\leq \left(\sum_{k=1}^n \|x_k\|^2 \right)^{1/2} \leq (c(1+\ln n))^{1/2} \| \sum_{k=1}^n g_k x_k \|_{L_2(\mu, A_2)}$$

(This statement is essentially a reformulation of lemma 11 in [PSX] with the use of theorem 8, see also the remark before theorem 15.)

Now we can prove the main result of this section.

3.5. Proposition: A_2 *is not a weak Hilbert space, but there is a constant* $c \geq 0$ *such that for all* $1 < r \leq 2$ *each operator* $S \in \mathfrak{L}(l_1, A_2)$ *is (r,1)-summing and*

$$\pi_{r,1}(S) \leq c \frac{r}{r-1} \|S\| .$$

Proof: By the definition of A_2 as an intermediate space of v_1 and l_∞ the sum operator factors through A_2 . This implies that A_2 can not be reflexive (see [LIP]) . But a weak Hilbert space is reflexive by [PS1].

On the other hand we note that by (3.3.) the inequaltity (3.4.) holds not only for A_2 but also for his dual space A_2^* for some universal constant $c > 0$. With (6) we deduce for $X \in \{A_2 , A_2^*\}$, $n \in \mathbf{N}$ and an operator $u \in \mathfrak{L}(l_2^n, X)$

$$(*) \qquad \pi_2(u) \leq 2\, \pi_2^{(n)}(u) = 2 \sup\{ (\sum_{k=1}^{n} \|uA(e_k)\|^2)^{1/2} \mid \|A: l_2^n \to l_2^n\| \leq 1 \}$$

$$\leq 2\, (c(1+\ln n))^{1/2} \sup\{ \|\sum_{k=1}^{n} g_k x_k\|_{L_2(\mu,X)} \mid \|A: l_2^n \to l_2^n\| \leq 1 \}$$

$$\leq 2\, (c(1+\ln n))^{1/2} \pi_\gamma(u) .$$

By trace duality this implies for every $v \in \mathfrak{L}(X, l_2^n)$

$$(**) \qquad \pi_\gamma^*(v) \leq 2\,(c(1+\ln n))^{1/2} \pi_2^*(v) = 2\,(c(1+\ln n))^{1/2} \pi_2(v) .$$

So we deduce with (*) for A_2 , the K-convexity of A_2 and (**) for A_2^* that for all $n \in \mathbf{N}$ and $u \in \mathfrak{L}(l_2^n, A_2)$

$$\pi_2(u) \leq 2\,(c(1+\ln n))^{1/2} \pi_\gamma(u) \leq 2\,(c(1+\ln n))^{1/2} K(A_2)\, \pi_\gamma^*(u^*)$$

$$\leq 2\,(c(1+\ln n))^{1/2} K(A_2)\; 2\,(c(1+\ln n))^{1/2} \pi_2(u^*)$$

$$= 4c\, K(A_2)\, (1+\ln n)\, \pi_2(u^*) .$$

By proposition 3.1. the assertion is proved. □

References:

[DJ] Defant, M. / Junge, M. : *On weak (r,2)-summing operators and weak Hilbert spaces,* (1988), Studia Math. , to appear.

[GAR] Garling, D.J.H. : *A class of reflexive symmetric BK-spaces,* Canad. J. Math. 21 (1969) , 602 - 608 .

[KW1] Kwapień, S. : *Some remarks on (p,q)-absolutely summing operators in l_p-spaces* , Studia Math. 29 (1968) , 327-337 .

[KW2] Kwapień, S. : *A linear topological characterization of inner product spaces,* Studia Math. 38 (1970) , 277-278 .

[KWP] Kwapień, S. / Pelczyński, A. : *The main triangle projection in matrix spaces and its applications,* Studia Math. 34 (1970) , 43-68 .

[LPI] Linde, W. / Pietsch, A. : *Mapping of Gaussian cylindrical measures in Banach spaces,* Theory of Prob. and its Appl. XIX (1974) , 445-460 .

[LIP] Lindenstrauss, J. / Pelczyński, A. : *Absolutely summing operators in L_p-spaces and their applications,* Studia Math. 29 (1968) , 275-326.

[PI1] Pietsch, A. : *Operator ideals,* Deutscher Verlag Wiss. , Berlin , 1978 and North-Holland , Amsterdam - New York - Oxford , 1980 .

[PI2] Pietsch, A. : *Eigenvalue distribution and geometry of Banach spaces,* (1980) , preprint .

[PS1] Pisier, G. : *Weak Hilbert spaces,* Proc. London Math. Soc. 56 (1988), 547-579.

[PS2] Pisier, G. : *Volume of convex bodies and geometry of Banach spaces,* forthcomming book .

[PSX] Pisier, G. / Xu, Q. : *Random series in the real interpolation spaces between the spaces v_p* , Lecture Notes in Math. 1267 , 185-209 .

[TOJ] Tomczak-Jaegermann, N. : *Banach Mazur distances and finite dimensional operator ideals,* Harlow, Longman and New York , Wiley 1988 .

1980 Mathematics Subject Classification (1985 Revision) : 47A30 , 47A65 , 47B10
Key words : Grothendieck inequality, weak Hilbert space, absolutely summing operators.

Martin Defant
Marius Junge

Mathematisches Seminar
Universität Kiel

Ohlshausenstr. 40-60

D-2300 Kiel 1 (West Germany)

A weak topology characterization of $\ell_1(m)$

S. J. Dilworth[1]

Abstract. Let m be any cardinal. The main result characterizes $\ell_1(m)$ as the only Banach lattice whose positive cone is metrizable in the weak topology. Two related theorems on ℓ_1-sums of Banach spaces are also proved.

1. Introduction

The three theorems of the paper concern ℓ_1-sums of Banach spaces. The first, which states that the Banach space $\ell_\infty(m)$ contains the ℓ_1-sum of 2^m copies of itself, is essentially a translation into Banach space terms of a theorem of Pondiczery on the product topology [10]. The case $m = \aleph_0$ of this result is reminiscent of the general theorem [9] that if X is any separable Banach space containing ℓ_1, then X^* contains $M(0,1)$, the space of finite Borel measures on [0,1]: the connection resides in the observation that $M[0,1]$ is linearly isomorphic to the ℓ_1-sum of c copies of $L_1(0,1)$. This observation is used to prove Theorem 2, which says that if X is as above then X^* contains 2^c mutually non-isomorphic closed linear subspaces.

Recall that the unit ball of a Banach space X is metrizable in the weak topology if X^* is separable, and recall too the consequence of the Baire category theorem that if X is infinite-dimensional then the weak topology on X and the weak-star topology on X^* are not metrizable. However, there are still some interesting *unbounded* subsets of Banach spaces that are metrizable in the weak or the weak-star topology. Recall, for example, that the positive cone of the dual of $C(K)$ (that is, the space of continuous functions on a compact metric space K), consisting of the non-negative finite Borel measures on K, is metrizable in the weak-star topology. It is not difficult, too, to see that the positive cone of $\ell_1(m)$ is metrizable in the weak topology for any cardinal m. Theorem 3 below - the paper's main result - says that the latter characterizes $\ell_1(m)$ among Banach lattices. Notation and terminology are mostly standard. For every cardinal number m, $\ell_\infty(m)$ denotes the space of real-valued bounded functions $(x_a)_{a\epsilon A}$ on a set of cardinality m with the norm $sup\{| x_a |: a\epsilon A\}$, and $\ell_1(m)$ denotes the space of absolutely summable functions with the norm $\sum_{a\epsilon A} | x_a |$. Given an indexed family of Banach spaces, $\{X_a : a\epsilon A\}$, the ℓ_1-sum, $(\sum_{a\epsilon A} \oplus X_a)_1$, consists of all vectors $x = (x_a)_{a\epsilon A}$, where $x_a \epsilon X_a$, equipped with the norm $||x|| = \sum_{a\epsilon A} ||x_a||$. The ℓ_1-sum of m copies of a Banach space X is denoted by $(\sum_m \oplus X)_1$, or by $(\sum_A \oplus X)_1$ when A is a set of cardinality m.

As usual, $\aleph_0$ and c denote the cardinality of the natural numbers and the real numbers respectively. The positive cone X^+ of a Banach lattice X is the weakly-closed cone of non-negative vectors in X. The sequence spaces $\ell_p (1 \le p < \infty)$ and the Lebesgue spaces $L_1(0,1)$ and $L_\infty(0,1)$ of real-valued functions on [0,1] are defined in the usual way. Finally, the Lebesgue measure and the indicator function of a measurable set $A \subset [0,1]$ are denoted by $|A|$ and $I(A)$ respectively.

[1] Research supported in part by the National Science Foundation under grant number DMS 8801731.

2. RESULTS

THEOREM 1. *Let m be an infinite cardinal. Then $\ell_\infty(m)$ contains a subspace isometric to the ℓ_1-sum of 2^m copies of $\ell_\infty(m)$.*

PROOF: $\ell_\infty(m)$ has a norming set D of linear functionals (i.e. $||x|| = sup\{d(x) : d\epsilon D\}$ for all $x\epsilon\ell_\infty(m)$) of cardinality m. For each $a\epsilon A$, where A is a set of cardinality 2^m, let D_a be a copy of D supplied with the discrete topology. By a theorem of Pondiczery [10] (also [5,7]) the topological product $\Pi_{a\epsilon A} D_a$ has a dense set $\{y^b = (y^b_a)_{a\epsilon A} : b\epsilon B\}$ of cardinality m. This set can be identified under the pairing $(y,x) = \sum_{a\epsilon A} y_a(x_a)$ with a collection of m linear functionals on $(\sum_A \oplus\ell_\infty(m))_1$. Let $x = (x_a)_{a\epsilon A}$ be a vector of norm one in $(\sum_A \oplus\ell_\infty(m))_1$ and let $\varepsilon > 0$. Select $a_1, a_2, \ldots, a_n$ in A such that $\sum_{k=1}^n ||x_{a_k}|| \geq 1-\varepsilon$. By the definition of the product topology there exists b in B such that $y^b_{a_k}(x_{a_k}) \geq ||x_{a_k}|| - \frac{\varepsilon}{n}$ for $k = 1,2,\ldots,n$. Thus $(y^b, x) \geq 1 - 3\varepsilon$. It follows that the mapping $x \to ((y^b, x))_{b\epsilon B}$ defines an isometric embedding from $(\sum_A \oplus\ell_\infty(m))_1$ into $\ell_\infty(m)$.

REMARK: Further applications of Pondiczery's theorem to Banach spaces have been noticed. In [11, p. 236], for example, it was observed that $C(\{0,1\}^{2^m})$ (that is, the space of all continuous functions on the Cantor space $\{0,1\}^{2^m}$) embeds isometrically into $\ell_\infty(m)$ (since $\{0,1\}^{2^m}$ has a dense set of cardinality m).

THEOREM 2. *Let X be a separable Banach space containing ℓ_1. Then X^* contains precisely 2^c mutually non-isomorphic closed subspaces.*

PROOF: Since X^* is isometric to a subspace of $\ell_\infty(\aleph_0)$ it has at most 2^c subspaces. By the observation made in the first paragraph of the introduction, to establish the converse it suffices to show that $Y = (\sum_c \oplus L_1(0,1))_1$ contains 2^c mutually non-isomorphic subspaces. Recall that $L_1(0,1)$ contains a subspace isometric to ℓ_p for each $1 < p < 2$ (for example, the closed linear span of a sequence of independent p-stable random variables). It follows that Y contains a subspace isometric to $(\sum_{p\epsilon A} \oplus\ell_p)_1$ for every subset A of (1,2). The proof will be complete once it is shown that distinct subsets of (1,2) do not give rise in this way to isomorphic Banach spaces. To see that that is so, suppose that $T : \ell_q \to (\sum_{p\epsilon A} \oplus\ell_p)_1$ is an isomorphic embedding, where $q > 1$ and $A \subset (1,2)$. It has to be shown that $q\epsilon A$. We may assume that $\delta||x|| \leq ||Tx|| \leq ||x||$ for some $\delta > 0$ and for all $x\epsilon\ell_q$. Since the range of T is separable we may also asume that A is countable. Let $p_1, p_2, \ldots$ be an enumeration of A, let P_n be the projection from $(\ell_{p_1} \oplus \ell_{p_2} \oplus \ldots)_1$ onto its first n factors, and let Q_n be the projection of ℓ_q onto the linear span of the first n basis vectors of its usual basis. Since $q > 1$ a routine gliding hump argument proves that $||(I - P_n)T(I - Q_n)|| \to 0$ as $n \to \infty$. Select n so that $||(I - P_n)T(I - Q_n)|| < \frac{1}{2}\delta$. Then P_nT restricts to an isomorphism from $(I - Q_n)\ell_q$ (which is isometric to ℓ_q) into $\ell_{p_1} \oplus \ell_{p_2} \oplus \cdots \oplus \ell_{p_n}$. By standard arguments (e.g. [1, pp. 194, 205]) this implies that $q\epsilon\{p_1, p_2, \ldots, p_n\}$, which completes the proof.

Recall the following useful concept due to Banach [1, p. 193]: two Banach spaces X and Y are said to have *incomparable linear dimensions* if there are no linear isomorphisms from X into Y or from Y into X. Theorem 2 allows the following refinement.

THEOREM 2*. *Let X be a separable Banach space containing ℓ_1. Then X^* contains 2^c closed subspaces with mutually incomparable linear dimensions.*

PROOF: Let S be acollection of 2^c subsets of (1,2) having the property that $A\backslash B$ and $B\backslash A$ are non-empty sets for all members A and B of S. To see that this is indeed possible, recall that the Stone-Cech compactification $\beta\mathbf{N}$ has cardinality 2^c, and that the points of $\beta\mathbf{N}$ are in bijective correspondence with the ultrafilters of $\mathbf{N}$. When regarded merely as sets of subsets of $\mathbf{N}$, the ultrafilters may be put in bijective correspondence with a collection S of subsets of (1,2) having the required property. The proof of Theorem 2 now shows that the spaces $(\sum_{p\epsilon A}\oplus\ell_p)_1$, for $A\epsilon S$, have mutually incomparable linear dimensions.

REMARK: Let $m \geq c$ be an infinite cardinal. In [6] it is proved that there exist exactly 2^m topologically distinct compact Hausdorff spaces of weight m. The corresponding spaces of continuous functions have density character m (that is, they have a dense set of cardinality m), and by the Banach-Stone theorem they are mutually non-isometric. Since every Banach space of density character m is isometric to one of the 2^m quotient spaces of $\ell_1(m)$, it follows that there are exactly 2^m mutually non-isometric Banach spaces of density character m. This argument does not say, however, how many mutually non-isomorphic spaces there are of a given density character.

We turn now to a proof of the following theorem characterizing the Banach lattice $\ell_1(m)$.

THEOREM 3. *Let X be a Banach lattice. Then the positive cone of X is metrizable in the weak topology if and only if X is lattice isomorphic to $\ell_1(m)$ for some cardinal m.*

Two preliminary results are required. The main idea in the first result below generalizes a fact about Hilbert space given in [8, p. 380]. I do not know a reference for Proposition 2: this result is of some interest in its own right and perhaps parts of it are known. Finally, recall that a topological space S is said to be a Fréchet space [4, p. 78] if, for every $A \subset S$ and for every x in the closure A, there exists a sequence $(x_n)_{n=1}^\infty$ in A which converges to x. In particular, metrizable and first countable spaces are Fréchet spaces.

PROPOSITION 1. *Let X be a Banach space with a normalized unconditional basis $(e_n)_{n=1}^\infty$, and suppose that X^+ is a Fréchet space is the weak topology. Then $(e_n)_{n=1}^\infty$ is equivalent to the standard basis of ℓ_1.*

PROOF: By renorming X, if necessary, we may assume that $(e_n)_{n=1}^\infty$ is a normalized basis such that

$$||\sum_{n=1}^{\infty} a_n e_n|| = ||\sum_{n=1}^{\infty} | a_n | e_n||$$

for all scalars $(a_n)_{n=1}^\infty$. First suppose that $\liminf_{n\to\infty} | f(e_n) | = 0$ for all $f\epsilon X^*$, and let $S = \{e_1 + e_m + me_n : 2 < m < n < \infty\}$. Every weak neighborhood of e_1 in X^* contains a neighborhood of the form

$$U = \{x\epsilon X^+ : |f_k(x - e_1)| < 1, 1 \leq k \leq N\},$$

where $(f_k)_{k=1}^N$ is a collection of positive linear functionals in X^*. Let $f = f_1 + f_2 + \cdots + f_N$. By assumption there exist $m \geq 2$ and $n > m$ such that $f(e_m) < \frac{1}{2}$ and $f(e_n) < \frac{1}{2m}$. For $1 \leq k \leq N$, it follows from the positivity of the f_k's that $| f_k(e_m + me_n) | \leq f(e_m + me_n) <$

1, and so $e_1 + e_m + me_n \epsilon U \cap S$, whence e_1 lies in the weak closure of S. Using the fact that weakly convergent sequences are norm bounded it is easy to see that e_1 is not the weak limit of a sequence from S, and it follows that X^+ is not a Fréchet space. This contradiction implies that there exists $f \epsilon X^*$ of norm one and $\delta > 0$ such that $f(e_n) > \delta$ for all n. Then

$$\begin{aligned}||\sum_{n=1}^{\infty} a_n e_n|| &= ||\sum_{n=1}^{\infty} | a_n | e_n|| \\ &\geq f(\sum_{n=1}^{\infty} | a_n | e_n) \\ &\geq \delta(\sum_{n=1}^{\infty} | a_n |),\end{aligned}$$

and so $(e_n)_{n=1}^{\infty}$ is equivalent to the standard basis of ℓ_1.

PROPOSITION 2. *(cf. [2, Prop. 4.1]). Let* $P = \{f \geq 0 : \int f(t)dt = 1\}$ *and let*

$Q = \{\frac{I(E)}{|E|} : E \subset (0,1), | E |> 0\}$.

(a) P is the weak closure of Q in $L_1(0,1)$.

(b) If $f_n \in Q$ and $f_n \to f$ weakly, then $f \in L_\infty(0,1)$.

(c) P is not a Fréchet space in the weak topology.

PROOF: (a) Let $f = \sum_{k=1}^{n} a_k I(E_k)$ be a simple function in P, where $E_1, E_2, \ldots, E_n$ are measurable disjoint subsets of (0,1). Suppose that $g_1, g_2, \ldots, g_m$ belong to $L_\infty(0,1)$. For each $1 \leq k \leq n$, the mapping $\phi_k(E) = (| E |, (\int I(E) g_j(t)dt)_{j=1}^{m})$ is a vector measure on the measurable subsets of E_k whose values lie in $\mathbf{R}^{m+1}$. By the Liapounoff convexity theorem (see e.g. [12, p. 114]) there exists sets $F_k \subset E_k (1 \leq k \leq n)$ such that $\phi_k(F_k) = \lambda a_k \phi_k(E_k)$ provided $0 < \lambda < \min[\frac{1}{a_k} : 1 \leq k \leq n]$. Let $F = U_{k=1}^{n} F_k$.

Then $| F |= \sum_{k=1}^{n} \lambda a_k | E_k |= \lambda$, and

$$\begin{aligned}\int I(F) g_j(t)dt &= \sum_{k=1}^{n} (\int I(F_k) g_j(t)dt) \\ &= \sum_{k=1}^{n} \lambda a_k (\int I(E_k) g_j(t)dt) \\ &= \lambda \int f(t) g_j(t)dt\end{aligned}$$

for $1 \leq j \leq m$. So

$$\int \frac{I(F)}{|F|} g_j(t)dt = \int f(t) g_j(t)dt \quad (1 \leq j \leq m),$$

and it follows that f lies in the weak closure of Q. Since the simple functions in P are norm dense in P, the desired conclusion follows. (b) If $f_n \to f$ weakly, then the set $S = \{f_n : n \geq 1\} \cup \{f\}$ is a weakly compact subset of $L_1(0,1)$. By the Dunford-Pettis criterion [3, p. 274] S is uniformly integrable. It follows that there exists M such that $||f_n||_\infty \leq M$ for all n, whence $||f||_\infty \leq M$ since $f_n \to f$ weakly. (c) This is an immediate consequence of (a) and (b).

REMARK: The converse of (b) is also true: f is a weak limit of a sequence from Q if and only if f belongs to $P \cap L_\infty(0,1)$.

PROOF OF THEOREM 3. It is easily verified that the weak and norm topologies coincide on the positive cone of $\ell_1(m)$: in particular, the weak topology is metrizable. To prove the converse, let X be a Banach lattice such that X^+ is metrizable in the weak topology. By Proposition 1 the closed linear span of every sequence of pairwise disjoint vectors in X is isomorphic to ℓ_1, and so by a result of Tzafriri [13. Proposition 1] X is lattice isomorphic to $L_1(\mu)$ for some measure space $(M, \sum, \mu)$. If μ is not purely atomic then $L_1(0,1)$ is lattice isomorphic to a sublattice of $L_1(\mu)$, but by Proposition 2 this would contradict the assumption that X^+ is metrizable in the weak topology. So μ is purely atomic and hence X is lattice isomorphic to $\ell_1(m)$ for some cardinal m.

The above arguments have in fact established the following stronger version of Theorem 3.

THEOREM 3*. *Let X be a Banach lattice. Then the following are equivalent:*

(1) *X^+ is a Fréchet space in the weak topology;*
(2) *X^+ is metrizable in the weak topology;*
(3) *the weak and norm topologies on X^+ are identical;*
(4) *X is lattice isomorphic to $\ell_1(m)$ for some cardinal m.*

REFERENCES

1. S. Banach, "Théorie des Opérations Linéaires (Second Edition)," Chelsea, New York, 1978.
2. N. L. Carothers, S. J. Dilworth and D. A. Trautman, *A theorem of Radon-Riesz type for the Lorentz space $L_{p,1}$. Preprint.*
3. N. Dunford and J. T. Schwartz, "Linear Operators Part I," Interscience, New York, 1967.
4. R. Engelking, "General Topology," PWN-Polish Scientific Publishers, Warsaw, 1977.
5. E. Hewitt, *A remark on density characters*, Bull. Amer. Math. Soc. **52** (1946), 641–643.
6. F. W. Lozier and R. H. Marty, *The number of Continua*, Proc. Amer. Math. Soc. **40** (1973), 271–273.
7. E. Marzcewski, *Séparabilité et multiplication cartésienne des espaces topologiques*, Fund. Math. **34** (1947), 127–143.
8. J. von Neumann, *Zur algebra der funktionaloperationen und theorie der normalen operatoren*, Math. Ann. **102** (1930), 370–427.
9. A. Pelczyński, *On Banach spaces containing $L_1(\mu)$*, Studia Math **30** (1968), 231–246.
10. E. S. Pondiczery, *Power problems in abatract spaces*, Duke Math. J. **11** (1944), 835–837.
11. H. P. Rosenthal, *On injective Banach spaces and the spaces $L_\infty(\mu)$ for finite measures μ*, Acta Math. **124** (1970), 205–248.
12. W. Rudin, "Functional Analysis," McGraw-Hill, New York, 1973.
13. L. Tzafriri, *An isomorphic characterization of L_p and c_0 spaces II*, Mich. Math. J. **18** (1971), 21–31.

Department of Mathematics
University of South Carolina
Columbia, SC 29208
U.S.A.

Singular integral operators: a martingale approach

Tadeusz Figiel

Introduction. Our purpose is to sketch a new approach to proving the boundedness of a vast class of linear operators which includes, e.g., the generalized Calderón-Zygmund operators discussed in [M]. The aproach is based on estimates of operator norms which come from applying recent results concerning the L_p-boundedness of martingale transforms.

In fact, the incentive for this work was the desire to extend some previously known boundedness results for operators acting in L_p-spaces of scalar-valued functions to the case of analogous spaces of X-valued Bochner measurable functions, where X is a Banach space. The recent results, due mainly to D. Burkholder and J. Bourgain, indicated that the class of the so-called *UMD*-spaces may be exactly the domain in which all results concerning Calderón–Zygmund integral operators and their generalizations remain valid. (Many Banach spaces which are important in classical analysis belong to that class.) Even the simplest singular integral operator, i.e., the Hilbert transform on the real line $\mathbf{R}$, has the property that its natural extension to an operator acting on $L_p(\mathbf{R}, X)$ where $1 < p < \infty$ is a bounded linear map if and only if the Banach space X is a *UMD*-space (cf. [Bu2] and [Bo]).

In order to obtain this extension it was necessary to find such proofs which make no use of any result that does not hold in the *UMD*-setting (for instance, the Fourier transform should be avoided, because it is not bounded in $L_2(\mathbf{R}, X)$, unless X is isomorphic to a Hilbert space). Thus it was a nice surprise that such austere methods could in fact lead to some results which were not less general than their counterparts established earlier with no restrictions on the range of admissible methods.

Our approach is indirect in the following sense. Rather than trying to prove that some "classical" operators are bounded, we started from considering certain rather new operators (cf. [F]), which in our opinion have a basic nature. (All the "singularities" which can occur in our context are neatly packaged inside the basic operators.) Having established precise estimates for the norms of those basic operators, we can take up the "general case". We just look at the class of those operators which can be realized as the sum of an absolutely convergent (in the operator norm) operator series whose summands are simple compositions of our basic operators. Then it turns out that the choice was sufficiently efficient for that class to contain so-called generalized Calderón-Zygmund operators (cf. [DJ], [M]) and much more. We illustrate the latter fact in the case of the famous $T(1)$-

theorem of David–Journé. In that case our aproach is applicable to kernels which (instead of being Hölder continuous off the diagonal) only have locally bounded mean oscillation which is dominated by an apropriate function of the distance to the diagonal of $\mathbf{R}^d \times \mathbf{R}^d$.

This short account cannot give full justice to all the important work done earlier in this area. The reader is advised to look into [M] to see some things which are not mentioned here and which could have influenced the present author.

We cannot adequately cover here all possibilities given by this aproach. Most of this exposition is devoted to stating the appropriate definitions and sketching the main steps.

The abstract scheme. Let us present first in abstract terms the general setting in which we shall be working.

Fix $d \geq 1$ and let $\mathcal{B}$ denote the space of bounded linear operators from the Hilbert space $L_2(\mathbf{R}^d)$ into itself. We shall consider a triple (CZ, WB, Υ), where WB and CZ are linear spaces with $CZ \subset WB$, while $\Upsilon : \mathcal{B} \to WB$ is a linear injection. We want to characterize those operators $T \in \mathcal{B}$ such that $\Upsilon(T) \in CZ$ and we shall do this entirely in terms of the class CZ. Thus we shall produce a condition, say $\mathcal{C}$, which is formulated in terms of CZ only, so that, if $J \in CZ$, then J satisfies $\mathcal{C}$ iff $J = \Upsilon(T)$ for some bounded operator $T \in \mathcal{B}$. Let us write $CZO = \Upsilon(\mathcal{B}) \cap CZ$.

The main advantage of our introducing the space CZ stems from a rather simple and transparent structure of that space. This is what makes CZ a good intermediary between the class of L_2-bounded operators and the class, denoted below by CZK, of appropriate singular kernels. (The class CZK is generated from CZ by means of a linear map Ψ which will be specified later on.) However, there may be something more intrinsic involved in this construction. In fact (besides the models based on the Haar system which we are going to describe in this exposition) there are other natural models for the triple (CZ, WB, Υ) and the just mentioned map Ψ, which share a very similar structure. In those models (which are defined in terms of other natural systems of functions, like the spline bases (cf. [CF]) or the wavelets (cf. [M]), the maps Υ and Ψ are replaced by their relatives, but the linear spaces WB, CZ and CZO remain essentially the same.

The main result in this exposition is a relative of the $T(1)$ theorem of David–Journé, because for those triples which we define below the condition $\mathcal{C}$ turns out to be analogous to the condition that $T(1)$, $T^*(1)$ are both in BMO, where $T = \Upsilon^{-1}(J)$. Namely, our condition $\mathcal{C}$ says that certain two objects (defined in terms of J only) should belong to a linear space closely related to BMO.

Actually, for each $J \in CZO$ the operator $T = \Upsilon^{-1}(J)$ will be associated in the usual sense with the kernel $\Psi(J)$ defined on $\Omega = \mathbf{R}^d \times \mathbf{R}^d \setminus \{(x,x) : x \in \mathbf{R}^d\}$, where Ψ will be a natural linear map $\Psi : CZ \to \mathcal{K}$ (and $\mathcal{K}$ stands for the set of all measurable "kernels" $K(x,y)$ on Ω). (In the case of Haar-type systems the resulting kernels are absolutely integrable on every diadic cube $I \subset \mathbf{R}^d \times \mathbf{R}^d$ whose interior is a subset of Ω.) It should be mentioned that the set $CZK = \Psi(CZ)$ of all kernels covered by our result will be rather rich.

The following remark is meant to clarify the relationship between our class CZK of all kernels and its subset $\Psi(CZO)$ consisting of those kernels that are associated with some members of the set $\Upsilon^{-1}(CZO)$ of our generalized Calderón-Zygmund operators.

In our models the kernel, i.e., the null space $\Psi^{-1}(0)$, of the map Ψ is always a subset of *CZO*. More precisely, for $J \in CZ$ the property $\Psi(J) = 0$ is equivalent to $J = \Upsilon(M_\phi)$ for some $\phi \in L_\infty(\mathbf{R}^d)$. (The operator $M_\phi \in \mathcal{B}$ is defined by $M_\phi(f) = \phi f$ for $f \in L_2(\mathbf{R}^d)$.) Thus, the kernel $K(x,y)$ (i.e., an element of *CZK*) cannot determine a unique operator $T \in \mathcal{B}$ such that $K = \Psi(\Upsilon(T))$, however, it carries enough information in order to find out whether or not it corresponds to such a T. Namely, if $K \in \Psi(CZO)$ and $J_1 \in CZ$, then $\Psi(J_1) = K$ implies $J_1 \in CZO$, i.e., the answer does not depend on making a particular choice of J_1 in $\Psi^{-1}(K)$.

Actually, the line of our argument has many points in common with the approach used in [DJ] and especially in [M]. The fact that the differences are not merely superficial becomes clear when one considers the problem of the boundedness of the Calderón–Zygmund operators in the L_p-spaces of X-valued Bochner measurable functions. The earlier approach, based e.g., on Cotlar's lemma, works only if X is isomorphic to a Hilbert space. However, the work of D. Burkholder and J. Bourgain has shown that in this context the proper class of Banach spaces is that of *UMD*-spaces. On the one hand, it is known (cf. [Bu]) that Calderón–Zygmund singular integral operators are bounded in the vector–valued *UMD* case, on the other if the Hilbert transform H is a bounded operator in $L_p(X)$ for some value of p, then the Banach space X must be a *UMD* space [Bo].

The estimates on which our proof is based depend only on martingale transform techniques and hence our results are applicable to the spaces of X-valued functions whenever X is a *UMD* space. For the sake of simplicity we shall skip the information about the X-valued versions in most statements given below.

The Haar system in $\mathbf{R}^d$ and related structures. The orthonormal system of Haar functions has been first introduced on the unit interval $[0,1]$. Later the analogues of that system have been developped in the multi-dimensional case (cf. [C]). In many applications, including the present context, it is not good enough to take tensor products of Haar functions on $\mathbf{R}$ – the supports of the functions in our systems on $\mathbf{R}^d$ should look like cubes rather than like parallelepipeds.

Below we recall the definitions and introduce a notation which will be convenient for our purposes.

First let $d = 1$. If $I \subset \mathbf{R}$ is an interval, then by $|I|$ we denote its length. By $\mathbf{Z}$ or $\mathbf{Z}^1$ we denote the set of integers. Diadic intervals in $\mathbf{R}^1$ are sets of the form $I = [k\,2^n, (k+1)2^n)$, where $k,\ n \in \mathbf{Z}$. We let $\mathcal{I}(1)$ denote the set of all diadic intervals in $\mathbf{R}^1$. It will be important for us that the group $\mathbf{Z}^1$ acts in a natural way on $\mathcal{I}(1)$. Given $m \in \mathbf{Z}$, every interval $I \in \mathcal{I}(1)$ is moved by m units of its own size, i.e., the image of I is the diadic interval $I + m|I| = [\inf I + m|I|, \sup I + m|I|)$.

The Haar functions on $\mathbf{R}^1$ are indexed by the set $\mathcal{I}(1)$, i.e., $HAAR(1) = \{\chi_I : I \in \mathcal{I}(1)\}$ (and each χ_I is supported on the set I). All of them can be in a simple way generated in terms of one of them. Namely, $\chi_{[0,1)} = 1_{[0,\frac{1}{2})} - 1_{[\frac{1}{2},1)}$ and for $I \in \mathcal{I}(1)$ one has

$$\chi_I(x) = |I|^{-\frac{1}{2}}\, \chi_{[0,1)}\Big(\frac{x - \inf I}{|I|}\Big).$$

The system $HAAR(1)$ is a complete orthonormal set in $L_2(\mathbf{R})$. Let us put for $I \in \mathcal{I}(1)$

$$\chi_I^1 = \chi_I, \qquad \chi_I^0 = |I|^{-\frac{1}{2}} 1_I = |\chi_I|.$$

Now let $d \in \mathbf{Z}$ be greater than 1. Diadic cubes in $\mathbf{R}^d$ are sets of the form

$$I = (I_0 + n_1|I_0|) \times (I_0 + n_2|I_0|) \times \ldots \times (I_0 + n_d|I_0|)$$

where $\mathbf{n} = (n_1, n_2, \ldots, n_d) \in \mathbf{Z}^d$ and $I_0 \in \mathcal{I}(1)$ is a diadic interval (which can be taken to be $[0, 2^n)$ for some $n \in \mathbf{Z}$). We let $\mathcal{I}(d)$ denote the set of all diadic cubes in $\mathbf{R}^d$. Given $I = I_1 \times \ldots \times I_d \in \mathcal{I}(d)$ and $\varepsilon = (\varepsilon_1, \ldots, \varepsilon_d) \in \{0,1\}^d$, we put

$$\chi_I^\varepsilon(x_1, \ldots, x_d) = \chi_{I_1}^{\varepsilon_1}(x_1) \cdot \ldots \cdot \chi_{I_d}^{\varepsilon_d}(x_d).$$

Clearly, the function χ_I^ε is supported on the set I. Let us we put

$$NS(d) = \{0,1\}^d \setminus \{(0,0,\ldots,0)\}.$$

The Haar system on $\mathbf{R}^d$ is defined by the formula

$$HAAR(d) = \{\chi_I^\varepsilon : I \in \mathcal{I}(d), \varepsilon \in NS(d)\}.$$

It is a complete orthonormal set in $L_2(\mathbf{R}^d)$.

Again, it is important that the group $\mathbf{Z}^d$ acts in a natural way on the set $\mathcal{I}(d)$. Namely, given $\mathbf{m} = (m_1, \ldots, m_d) \in \mathbf{Z}^d$, every diadic cube $I \in \mathcal{I}(d)$, say $I = I_1 \times \ldots \times I_d$, is transformed by $\mathbf{m}$ so that the image of I is the diadic cube $(I_1 + m_1|I_1|) \times \ldots \times (I_d + m_d|I_d|)$, which we again denote by $I + \mathbf{m}|I|$ (here $|I|$ refers to the size of I rather than to its volume).

Now we shall explain the notation which we use when dealing with functions on the space $\mathbf{R}^{2d}$ regarded as the product $\mathbf{R}^d \times \mathbf{R}^d$ (a pair $(\mathbf{x}, \mathbf{y}) \in \mathbf{R}^d \times \mathbf{R}^d$ being identified with the point $(x_1, \ldots, x_d, y_1, \ldots, y_d) \in \mathbf{R}^{2d}$).

We shall often represent elements $\sigma \in NS(2d)$ as pairs (ε, η), where ε, $\eta \in \{0,1\}^d$. Note that every element $h = \chi_I^\sigma$ of $HAAR(2d)$ can be written as $h = h^{(x)} \otimes h^{(y)}$ (i.e., $h(\mathbf{x}, \mathbf{y}) = h^{(x)}(\mathbf{x}) h^{(y)}(\mathbf{y})$ for each $(\mathbf{x}, \mathbf{y}) \in \mathbf{R}^d \times \mathbf{R}^d$). Indeed, it is clear that we can take $h^{(x)} = \chi_{I_1}^\varepsilon$ and $h^{(y)} = \chi_{I_2}^\eta$, where I_1, $I_2 \in \mathcal{I}(d)$ are apropriate cubes and not both ε and η are equal to $0 = (0, \ldots, 0)$.

Finally, we can specify the space WB of weakly bounded elements. We shall take $WB = l_\infty(HAAR(2d))$, the space of bounded functions on the countable set $HAAR(2d)$ equipped with the usual $\|\cdot\|_\infty$ norm.

The map $\Upsilon : \mathcal{B} \to WB$ from the space of bounded linear operators on $L_2(\mathbf{R}^d)$ into WB is defined by the formula

$$\Upsilon(T)(h) = (T(h^{(y)}), h^{(x)})$$

in which $(\cdot, \cdot)$ stands for the natural inner product on $L_2(\mathbf{R}^d)$.

Before we describe our subspace $CZ \subset WB$, let us mention that it will contain all finitely supported elements of WB. Every such element, interpreted as the corresponding linear combination of elements of $HAAR(2d)$, obviously determines a "kernel" on $\mathbf{R}^{2d}$. The map Ψ will be the natural extension of the latter correspondence.

Our further work with the kernels of operators in terms of $HAAR(2d)$ will be made more transparent by passing to a different parametrization of that set. Consider the set $IS(d) = \mathcal{I}(d) \times \mathbf{Z}^d \times NS(2d)$ as a new set of indices. For $i \in IS(d)$, say $i = (I, \mathbf{m}, (\varepsilon, \eta))$, let $h_i = \chi^{\varepsilon}_{I+\mathbf{m}|I|} \otimes \chi^{\eta}_{I}$. Obviously, h_i coincides with an element of $HAAR(2d)$, namely

$$h_{(I,\mathbf{m},(\varepsilon,\eta))} = \chi^{(\varepsilon,\eta)}_{(I+\mathbf{m}|I|)\times I} \, .$$

It is clear that the map $i \to h_i$ establishes a one-to-one correspondence between the index set $IS(d)$ and the Haar system $HAAR(2d)$. The latter map induces an isomorphism between WB and the space $l_\infty(IS(d))$. Formally, to each $J \in WB$ there corresponds the element $\tilde{J} \in l_\infty(IS(d))$, defined by

$$\tilde{J}(I, \mathbf{m}, \varepsilon, \eta) = J(h_{(I,\mathbf{m},(\varepsilon,\eta))}\,).$$

Thus $l_\infty(IS(d))$ is just another incarnation of WB. We often prefer to work with $\tilde{J}$ rather then with J, we hope that it will not do much harm if we sometimes skip the " $\tilde{}$ " in our notation.

The choice of the subspace which will be our CZ is not canonical. Knowing how the proof goes and which estimates of the basic operators can be improved (perhaps by taking advantage of some special features of the case that one is interested in), one can modify the choice of CZ so as to obtain a somewhat bigger subspace for which the arguments still work. At this moment it is preferable to choose a simple variant.

For $\mathbf{n} \in \mathbf{Z}^d$, we let $\delta_{\mathbf{n}}$ denote the indicator function of the subset of $IS(d)$ consisting of those $(I, \mathbf{m}, (\varepsilon, \eta))$ such that $\mathbf{m} = \mathbf{n}$.

Later on we shall specify a real function ω on $\mathbf{Z}^d$ with $\omega(\mathbf{n}) \geq 1$, in terms of which we define the space CZ and the norm $\|\cdot\|_{CZ}$, using the formula

$$\|J\|_{CZ} = \sum_{\mathbf{n}\in\mathbf{Z}^d} \omega(\mathbf{n}) \|\delta_{\mathbf{n}} \tilde{J}\|_\infty \, .$$

the space CZ will, of course, consist of those $J \in WB$ such that $\|J\|_{CZ} < \infty$.

In fact, some properties of CZ are true under the minimal assumption $\inf \omega(\mathbf{Z}^d) \geq 1$.

For $\mathbf{n} \in \mathbf{Z}^d$, we let $|\mathbf{n}| = \|\mathbf{n}\|_\infty = \max_i |n_i|$.

Finally, let us recall a definition of the spaces $H_1^{diad}(\mathbf{R}^d)$ and $BMO^{diad}(\mathbf{R}^d)$. (Both definitions work also in the case of X-valued functions, where X is a Banach space.)

A function $f \in L_\infty(\mathbf{R}^d)$ is said to be a diadic atom, if there is a diadic cube $I \in \mathcal{I}(d)$ such that $1_I f = f$, $\|f\|_\infty \leq |I|^{-1}$ and $(f, 1) = 0$. The space $H_1^{diad}(\mathbf{R}^d)$ consists of those functions $f \in L_1(\mathbf{R}^d)$ which can be represented as $\sum_i t_i f_i$, where the f_i's are diadic atoms and $\sum_i |t_i| < \infty$. Then $\|f\|_{H_1^{diad}(\mathbf{R}^d)}$ is the infimum of all those sums $\sum_i |t_i| < \infty$.

A locally integrable X-valued function g on $\mathbf{R}^d$ can be integrated against any diadic atom f on $\mathbf{R}^d$. It is easy to verify that

$$\sup\{ \int_{\mathbf{R}^d} fg\,dx : f \text{ is a diadic atom on } \mathbf{R}^d \} = \sup_{I\in\mathcal{I}(d)} |I|^{-1} \inf_c \|1_I(g-c)\|_1 .$$

Those functions g for which the latter expression is finite are said to have bounded mean oscillation on diadic cubes. The value of above expression is denoted by $\|g\|_{BMO^{diad}(\mathbf{R}^d)}$. The space $BMO^{diad}(\mathbf{R}^d)$ consists of all those functions (modulo the constant functions on $\mathbf{R}^d$). It is well known that the space $BMO^{diad}(\mathbf{R}^d)$ can be naturally identified with the dual space of $H_1^{diad}(\mathbf{R}^d)$ in the way we have described (say, in the scalar case).

Let us recall a characterization of elements of $BMO^{diad}(\mathbf{R}^d)$ in terms of their coefficients with respect to $HAAR(d)$ (in the case of scalar-valued functions). We put, for a scalar valued function a defined on the set $\mathcal{I}(d) \times NS(d)$,

$$\|a\|_{bmo(\mathbf{R}^d)} = \sup_I \Big(|I|^{-1} \sum_{J \in \mathcal{I}(d),\, J \subseteq I} \ \sum_{\varepsilon \in NS(d)} |a(J,\varepsilon)|^2 \Big)^{\frac{1}{2}}.$$

The condition $\|a\|_{bmo(\mathbf{R}^d)} < \infty$ is necessary and sufficient for a to be the family of Haar coefficients of a function $g \in BMO^{diad}(\mathbf{R}^d)$. Let us mention that the latter statement should be formulated in a different way, if one considers the X-valued case.

We let $bmo(\mathbf{R}^d)$ denote the space of those a such that $\|a\|_{bmo(\mathbf{R}^d)} < \infty$.

Main result. We start with two heuristic formulae which provide some motivation for the definition of Θ_x and Θ_y which is given below. It does not matter that the derivation of these formulae makes sense only for very special operators $T \in \mathcal{B}$, because after reading the statement of the theorem, it will become clear that for the operators $T \in \Upsilon^{-1}(CZO)$, all steps in the derivation of the two formulae are justified, and this is the class in which we are interested.

Our purpose is to identify the elements $T1$ and T^*1 (if they make sense) by determining their coefficients with respect to all the members χ_I^ε of the Haar system $HAAR(d)$. First we write

$$\begin{aligned}(T1, \chi_I^\varepsilon) = (1, T^*\chi_I^\varepsilon) &= \sum_{\mathbf{n} \in \mathbf{Z}^d} (1_{I+\mathbf{n}|I|}, T^*\chi_I^\varepsilon) = |I|^{\frac{1}{2}} \sum_{\mathbf{n} \in \mathbf{Z}^d} (T\chi^0_{I+\mathbf{n}|I|}, \chi_I^\varepsilon) \\ &= |I|^{\frac{1}{2}} \sum_{J \in \mathcal{I}(d),\, |J|=|I|} (\Upsilon T)(\chi_I^\varepsilon \otimes \chi_J^0).\end{aligned}$$

Then to obtain the second formula we just apply the first one to T^*

$$(T^*1, \chi_I^\varepsilon) = |I|^{\frac{1}{2}} \sum_{J \in \mathcal{I}(d),\, |J|=|I|} (\Upsilon T^*)(\chi_I^\varepsilon \otimes \chi_J^0) = |I|^{\frac{1}{2}} \sum_{J \in \mathcal{I}(d),\, |J|=|I|} \overline{(\Upsilon T)(\chi_J^0 \otimes \chi_I^\varepsilon)}.$$

Observe that in both formulae each term on the right-hand side is well defined but the infinite sums are in general case divergent. Since those terms make sense even if ΥT is replaced by any element of WB, and since elements of CZ are so nice, it is not unnatural to introduce two linear operators, say Θ_x and Θ_y, which map the space CZ into scalar functions on the set $HAAR(d)$. We simply put, for each $(I, \varepsilon) \in \mathcal{I}(d) \times NS(d)$,

$$\Theta_x(J)(\chi_I^\varepsilon) = |I|^{\frac{1}{2}} \sum_{J \in \mathcal{I}(d),\, |J|=|I|} J(\chi_I^\varepsilon \otimes \chi_J^0),$$

$$\Theta_y(J)(\chi_I^\varepsilon) = |I|^{\frac{1}{2}} \sum_{J \in \mathcal{I}(d),\, |J|=|I|} J(\chi_J^0 \otimes \chi_I^\varepsilon).$$

Since $\omega(\mathbf{n}) \geq 1$, the above sums are absolutely convergent for each $J \in CZ$. In fact, one has $|\Theta_z(J)(\chi_I^\varepsilon)| \leq |I|^{\frac{1}{2}}\|J\|_{CZ}$ for $z \in \{x, y\}$ and for each $(I, \varepsilon) \in \mathcal{I}(d) \times NS(d)$.

Now we can state the main result.

Theorem. *Suppose that the sequence ω in the definition of CZ satisfies the estimate $\omega(\mathbf{n}) \geq c\log(2 + |\mathbf{n}|)$ for some $c > 0$ and every $\mathbf{n} \in \mathbf{Z}^d$. Then, for $J \in CZ$,*

$$J \in \Upsilon(\mathcal{B}) \qquad \text{if and only if} \qquad \Theta_x(J),\, \Theta_y(J) \in bmo(\mathbf{R}^d).$$

Moreover, if this is the case (i.e., if $J \in CZO$), then the operators $T = \Upsilon^{-1}(J)$ and T^* are both bounded from $H_1^{diad}(\mathbf{R}^d)$ to $L_1(\mathbf{R}^d)$. Consequently, T and T^* are bounded as maps from $L_\infty(\mathbf{R}^d)$ into $BMO^{diad}(\mathbf{R}^d)$ and hence from $L_p(\mathbf{R}^d)$ into $L_p(\mathbf{R}^d)$ for $1 < p < \infty$.

Furthermore, if $\Upsilon(T) \in CZO$ and X is a Banach space with the UMD-property, then the statements made above concerning the boundedness of T in various function spaces carry over to the case of the spaces of X-valued Bochner measurable functions on $\mathbf{R}^d$.

Remark 1. Observe that the theorem implies in particular that, if $T \in \mathcal{B}$ satisfies $J = \Upsilon(T) \in CZ$, then T has all the boundedness properties listed in the theorem (there is no need to verify that $T1$, T^*1 are BMO^{diad}-functions, because this is assured by the theorem).

Remark 2. The BMO^{diad}-condition is weaker than the classical BMO-condition, hence our assertion concerning the L_p-boundedness covers a wider range of operators in $\mathcal{B}$. The elements of that class (i.e., of the set $\Upsilon^{-1}(CZO)$) usually are not L_∞-BMO and H_1-L_1 bounded, since the latter implies a stronger property, namely that both $T(1)$ and $T^*(1)$ (which, as we know, are well-defined elements of $BMO^{diad}(\mathbf{R}^d)$) belong to $BMO(\mathbf{R}^d)$. Now, for a properly chosen subclass CZ_{reg} of CZ (to be specified below), our main result implies easily that the condition $T(1), T^*(1) \in BMO(\mathbf{R}^d)$ is necessary and sufficient for the L_∞-BMO and H_1-L_1 boundedness. (The trick is that a function g on $\mathbf{R}^d$ is in $BMO(\mathbf{R}^d)$ iff the translates $\{g_z : z \in \mathcal{S}\}$ form a bounded subset of $BMO^{diad}(\mathbf{R}^d)$ for a sufficiently big subset $\mathcal{S} \in \mathbf{R}^d$ (we let $g_z(x) = g(x - z)$ for $x, z \in \mathbf{R}^d$).) The relevant property of the set $\mathcal{S}$ is that for every cube $Q \in \mathbf{R}^d$ there is $z \in \mathcal{S}$ and a diadic cube $I \in \mathcal{I}(d)$ such that $Q + z \subseteq I$ and the ratio of volumes of I and Q is bounded.)

In our setting, the class CZ_{reg} can be chosen to consist of those $T \in CZO$ whose all translates T_z by elements of a subset $\mathcal{S}$ of $\mathbf{R}^d$ have their images in WB, contained in a bounded subset of CZ, i.e., $\sup_{z\in\mathcal{S}} \|\Upsilon(T_z)\|_{CZ} < \infty$. A good choice for $\mathcal{S}$ is, for instance, the set $\bigcup_{n\in\mathbf{Z}}\{-2^n, 2^n\}^d$.

The above criterion for L_∞-BMO and H_1-L_1 boundedness of operators can be significantly improved, yet the proper setting for such results seems to be in the context of other systems of functions on $\mathbf{R}^d$ which we have already mentioned. The corresponding results for those systems can be deduced from the results presented in this exposition with the use of the equivalence established in [F].

Application to the Hilbert transform. Now let us see how this result can be applied in the simplest non-trivial case of a Calderón-Zygmund integral operator, i.e., to the Hilbert transform H. A thorough investigation of this special case will shed some light on typical problems arising when one studies such operators. So let us pretend, for the time being, that we do not know that H is L_2-bounded (or that we want to obtain, using our main result, the known fact that H is bounded in a space $L_p(\mathbf{R}, X)$ of vector-valued functions, where $1 < p < \infty$ and X is a *UMD*-space).

We want to construct a bounded operator on $L_2(\mathbf{R})$, say T, which is associated with the kernel K defined as $K(x, y) = \frac{1}{x-y}$ for $x, y \in \mathbf{R}$, $x \neq y$. As we know, such an operator (if it exists) is not unique, but our T should also be ***anti-symmetric*** (which means that $(Tf, g) = -(f, Tg)$ for $f, g \in L_2(\mathbf{R}^d)$) and ***real*** (i.e, it should take real-valued functions into real-valued functions).

We first find a suitable element of *CZ*, so that we can apply of the results we stated above. To do this we consider the expressions $a(I, n)$, where $I \in \mathcal{I}(1)$ is a diadic interval and n is an integer, defined (if $n \neq 0$) as

$$a(I, n) = \int_{(I+n|I|)\times I} K(x, y)dxdy,$$

because the *HAAR*(2)-coefficients of K can be easily expressed as soon as we know the $a(I, n)$'s. Namely, if I', $I'' \in \mathcal{I}(1)$ satisfy $I = I' \cup I''$ and $I'' = I' + |I'|$, then for $n \neq 0$

$$\begin{aligned}(K, \chi^{\varepsilon,\eta}_{(I+n|I|)\times I}) = |I|^{-1} \big(&a(I', 2n) + (-1)^{\varepsilon} a(I', 2n + 1) \\ &+ (-1)^{\eta} a(I'', 2n - 1) + (-1)^{\varepsilon+\eta} a(I'', 2n)\big).\end{aligned}$$

Of course, for $n = 0$ the above definition of $a(I, n)$ would not make much sense, hence for the time being let us just take $a(I, 0)$ to be 0 for each $I \in \mathcal{I}(1)$ and examine the consequences. Reasons for making that particular choice are made clear later on.

If $n \neq 0$, then the $a(I, n)$'s are well-defined, because K is absolutely integrable on the set $(I + n|I|) \times I$. We could compute the numerical value of each those $a(I, n)$'s exactly, but now it will be enough for us to observe that $a(I, n) = b(n) \cdot |I|$ for $n \neq 0$, where the sequence $b(n)$ satisfies $b(-n) = b(n)$, and $b(n) = n^{-1} + O(n^{-2})$ as n tends to ∞. Because of our earlier choice, letting $b(0) = 0$ we have now $a(I, n) = b(n) \cdot |I|$ for all $n \in \mathbf{Z}$.

Using the latter relation we obtain simple formulas for all $I \in \mathcal{I}(1)$ and $n \in \mathbf{Z} \setminus \{0\}$

$$\begin{aligned}(K, \chi^{1,0}_{(I+n|I|)\times I}) &= -(K, \chi^{0,1}_{(I+n|I|)\times I}) = \tfrac{1}{2}(b(2n - 1) - b(2n + 1)), \\ (K, \chi^{1,1}_{(I+n|I|)\times I}) &= b(2n) - \tfrac{1}{2}\left(b(2n - 1) + b(2n + 1)\right).\end{aligned}$$

Now we define $J \in WB$ by letting for $h \in HAAR(2)$, say $h = \chi^{\varepsilon,\eta}_{(I+n|I|)\times I}$ where $n = 0$ is now a legitimate choice,

$$J(h) = \tfrac{1}{2}\left((1 + (-1)^{\varepsilon+\eta})b(2n) + (-1)^{\varepsilon} b(2n + 1) + (-1)^{\eta} b(2n - 1)\right).$$

Observe that $J \in CZ$, because one has $|J(\chi^{\varepsilon,\eta}_{(I+n|I|)\times I})| = O(n^{-2})$ as $|n|$ tends to ∞.

It is very easy to check that our J satisfies $\Theta_x(J) = \Theta_y(J) = 0$, thus by our main result we have shown that $J \in CZO$, i.e. $J = \Upsilon(T)$ for some $T \in \mathcal{B}$. It is also easy to check that the operator T is anti-symmetric and real, as we requested (we started with a real-valued anti-symmetric kernel).

Now, if we just tried to choose any real values for the $a(I,0)$'s appearing in our recipe for J, so as to obtain an anti-symmetric element, say $J_1 \in WB$, then for each $I \in \mathcal{I}(1)$ there would be two equations to be satisfied, namely $J_1(\chi_I^1 \otimes \chi_I^1) = 0$ and $J_1(\chi_I^1 \otimes \chi_I^0) = -J_1(\chi_I^0 \otimes \chi_I^1)$. Using the formulae obtained earlier we can conclude that $a(I',0) = a(I'',0) = 0$. Thus our choice of the $a(I,0)$'s is that only one which yields a real-valued and anti-symmetric J.

It should be quite clear that, if one knows beforehand about H that it is an L_2-bounded, anti-symmetric, real operator which corresponds to the kernel cK, where $c = -\frac{1}{\pi}$, and hence its $HAAR(2)$-coefficients (i.e., the element $\Upsilon(H) \in WB$) coincide with those of cJ, then it now follows that $T = \frac{1}{c}H$, because Υ is a one-to-one mapping.

On the other hand, if we insist on not using anything from the classical theory of H (which is necessary if we want to obtain a new proof), then we still have to find the "kernel" of T, i.e., the element $\Psi(\Upsilon(T)) = \Psi(J)$, and show that it is equal almost everywhere on $\mathbf{R}^2$ to the function $K(x,y)$ with which we started. We also should establish the formula

$$(Tf,g) = \int K(x,y)\, f(y)\, g(x)\, dy\, dx,$$

say for those f, $g \in \mathrm{L}_2(\mathbf{R})$ which are supported on non-overlapping diadic intervals.

To do this we consider three sets of functions. We let $\mathcal{S}_a = HAAR(2) \setminus \{\chi_{I\times I}^{\epsilon,\eta} : I \in \mathcal{I}(1), (\epsilon,\eta) \in NS(2)\}$. Then let $\mathcal{S}_b$ be the set of the indicator functions of those $Q \in \mathcal{I}(2)$ such that Q is on one side of the diagonal, and no Q' stricly containing Q has that property. Finally, let $\mathcal{S}_c$ consist of the indicator functions of those $Q \in \mathcal{I}(2)$ which are either of the form $[0,2^n) \times [-2^n,0)$ or of the form $[-2^n,0) \times [0,2^n)$. We check easily that for $f \in \mathcal{S}_a \cup \mathcal{S}_b \cup \mathcal{S}_c$, one has $\int \Psi(J)f = \int Kf$. Of course, this is enough to show that the two functions are equal almost everywhere. The integral formula for $T(f,g)$ is then an easy consequence.

Comparison of kernels in *CZK* with standard kernels. A typical class of generalized Calderón-Zygmund operators is that which corresponds to the so called standard kernels (cf. [DJ], [M]). In the case of the d-dimensional space $\mathbf{R}^d$, where $d \geq 1$, the latter class consists of those functions K defined on the set $\Omega = \mathbf{R}^d \times \mathbf{R}^d \setminus \{(x,x) : x \in \mathbf{R}^d\}$ such that, for some $\delta > 0$ and $B < \infty$, one has

$$|K(x,y)| \leq B|x-y|^{-d},$$
$$|K(x,y) - K(x',y)| \leq B|x-x'|^{\delta}|x-y|^{-d-\delta},$$
$$|K(x,y) - K(x,y')| \leq B|y-y'|^{\delta}|x-y|^{-d-\delta},$$

for all (x,y), (x',y), $(x,y') \in \Omega$ such that $|x-x'| \leq \frac{1}{2}|x-y|$ and $|y-y'| \leq \frac{1}{2}|x-y|$.

Given a standard kernel K, does it belong to the class *CZK*?

The answer is yes, if K is associated to a bounded operator T on $L_2(\mathbf{R}^d)$ (which is then said to be a generalized Calderón-Zygmund operator). In that case we expect K

to coincide with $\Psi(J)$ where $J = \Upsilon(T)$. Of course, $J \in WB$, and we should check that $J \in CZ$ to make sure that $\Psi(J)$ is defined, showing at the same time that $\Psi(J) \in CZK$. (Then it is easy to verify that $\Psi(J) = K$ almost everywhere on $\mathbf{R}^{2d}$, and hence $K \in CZK$.) It will suffice if we get an estimate of the form

$$|J(\chi^{\varepsilon,\eta}_{(I+\mathbf{n}|I|)\times I})| \leq cB|\mathbf{n}|^{-d-\delta},$$

for all $I \in \mathcal{I}(d)$, $(\varepsilon, \eta) \in NS(2d)$ and all $\mathbf{n} \in \mathbf{Z}^d$ such that $|\mathbf{n}|$ is sufficiently big, say $\geq n_0$. (The subset of CZ determined by the above condition will be denoted briefly by CZ_δ.) Such an estimate follows rather easily from the second and third condition in the definition of standard kernels. Thus the class CZK described in our main result contains the kernels of all generalized Calderón-Zygmund operators in the sense of [DJ] or [M].

The answer may sometimes be negative (consider the kernel $K(x,y) = |x-y|^{-d}$ on Ω, the case of dimension $d = 1$ being rather easy).

On the other hand, in case of a general standard kernel $K(x,y)$, the question is harder. In order to construct a $J \in CZ$ such that $\Psi(J) = K$ we should determine all the $HAAR(2d)$-coefficients of J, i.e., the values $J(\chi^{\varepsilon,\eta}_{(I+\mathbf{n}|I|)\times I})$ for $I \in \mathcal{I}(d)$, $\mathbf{n} \in \mathbf{Z}^d$ and $(\varepsilon, \eta) \in NS(2d)$, in a manner consistent with the values obtained by integrating K against the Haar functions. Those coefficients whose index $(I, \mathbf{n}, \varepsilon, \eta)$ satisfies $\mathbf{n} \neq 0$ are defined uniquely by K and their absolute values are dominated by $cB|\mathbf{n}|^{-(d+\delta)}$. (We know this already for $|\mathbf{n}| \geq n_0$, and if $1 \leq |\mathbf{n}| < n_0$, this estimate follows from the first condition on the standard kernel.)

Thus a suitable J (with a finite CZ-norm) can be found for our K if and only if the remaining values of J (those taken on the remainder of the set $HAAR(2d)$) can be chosen to be bounded and so that $\Psi(J) = K$. Those remaining values correspond to the diagonal elements of $HAAR(2d)$, i.e., they form the element $\delta_0 J$, which is not determined uniquely by $\Psi(J)$, hence *a fortiori* by K. However, the condition $\Psi(J) = K$ is in this way reduced to a countable system of linear equations. (There is one equation for each diadic cube which is maximal in the family of all diadic cubes whose interior is a subset of Ω and there are also equations corresponding to diadic cubes which are located as those in the definition of the family $\mathcal{S}_c$.) Finding a bounded solution of that system is the heart of the matter. We are going to say more on this subject in another paper.

Elements of our class CZK need not be Hölder continuous off the diagonal. Their definition given in terms of CZ is at the same time a description in the sense of constructive theory of real functions. Here we shall describe only the subclasses of CZK that are induced by the classes CZ_δ, which generalize the standard kernels. Recall that, if $\delta > 0$ then $CZ_\delta = \{J \in CZ : |\tilde{J}(I, \mathbf{n}, \varepsilon, \eta)| = O(|\mathbf{n}|^{-d-\delta})\}$. In view of the obvious estimate $|J(\chi^{\varepsilon,\eta}_Q)| \leq \|\chi^{\varepsilon,\eta}_Q\|_{H^{diad}_1(Q)} \|1_Q \Psi(J)\|_{BMO^{diad}(Q)}$, valid for every Haar function $\chi^{\varepsilon,\eta}_Q \in HAAR(2d)$ such that $Q = (I + \mathbf{n}|I|) \times I$, $I \in \mathcal{I}(d)$, $\mathbf{n} \in \mathbf{Z}^d$ $|\mathbf{n}| > 1$, the class CZ_δ contains the set

$$\{J \in CZ : \sup_{I \in \mathcal{I}(d),\ \mathbf{n} \in \mathbf{Z}^d,\ |\mathbf{n}|>1} |\mathbf{n}|^{d+\delta} |I|^d \|1_{(I+\mathbf{n}|I|)\times I} \Psi(J)\|_{BMO^{diad}((I+\mathbf{n}|I|)\times I)} < \infty\}.$$

It turns out that the latter set equals CZ_δ (this can be checked easily using the characterization of $BMO^{diad}(Q)$, in terms of the $HAAR(2d)$-coefficients), i.e., for $K \in \Psi(CZ)$ the condition $K \in \Psi(CZ_\delta)$ is equivalent to the following estimate

$$\|1_{(I+\mathbf{n}|I|)\times I}K\|_{BMO^{diad}((I+\mathbf{n}|I|)\times I)} \le B|\mathbf{n}|^{-d-\delta}|I|^{-d}, \qquad I \in \mathcal{I}(d),\ \mathbf{n} \in \mathbf{Z}^d,\ |\mathbf{n}| > 1,$$

which should be compared with the second and third conditions on the standard kernels.

Basic operators and the decomposition of *CZO*. Let us present our basic operators in $L_2(\mathbf{R}^d)$.

Lemma 1. *For each $\mathbf{n} \in \mathbf{Z}^d$ there are operators $T_\mathbf{n}$ and $U_\mathbf{n} \in \mathcal{B}$ such that for each $\chi_I^\varepsilon \in HAAR(d)$ one has*

$$T_\mathbf{n}(\chi_I^\varepsilon) = \chi_{I+\mathbf{n}|I|}^\varepsilon, \qquad U_\mathbf{n}(\chi_I^\varepsilon) = \chi_{I+\mathbf{n}|I|}^0 - \chi_I^0.$$

Moreover, the operators $T_\mathbf{n}$ and $U_\mathbf{n}$ are both bounded from $H_1^{diad}(\mathbf{R}^d)$ to $L_1(\mathbf{R}^d)$ and from $L_\infty(\mathbf{R}^d)$ into $BMO^{diad}(\mathbf{R}^d)$ and hence from $L_p(\mathbf{R}^d)$ into $L_p(\mathbf{R}^d)$ for $1 < p < \infty$. If X is a Banach space with the UMD-property, then the operators $T_\mathbf{n}$ and $U_\mathbf{n}$ are both bounded in the corresponding spaces of X-valued Bochner measurable functions on $\mathbf{R}^d$. In each of those cases the norms of $T_\mathbf{n}$ and $U_\mathbf{n}$ are $\le C\log(2+|n|)$, where $C < \infty$ depends only on p and X.

This lemma essentially appears in [F], where it is proved in terms of the Haar system on the interval $[0,1]$ (cf. Theorem 1) and it is explained how to deduce the analogous fact for the system $HAAR(1)$ using a simple rescaling procedure. It is observed in [F] that there is no basic difficulty in extending the martingale method used to prove Theorem 1 to the case of $\mathbf{R}^d$ and we shall do that in detail elsewhere.

In the next lemma we introduce the operators (namely P_a and P_a^*) which are the diadic version of the operators known as paraproducts (cf. [DJ], [M]).

Lemma 2. *For each $a \in bmo(\mathbf{R}^d)$ there is $P_a \in \mathcal{B}$ such that for each $\chi_I^\varepsilon \in HAAR(d)$ one has*

$$P_a(\chi_I^\varepsilon) = a(I,\varepsilon)|I|^{-1}1_I.$$

Moreover, the operators P_a and P_a^* are both bounded in the same spaces as the operators $T_\mathbf{n}$ and $U_\mathbf{n}$ from Lemma 1 and in each of those cases the norm of P_a and of P_a^* is $\le C\|g\|_{bmo(\mathbf{R}^d)}$, where $C < \infty$ depends only on p and X.

In the case of spaces of scalar-valued functions Lemma 2 follows from standard facts (we give a proof in the final section — the case where $d = 1$ contains all the ingredients needed in the general case). The proof in the case of X-valued functions, which relies on an estimate due to Jean Bourgain (October 1987, unpublished) and is more technical, will be presented elsewhere.

We start with more notation. (The reader who is not familiar with bases in function spaces on $\mathbf{R}^d$ is advised to think first of the case of dimension $d = 1$.) Recall that the set $IS(d) = \mathcal{I}(d) \times \mathbf{Z}^d \times NS(2d)$ can be regarded as a convenient set of indices for the system $HAAR(2d)$ and that there is a natural isomorphism between WB and $l_\infty(IS(d))$.

We need more $\{0,1\}$-valued functions on $IS(d)$. The three new ones will be denoted by δ^{NN}, δ^{ZN} and δ^{NZ}. They are the indicator functions of the following three subsets of $IS(d)$: $\{(I,\mathbf{n},\varepsilon,\eta):\varepsilon\neq 0,\ \eta\neq 0\}$, $\{(I,\mathbf{n},\varepsilon,\eta):\varepsilon=0\}$ and $\{(I,\mathbf{n},\varepsilon,\eta):\eta=0\}$.

We shall show that the elements of the space CZ admit a nice decomposition. Each $J\in CZ$ can be written in the form $J=J_0+J_{\text{far}}$, where $J_{\text{far}}\in CZO$ and J_0 satisfies $\delta_0 J=J$, $\delta^{NN}J=0$. (Let us remark that J_0 contains the same information as the pair $(\Theta_x(J),\Theta_x(J))$ which appears in the main result and this shows a link with Lemma 2.) The element J_{far} is constructed below in terms of J, from the operators appearing in Lemma 1 and other operators which are diagonal with respect to $HAAR(d)$.

For elements $J\in CZ$ we shall use the notation $\|J\|_{\text{op}}$ which will denote either $+\infty$ or the norm of $\Upsilon^{-1}(J)$ in the sense of the space of operators which is being considered at the moment. (We prove first that some expressions define bounded operators on L_2 and later want to use the same formulae again in order to estimate, for instance, the norm from $L_\infty(\mathbf{R}^d)$ to $BMO^{diad}(\mathbf{R}^d)$.)

The component $\delta^{NN}J_{\text{far}}$ is defined as the sum of the series $S^{NN}=\sum_{\mathbf{n}\in\mathbf{Z}^d}\delta_{\mathbf{n}}\delta^{NN}J$. We shall estimate the $\|\cdot\|_{\text{op}}$ of the terms of S^{NN} and show that their sum is $\leq C\|J\|_{CZ}$. (This argument utilizes the operators $T_{\mathbf{n}}$ from Lemma 1.) It will follow that the element $\delta^{NN}J$, which is the sum of that series, belongs to CZO and is bounded in all the norms which we consider.

Then a similar trick can be done in order to express the component $\delta^{ZN}J_{\text{far}}$ as the sum of a series of the form $S^{ZN}=\sum_{\mathbf{n}\in\mathbf{Z}^d\setminus\{0\}}(\delta_{\mathbf{n}}\delta^{ZN}J-J^{ZN}_{\mathbf{n}})$. (We let the expressions $J^{ZN}_{\mathbf{n}}$ stand for suitable compensating summands (defined below) which satisfy $\delta_0\delta^{ZN}J^{ZN}_{\mathbf{n}}=J^{ZN}_{\mathbf{n}}$. Hence the sum of the series S^{ZN} vanishes where $\delta^{ZN}=0$ and it mimics J where $\delta_0=0$ and $\delta^{ZN}=1$.) This part uses the operators $U_{\mathbf{n}}$ from lemma 1.

The third component of J_{far}, i.e., $\delta^{NZ}J_{\text{far}}$, is obtained in an analogous way as $\delta^{ZN}J_{\text{far}}$. We use a series of the form $S^{NZ}=\sum_{\mathbf{n}\in\mathbf{Z}^d\setminus\{0\}}(\delta_{\mathbf{n}}\delta^{NZ}J-J^{NZ}_{\mathbf{n}})$, etc. (just swap the Z's with the N's). This case is dual to the previous one, the relevant operators being now the $U^*_{\mathbf{n}}$'s.

A fundamental estimate is the following simple inequality which reflects the unconditionality of the Haar system on $\mathbf{R}^d$ in various function spaces. This inequality is valid for $J\in WB$ (with $\|\cdot\|_{\text{op}}$ denoting the norm in the sense of the space of operators acting either on $L_p(\mathbf{R}^d)$ with $1<p<\infty$, or on $BMO^{diad}(\mathbf{R}^d)$, or $H_1^{diad}(\mathbf{R}^d)$, or the X-valued version of one of the above, X being a UMD-space)

$$\|\delta_0\delta^{NN}J\|_{\text{op}}\leq C\|\delta_0\delta^{NN}J\|_\infty,$$

and the constant $C<\infty$ may depend on the operator norm which is being considered, but it does not depend on J. For instance, if $\|\cdot\|_{\text{op}}$ denotes the norm of operators on L_2, then we can take $C=D^2$, where $D=2^d-1$. (Being more careful, one can get $C=D$ (rather than D^2) in the last estimate, the case of $d=1$ is really simpler.) The latter statement is quite obvious, because the operator T can be decomposed into D^2 components (corresponding to fixing the values of ε and η in the indices of J). Those components either are diagonal operators with respect to the complete orthonormal system $HAAR(d)$ in $L_2(\mathbf{R}^d)$ (if $\varepsilon=\eta$) or they can be written as compositions of such diagonal operators with certain permutations of $HAAR(d)$ (if $\varepsilon \neq \eta$), hence their operator norms in $L_2(\mathbf{R}^d)$ are dominated by the supremum of the coefficients. Thus $T = \Upsilon^{-1}(\delta_0\delta^{NN}J) \in \mathcal{B}$. A similar argument works for all the operator norms which are listed above, because the unconditionality of the Haar system in all those norms is known to be equivalent to X being a *UMD*-space (and the 1-dimensional Banach space $\mathbf{R}$ is a *UMD*-space).

Now, using the operators $T_{\mathbf{n}}$, we can reduce the estimates for $\delta_{\mathbf{n}}\delta^{NN}J$ to the fundamental estimate. It will be convenient for us to have a notation for the action of the group $\mathbf{Z}^d$ on the space WB (regarded as the space $l_\infty(IS(d))$). We put for $\mathbf{n} \in \mathbf{Z}^d$ and $J \in WB$

$$(\mathcal{T}_{\mathbf{n}}(\tilde{J}))(I, \mathbf{m}, \varepsilon, \eta) = \tilde{J}(I, \mathbf{m} - \mathbf{n}, \varepsilon, \eta).$$

For $J \in WB$ we have $\delta_{\mathbf{n}}\delta^{NN}J = \Upsilon(T_{\mathbf{n}} \circ \Upsilon^{-1}(\mathcal{T}_{-\mathbf{n}}(\delta_{\mathbf{n}}\delta^{NN}J)))$, hence we obtain

$$\|\delta_{\mathbf{n}}\delta^{NN}J\|_{\mathrm{op}} \leq \|T_{\mathbf{n}}\| \, \|\mathcal{T}_{-\mathbf{n}}(\delta_{\mathbf{n}}\delta^{NN}J)\|_{\mathrm{op}} \leq C\|T_{\mathbf{n}}\| \, \|\delta_{\mathbf{n}}\delta^{NN}J\|_\infty.$$

It should be clear now how to complete the estimate of $\|\delta^{NN}J_{\mathrm{far}}\|_{\mathrm{op}}$ in all the cases stated in the Theorem. (Here and also below the case of L_∞-BMO^{diad} and of H_1^{diad}-L_1 estimates may require some additional routine considerations which are better left to the diligent reader.)

Now we take up the second term, i.e., $\delta^{ZN}J_{\mathrm{far}}$. We have to describe those compensating summands which were denoted by $J_{\mathbf{n}}^{ZN}$ in the series for $\delta^{ZN}J_{\mathrm{far}}$. We simply take $J_{\mathbf{n}}^{ZN} = \mathcal{T}_{-\mathbf{n}}(\delta_{\mathbf{n}}\delta^{ZN}J)$. Consider the diagonal operator with respect to $HAAR(d)$, say Δ, defined by the formula $\Delta(\chi_I^\eta) = \tilde{J}(I, \mathbf{n}, 0, \eta)\chi_I^\eta$ and observe that

$$\Upsilon(U_{\mathbf{n}} \circ \Delta) = \delta_{\mathbf{n}}\delta^{ZN}J - \mathcal{T}_{-\mathbf{n}}(\delta_{\mathbf{n}}\delta^{ZN}J).$$

This allows us to show that the $\|\cdot\|_{\mathrm{op}}$ norm of the difference on the right-hand side is

$$\leq \|U_{\mathbf{n}}\| \, \|\Delta\|_{\mathrm{op}} \leq C\|U_{\mathbf{n}}\| \|\delta_{\mathbf{n}}\delta^{ZN}J\|_\infty,$$

which is again sufficient to complete the estimate. The third component of J_{far} is handled in an analogous way.

Having constructed the element $J_{\mathrm{far}} \in CZO$, one can evaluate the difference $J_0 = J - J_{\mathrm{far}}$, and note that it is closely related to our formulas for $\Theta_y(J)$ and $\Theta_x(J)$. Namely, if those two elements are both in $bmo(\mathbf{R}^d)$, then Lemma 2 yields two operators P_a and $P_{a'} \in \mathcal{B}$ so that $a = \Theta_y(J)$, $a' = \overline{\Theta_x(J)}$ and

$$\Upsilon(P_a) = \delta_0\delta^{ZN}J_0, \qquad \Upsilon(P_{a'}^*) = \delta_0\delta^{NZ}J_0.$$

Since $J_0 = \delta_0\delta^{ZN}J_0 + \delta_0\delta^{NZ}J_0$, one obtains that J_0, and hence also J, belongs to CZO. The converse is true too, i.e., one can show that, if $T \in \mathcal{B}$ satisfies the relation $\Upsilon(T) = J = \delta_0 J$, then both elements $\delta^{ZN}J$ and $\delta^{NZ}J$ are in $\Upsilon(\mathcal{B})$ and hence $\delta^{ZN}J = \Upsilon(P_a)$ and $\delta^{NZ}J = \Upsilon(P_{a'}^*)$, where a, $a' \in bmo(\mathbf{R}^d)$, $a = \Theta_y(J)$ and $a' = \overline{\Theta_x(J)}$.

This completes our sketch of the proof of the main result.

The boundedness of paraproduct operators. For $k \in \mathbf{Z}$ let $E_k = \mathbf{E}_{\Sigma_k}$ denote the operation defined in the space of locally integrable functions on $\mathbf{R}$, which transforms a function f into its conditional expectation with respect to the σ-field Σ_k generated by the diadic intervals of length 2^{-k}. We let $\Delta_k = E_k - E_{k-1}$.

Consider the bilinear operation P defined by the following formula

$$P(g, f) = \sum_k (E_k f)\, \Delta_{k+1} g,$$

where f,g are locally integrable functions on $\mathbf{R}$. (Throughout the argument one may assume, to keep the things simple, that g is in the linear span of $HAAR(1)$. The estimates proved for that special case would suffice for our purposes.)

For $1 < p < \infty$ and for suitable functions g, we would like to obtain an estimate of the form

$$\|P(g, f)\|_p \leq C \|f\|_p,$$

for all $f \in L_p(\mathbf{R})$, with a constant $C < \infty$ which depends only on g and p.

Using the basic estimate for the martingale transforms (cf. [Bu]), we can obtain the following. Let $I \in \mathcal{I}(1)$ be a diadic interval, $|I| = 2^{-m}$. If f, $\tilde{g}$ are scalar-valued and $f \in L_\infty(\mathbf{R})$, then for every $1 < p < \infty$ one has

$$\|1_I P(\tilde{g}, f)\|_p \leq C_p \|f\|_\infty \|1_I \tilde{g}\|_p.$$

(In fact, in the scalar case $C_p = \max\{p-1, (p-1)^{-1}\}$.) Indeed, when restricted to the probability space $(I, |I|^{-1} dx)$, equipped with the σ-fields induced by the sequence $(\Sigma_k)_{k>m}$, the sum in the definition of $P(\tilde{g}, f)$ can be regarded (if we skip the terms with $k \leq m$ which are constant on I) as the transform of the martingale difference sequence $(\Delta_{k+1}\tilde{g})$ by the adapted sequence $(E_k f)$.

Now, if $g \in BMO^{diad}(\mathbf{R})$ and $1 < p < \infty$, then

$$\inf_{c \in \mathbf{R}} \|1_I (g - c)\|_p \leq C'_p |I|^{\frac{1}{p}} \|g\|_{BMO^{diad}(\mathbf{R})}.$$

Let $\tilde{g} = 1_I (g - \tilde{c})$, where $\tilde{c}$ is chosen so as to minimize the left-hand side of the above formula. Observe that $1_I P(\tilde{g}, 1_I f) = 1_I (P(g, f) + c')$ for some c'. Hence we can estimate

$$\begin{aligned}\inf_c \|1_I (P(g, f) - c)\|_p &\leq \|P(\tilde{g}, 1_I f)\|_p \leq C_p \|f\|_\infty \|\tilde{g}\|_p \\ &\leq C_p |I|^{1/p} \|f\|_\infty \|g\|_{BMO^{diad}(\mathbf{R})}.\end{aligned}$$

If $p \in [1, \infty)$ is fixed, then the $BMO^{diad}(\mathbf{R})$-norm of a measurable function h on $\mathbf{R}$ is dominated by $\sup_{I \in \mathcal{I}(1)} |I|^{-1/p} \inf_c \|1_I (h - c)\|_p$, hence the last estimate implies that

$$\|P(g, f)\|_{BMO^{diad}(\mathbf{R})} \leq C_p \|g\|_{BMO^{diad}(\mathbf{R})} \|f\|_\infty.$$

Furthermore, if $f \in L_\infty(\mathbf{R})$ is a diadic atom supported by an interval $I \in \mathcal{I}(1)$ (i.e., $1_I f = f$, $\|f\|_\infty \leq |I|^{-1}$ and $(f, 1) = 0$), then we have $1_I P(g, f) = P(g, f)$ and also $(P(g, f), 1) = 0$. Therefore, using Hölder's inequality and the previous estimate, we obtain

$$\|P(g,f)\|_1 \leq |I|^{1/2} \|1_I P(g,f)\|_2 \leq 2 |I|^{1/2} \inf_c \|1_I (P(g,f) - c)\|_2 \leq |I| C \|f\|_\infty = C.$$

This shows that the operator $P(g, \cdot)$, where $P(g, \cdot)(f) = P(g, f)$, is bounded from $L_\infty(\mathbf{R})$ into $BMO^{diad}(\mathbf{R})$) and from $H_1^{diad}(\mathbf{R})$ into $L_1(\mathbf{R})$. By interpolation, it is also bounded on $L_p(\mathbf{R})$, for $1 < p < \infty$ (cf. [BS] for a proof of the interpolation formulae $(L_\infty, H_1)_{1/p,p} = L_p$ and $(BMO, L_1)_{1/p,p} = L_p$, which, after obvious modifications, works also in the diadic case and which is valid for spaces of vector-valued functions as well).

Thus we have proved that, for $1 < p < \infty$,

$$\|P(g,f)\|_p \leq C_p \|f\|_p \|g\|_{BMO^{diad}(\mathbf{R})}.$$

In particular $T = P(g, \cdot)$ is a bounded operator on $L_2(\mathbf{R})$. Computing $\Upsilon(T)$ we find that T is nothing else than P_a^*, where $a = (a_I)_{I \in \mathcal{I}(1)}$ is the family of the $HAAR(1)$-coefficients of g. Thus we have shown that P_a^* (and hence also P_a) has the required boundedness properties, at least in the case where X is finite dimensional.

It is not difficult to adjust the above argument to the case of $d > 1$. The argument needs a less obvious modification in order to become valid in the case where the function f takes values in an arbitrary UMD-space X. It depends on obtaining a new intermediate estimate for $P(g, f)$. That estimate, which we owe to Jean Bourgain, will be presented elsewhere.

Acknowledgment. I would to thank the organizers of the Strobl conference for their successful effort. My special thanks are due to Prof. Walter Schachermayer who was able to alleviate my reluctance to writing up these results.

References

[BS] C. Bennett and R. Sharpley, *Weak type inequalities for H^p and BMO,* Proc. Symp. in Pure Math. **35 (I)** (1979), 201–229.

[Bo] J. Bourgain, *Some remarks on Banach spaces in which martingale difference sequences are unconditional,* Ark. Mat. **21** (1983), 163–168.

Bo2] J. Bourgain, *Vector-valued singular integrals and the H^1 - BMO duality,* Probability Theory and Harmonic Analysis, J. A. Chao and W. A. Woyczyński, editors, Marcel Dekker (1986), New York 1–19.

[Bu] D. Burkholder, *A geometrical characterization of Banach spaces in which martingale difference sequences are unconditional,* Ann. Prob. **9** (1981), 997–1011.

Bu2] D. Burkholder, *Martingales and Fourier analysis in Banach spaces,* Springer-Verlag, Lecture Notes in Mathematics **1206** (1986), 61–108.

[C] Z. Ciesielski, *Haar orthogonal functions in analysis and probability,* Colloquia Mathematica Societatis Janos Bolyai, 49. Alfred Haar Memorial Conference, Budapest (1985), 25–56.

[CF] Z. Ciesielski and T. Figiel, *Spline bases in classical function spaces on compact C^∞ manifolds, II,* Studia Math., **76** (1983), 95–136.

[DJ] G. David and J.-L. Journé, *A boundedness criterion for generalized Calderón-Zygmund operators,* Annals of Math., **120** (1984), 371–397.

[F] T. Figiel, *On equivalence of some bases to the Haar system in spaces of vector-valued functions* Bull. Pol. Acad. Sci. **36** (1988), 119–131.

[M] Y. Meyer, *Wavelets and operators,* Proceedings of the Special Year in Modern Analysis at the University of Illinois, 1986–87, vol. 1, Edited by Earl R. Berkson, N. T. Peck & J. Uhl, Cambridge University Press, London Mathematical Society Lecture Note Series **137** (1989), 256–363.

Institute of Mathematics
Polish Academy of Sciences
ul. Abrahama 18
81-825 Sopot, Poland

Remarks about the interpolation of Radon-Nikodym operators

by

N. Ghoussoub, B. Maurey, W. Schachermayer

0 Introduction : A bounded linear operator T between Banach spaces X and Y is said to be *a Radon-Nikodym operator* if it maps bounded X-valued martingales into converging Y-valued martingales. In [G-J], an example is given of an R.N.P operator that does not factor through an R.N.P space (i.e a space where the identity is an R.N.P operator). However, the question is still open in the case of *a strong R.N.P operator* i.e. when the closure of the image of the unit ball by the operator is an R.N.P set in the range space.

This note consists of two parts. In the first, we shall prove that the interpolation method of [D-F-J-P] gives a positive answer to the above question if one considers what we call *a controllable R.N.P operator* (See the definition below). Most of the commonly known *strong R.N.P operators* are of this type. However, in the second part of this note, we shall give examples which are not. Actually, we shall construct *strong R.N.P operators* for which the [D-F-J-P] factorization scheme gives Banach spaces that contain c_0. On the other hand, these operators are not counterexamples to the general problem because we do not know whether they factor through an R.N.P space by another scheme.

It is well known that the above mentioned type of questions is equivalent to the following *interpolation problem*:

Suppose C is a closed circled convex bounded R.N.P subset of a Banach space

X. Can then one "embed" C in a Banach space with the R.N.P? In other words, if one considers the linear span Y of C equipped with the gauge $\|.\|_C$ of C. Can one factor the injection $j : Y \to X$ through a space with the R.N.P?

We shall also deal with the analogous problem of interpolating P.C.P sets: i.e those whose all closed subsets contain points of *weak to norm* continuity.

To justify the definitions below, we recall the following result proved in [G-M2]:

A closed convex bounded subset C of a separable Banach space X has the P.C.P (resp the R.N.P) if and only if it is a *strong* $w^* - G_\delta$ (resp *strong* $w^* - H_\delta$) in some dual space. (i.e there exists a separable Banach space Z such that X can be identified with a closed subspace of Z^*, in such a way that $\overline{C}^* \setminus C = \cup_n K_n$ with each K_n being w^*-compact (resp w^*-compact and convex) in Z^* and such that $\text{dist}(K_n, X) > 0$).

For two sets G and H, let us denote by

$$\text{dist}_m(G,H) = \inf\{\text{dist}(x,H); x \in G\} \quad (\text{resp, } \text{dist}_M(G,H) = \sup\{\text{dist}(x,H); x \in G\})$$

the minimal (resp, the "maximal") distance between the sets G and H.

We need the following definition:

Definition (1): A closed convex bounded subset of a Banach space X is said to be a *controllable P.C.P set* (resp a *controllable R.N.P set*) if in the above representation as a $w^* - G_\delta$-set (resp $w^* - H_\delta$-set) the K_n's can be chosen in such a way that for some constant $L > 0$, we have for each n:

$$\text{dist}_M(K_n, X) \leq L.\text{dist}_m(K_n, X).$$

In other words, the oscillations of the distance to X on each K_n should be uniformly bounded.

An operator T is then said to be *a controllable P.C.P* (resp *a controllable R.N.P*) operator if the closure of the image of the unit ball by T is a controllable P.C.P (resp R.N.P) set in the range.

I. A positive result:

In this section, we shall prove the following

Theorem (2): *If $T : X \to Y$ is a controllable P.C.P (resp R.N.P) operator between two Banach spaces X and Y, then T factors through a Banach space E with the P.C.P (resp the R.N.P). The space E being the D-F-J-P interpolation space associated to the operator T.*

Proof: Let $C = \overline{T(\text{Ball}(X)}$ and let $D = \cap_n[2^n C + 2^{-n}\text{Ball}(Y)]$. We need to show that D is also a P.C.P (resp R.N.P) subset of X.

Before proving that claim, let us show how it implies the above theorem. First recall that if E is the interpolation space given in [D-F-J-P], and if $T = SoR$ where $R : X \to E$ and $S : E \to Y$, then we have that S is a Tauberian embedding; that is: $S(\text{Ball}(E))$ is a closed subset of D and $S^*(\text{Ball}(Y^*)$ is norm dense in E^*. This necessarily implies that S is one to one.

Note that if D is a P.C.P (resp R.N.P) set, then E has P.C.P (resp R.N.P) since S is then a *nice G_δ embedding* [G-M1], (resp a *semi-embedding* [B-R]). Actually, the weak topologies on $\text{Ball}(E)$ and its image by S are identical.

Back to the claim, we shall prove that D is a $w^* - G_\delta$ (resp $w^* - H_\delta$) set in an appropriate dual space. For that, start by using the hypothesis on C to find a separable Banach space Z such that X is an isometric subspace of Z^* in such a way that $\overline{C}^* \backslash C = \cup_n K_n$ with each K_n being w^*-compact (resp w^*-compact and convex) and such that for some constant $L > 0$, we have $\text{dist}_M(K_n, X) \leq L.\text{dist}_m(K_n, X)$ for each n.

Let $\overline{D}^*$ be the w^*- closure of D in Z^*. For each $(n, k) \in \mathbf{N}^2$, define

$$L_{n,k} = 2^k[K_n + \frac{\text{dist}_m(K_n, X)}{2}.\text{Ball}(Z^*)] \cap \overline{D}^*.$$

Note that for each n, k the set $L_{n,k}$ is w^*-compact (resp w^*-compact and convex). It remains to prove that

$$\overline{D}^* \setminus D = \cup_n \cup_k L_{n,k}.$$

Note first that each $L_{n,k}$ is disjoint from X so that $\cup_n \cup_k L_{n,k} \subset \overline{D}^* \setminus D$. On the other hand, if $y^* \in \overline{D}^* \setminus D$, we let $\eta = \text{dist}(y^*, X) > 0$ and we choose k_0 such that $2^{-k_0} < \eta/4L$. We may write $y^* = y_1^* + y_2^*$ with $y_1^* \in 2^{k_0}\overline{D}^*$ and $\|y_2^*\| \leq 2^{-k_0}$. As $\text{dist}(y_1^*, X) \geq \eta - \frac{\eta}{4L} \geq \eta/2 > 0$, there is an integer n_0 such that $2^{-k_0} y_1^* \in K_{n_0}$.

This implies that $\text{dist}_M(K_{n_0}, X) \geq \text{dist}(2^{-k_0} y_1^*, X) > 2^{-k_0}\frac{\eta}{2}$ and by our assumption, we get that $\text{dist}_m(K_{n_0}, X) \geq 2^{-k_0}\frac{\eta}{2L}$.

Note now that

$$\|y_2^*\| \leq 2^{-k_0} \leq 2^{-k_0}.2^{k_0}.\frac{\eta}{4L} \leq 2^{k_0}\frac{\text{dist}_m(K_{n_0}, X)}{2}.$$

In other words

$$y^* = y_1^* + y_2^* \in 2^{k_0}K_{n_0} + 2^{k_0}\text{dist}_m(K_{n_0}, X)\text{Ball}(Y^*) = L_{n_0,k_0}$$

Thus proving the claim and the Theorem.

Remark 3): One can also prove directly that the interpolation space E has the P.C.P (resp the R.N.P) by showing that it is a $w^* - G_\delta$ (resp $w^* - H_\delta$) in the space F^* obtained by the same interpolation applied this time to the w^*-closure of C in Z^*.

II. The couterexample:

In this section, we establish the following:

Theorem (4): *There exists a closed absolutely convex R.N.P subset C of the unit ball of c_0 which is not a controllable R.N.P set. More precisely, the [D-F-J-P]-interpolation space associated to C contains an isomorphic copy of c_0 and hence fails to be an R.N.P space.*

Proof: Let $(a_n)_n$ and $(b_n)_n$ be two fixed positive sequences such that $a_n \uparrow +\infty$ and $b_n \downarrow 0$. For each pair of sequences $\tilde{\alpha} = (\alpha_n)_n$ and $\tilde{\beta} = (\beta_n)_n$ both going to zero, we shall construct a closed absolutely convex subset $C(\tilde{\alpha}, \tilde{\beta})$ of the unit ball of c_0 such that:

A) $C(\tilde{\alpha}, \tilde{\beta})$ is a $w^* - H_\delta$ in ℓ_∞ and therefore has the R.N.P.

B) If $0 < \beta_n \leq \alpha_n b_n$ and $1 \geq \alpha_n \geq a_n^{-1}$, then the set $D = \cap_n (a_n C(\tilde{\alpha}, \tilde{\beta}) + b_n \mathrm{Ball}(c_0))$ contains a subset that is affinely isometrically isomorphic to the unit ball of c_0 and therefore the [D-F-J-P]- factorization space associated to C fails the R.N.P.

First we introduce some notation. Let $\mathcal{T} = \cup_{k=1}^{\infty} \{-1, +1\}^k$ be the dyadic tree and let $\overline{\mathcal{T}} = \mathcal{T} \cup \Gamma$ where $\Gamma = \{-1, +1\}^{\mathbf{N}}$. Let $J = \cup_{k=1}^{\infty} J_k$ with each J_k being the power set of $\{-1, +1\}^k$. We shall denote by P_k an element of J that belongs to J_k. Let I be the countable index set $I = I_1 \cup I_2$ where $I_1 = \mathbf{N}^* = \mathbf{N} \setminus \{0\}$ and $I_2 = J \times \mathbf{N}^* \times \mathbf{N}^*$. For $i \in I$, we either write $i = (n)$ to indicate that it belongs to I_1 or we write $i = (P_k, l, n)$ with $P_k \in J$ and $(l, n) \in \mathbf{N}^{*2}$ to indicate that $i \in I_2$. The space $c_0 = c_0(I)$ will be defined over the index set I.

Fix now sequences $(\alpha_n)_n$ and $(\beta_n)_n$ both going to zero.

For $n \in \mathbf{N}^*$ and $(\varepsilon_1, ..., \varepsilon_m) \in \mathcal{T}$, define the following elements of $c_0(I)$:

$$\varphi^n(\varepsilon_1, ..., \varepsilon_m)(i) = \begin{cases} \alpha_n \varepsilon_j & \text{if } i \in I_1 \text{ and } i = (j) \text{ for some } 1 \leq j \leq m \\ \beta_n & \text{if } i = (P_k, l, n) \in I_2, 1 \leq l \leq m, 1 \leq k \leq m \text{ and } (\varepsilon_1, ..., \varepsilon_k) \in P_k \\ 0 & \text{elsewhere.} \end{cases}$$

For $n = 0$ and $(\varepsilon_1, ..., \varepsilon_m) \in \mathcal{T}$, define

$$\varphi^0(\varepsilon_1, ..., \varepsilon_m)(i) = \begin{cases} \varepsilon_j & \text{if } i \in I_1 \text{ and } i = (j) \text{ for some } 1 \leq j \leq m \\ 0 & \text{elsewhere.} \end{cases}$$

It is clear that the above defined elements are finitely supported elements in $c_0(I)$

and their norms are $\|\varphi^0(\varepsilon_1,...,\varepsilon_m)\| = 1$ while $\|\varphi^n(\varepsilon_1,...,\varepsilon_m)\| = \alpha_n$ for every $n \in \mathbf{N}^*$ and any $(\varepsilon_1,...,\varepsilon_m) \in \mathcal{T}$.

Define now for each $n \in \mathbf{N}$, the sets $A_n = \{\varphi^n(\varepsilon_1,...,\varepsilon_m); (\varepsilon_1,...,\varepsilon_m) \in \mathcal{T}\}$ and C_n =closed convex circled hull of A_n. Let then $A = \cup_{n\in\mathbf{N}^*} A_n \cup \{0\}$ and let $C = C(\tilde{\alpha}, \tilde{\beta})$ be the closed convex hull of $\cup_{n\in\mathbf{N}^*} C_n$. We will denote by $\tilde{A}$ the w^*-closure in ℓ_∞ of any subset A of c_0. We start by proving assertion (A). We divide the proof into four steps:

Claim 1): For each $n \geq 1$, there exists a homeomorphic embedding $\tilde{\varphi}^n : \bar{\mathcal{T}} :\to \ell^\infty(I)$ such that $\tilde{\varphi}^n(\bar{\mathcal{T}}) = \tilde{A}_n$ and $\varphi^n(\mathcal{T}) = A_n = \tilde{A}_n \cap c_0(I)$.

Indeed, it is enough to take for $\tilde{\varphi}^n$ the continuous extension of φ^n to $\bar{\mathcal{T}}$. Note that the action of $\tilde{\varphi}^n$ on an element $\gamma = (\varepsilon_j)_{j=1}^\infty \in \Gamma$ is given by

$$\tilde{\varphi}^n(\gamma)(i) = \begin{cases} \alpha_n \varepsilon_j & \text{if } i \in I_1 \text{ and } i = (j) \text{ for some } 1 \leq j < \infty \\ \beta_n & \text{if } i = (P_k, l, n) \in I_2 \text{ is such that } l \in \mathbf{N}^* \text{ and } (\varepsilon_1,...,\varepsilon_k) \in P_k \\ 0 & \text{elsewhere.} \end{cases}$$

The proof of the following claim is obvious and is left to the reader.

Claim 2): For $n \geq 1$, $P_k \in J$, and $x \in \tilde{A}_n$, the limit when $l \to \infty$ of $x((P_k, l, n))$ exists and is equal to either 0 or β_n. Moreover, the limit is β_n if and only if x is equal to $\tilde{\varphi}^n(\gamma)$ for some $\gamma = (\varepsilon_j)_{j=1}^\infty \in \Gamma$ with $(\varepsilon_1,...,\varepsilon_k) \in P_k$.

For a signed measure μ supported on $\tilde{A}$, we shall denote by bar(μ) the difference bar(μ^+) - bar(μ^-) of the barycenters of the positive and negative parts of μ.

Claim 3): a) We have:

$$\tilde{C}_n = \{\text{bar}(\mu); \mu \quad \text{is a signed measure on } (\tilde{A}_n, w^*) \text{ with } \|\mu\|_1 \leq 1\}.$$

b) If μ is a signed measure on $(\tilde{A}_n, w^*)$ with $\|\mu\| \leq 1$ and if $x = \text{bar}(\mu)$, then the following assertions are equivalent:

i) $x \in C_n = \tilde{C}_n \cap c_0$.

ii) $|\mu|(\tilde{\varphi}^n(\bar{\mathcal{T}} \setminus \mathcal{T})) = 0$.

iii) For all $P_k \in J$, $\liminf_{l\to\infty} x(P_k, l, n) \leq 0$ and $\liminf_{l\to\infty}(-x(P_k, l, n)) \leq 0$.

The proof of assertion a) of claim 3) as well as the implications $ii) \Rightarrow i) \Rightarrow iii)$ of b) are obvious.

To prove that $iii) \Rightarrow ii)$, assume that μ and $x = \mathrm{bar}(\mu)$ verify the condition iii). Suppose that ii) does not hold: that is $|\mu|(\tilde{\varphi}^n(\bar{\mathcal{T}} \setminus \mathcal{T})) = |\mu|(\tilde{\varphi}^n(\Gamma)) > 0$. To every $P_k \in J$, we associate the relatively clopen subset V_{P_k} of $\tilde{\varphi}^n(\Gamma) = \tilde{A}_n \setminus A_n$ defined by

$$V_{P_k} = \{\varphi^n((\delta_j)_{j=1}^{\infty}); (\delta_j)_{j=1}^{\infty} \in \Gamma \quad \text{and } (\delta_1, ..., \delta_k) \in P_k\}).$$

Note that $\{V_{P_k}; P_k \in J\}$ forms a Boolean algebra generating the Borel σ-algebra of $(\tilde{A}_n \setminus A_n, w^*)$. The assumption therefore implies the existence of some $P_k \in J$ such that $\eta = \mu(V_{P_k}) \neq 0$. We can also suppose that $\eta > 0$ by passing to $-\mu$ and $-x$ if necessary. Find now $l_0 \geq k$ such that $|\mu|(\{\tilde{\varphi}^n(t); t \in \mathcal{T}, |t| \geq l_0\}) < \eta/2$. Since x is the barycenter of μ, it follows that for every $l \geq l_0$, $x(P_k, l, n) > \eta/2$ which is a contradiction.

Claim 4): a) $\tilde{C} = \{\sum_{n=1}^{\infty} \lambda_n x_n; \sum_{n=1}^{\infty} |\lambda_n| \leq 1, x_n \in \tilde{C}_n\}$.

b) An element $x \in \tilde{C}$ is in C if and only if for every $n \geq 1$ and $P_k \in J$, we have $\liminf_{l\to\infty} x(P_k, l, n) \leq 0$ and $\liminf_{l\to\infty}(-x(P_k, l, n)) \leq 0$. In particular, C is a strong $w^* - H_\delta$ set in $\tilde{C}$ and therefore it is an R.N.P set.

Assertion a) follows from the fact that $\alpha_n = \sup\{\|x\|; x \in \tilde{C}_n\}$ tends to zero when $n \to \infty$.

For b), let $x \in \tilde{C}_n$ and write $x = \sum_{n=1}^{\infty} \lambda_n x_n$ with $\sum_{n=1}^{\infty} |\lambda_n| \leq 1$ and $x_n \in \tilde{C}_n$. If each x_n with $\lambda_n > 0$ is in C_n, then clearly $x \in C$. On the other hand, if there is n_0 such that $\lambda_{n_0} > 0$ and $x_{n_0} \in \tilde{C}_{n_0} \setminus C_{n_0}$, then by claim 3), we may find $P_k \in J$ such that $\liminf_{l\to\infty} x_{n_0}(P_k, l, n_0) > 0$ or $\liminf_{l\to\infty}(-x_{n_0}(P_k, l, n_0)) > 0$. Noting that $\lambda_{n_0} x_{n_0}(P_k, l, n_0) = x(P_k, l, n_0)$ for every $l \in \mathbf{N}^*$ we obtain that

$\liminf_{l\to\infty} x(P_k, l, n_0) > 0$ or $\liminf_{l\to\infty} - x(P_k, l, n_0) > 0$. This finishes the proof of b) and therefore part (A) of the Theorem.

To prove part (B), assume that $\alpha_n^{-1} \le a_n$ and $0 < \beta_n \le \alpha_n b_n$ for every n. We claim that the set $D = \cap_{n=1}^{\infty}(a_n C + b_n.\text{Ball}(c_0(I)))$ contains C_0 which in turn can be identified with the unit ball of c_0.

Indeed, we only need to show that $\varphi^0(\varepsilon_1, ..., \varepsilon_m) \in a_n.C + b_n\text{Ball}(c_0(I))$ for every $n \ge 1$ and any $(\varepsilon_1, ..., \varepsilon_m) \in \mathcal{T}$. Since $\alpha_n^{-1} \le a_n$ we have that $\alpha_n^{-1}\varphi^n(\varepsilon_1, ..., \varepsilon_m) \in a_n C_n \subset a_n C$ and since $\beta_n \le \alpha_n b_n$ we get

$$\|\alpha_n^{-1}\varphi^n(\varepsilon_1, ..., \varepsilon_m) - \varphi^0(\varepsilon_1, ..., \varepsilon_m)\| = \frac{\beta_n}{\alpha_n} \le b_n.$$

This clearly proves assertion (B).

To establish the theorem, it is enough to consider the sequences $a_n = 2^n$, $b_n = 2^{-n}$, $\alpha_n = 2^{-n}$, $\beta_n = 4^{-n}$ to conclude.

References

[B-R] J. Bourgain, H.P. Rosenthal: *Applications of the theory of semi-embeddings to Banach space theory.* J. Funct. Anal. **52** (1983), p 149-187.

[B] R.D. Bourgin: *Geometric aspects of convex sets with the Radon- Nikodym property.* Lecture notes in Mathematics, 993, Springer- Verlag(1983).

[D-F-J-P] W. J Davis, T. Figiel, W.B Johnson and A. Pelczynski: *Factoring weakly compact operators.* J. Funct. Analysis.17.(1974), p.311-327.

[G-M1] N. Ghoussoub, B. Maurey: *G_δ -embeddings in Hilbert space.* Journal of Functional Analysis.Vol.61, (1985) p.72-97

[G-M2] N. Ghoussoub, B. Maurey: *H_δ -embeddings in Hilbert space and Optimization on G_δ -sets.* Memoirs of the A.M.S. No.34 (1986).

[G-J] N. Ghoussoub, W.B. Johnson: *Counterexamples to several problems on the factorization of bounded linear operators.* Proc. A.M.S **92**, (1984), p 233-238.

N. Ghoussoub
Department of Mathematics
University of Bristish Columbia
Vancouver, B.C, V6T1Y4
Canada.

B. Maurey
U.F.R de Mathématiques
Université Paris VII
75251 Paris Cedex 05
France.

W. Schachermayer
Institut für Mathematik
J. Kepler Universität
A-4040 Linz
Austria.

SYMMETRIC SEQUENCES IN FINITE-DIMENSIONAL NORMED SPACES

W. T. Gowers

Trinity College, Cambridge,
CB2 1TQ, England

§1. Introduction

In 1982 and 1985 Amir and Milman published two important papers [2,3] showing that concentration of measure could be used to find large almost symmetric basic sequences in certain finite-dimensional normed spaces. In particular they exploited Azuma's inequality [4] and consequences of it due to Maurey [9] and Schechtman [11] in order to obtain the measure-concentration results they needed. In this survey we consider a number of properties of finite-dimensional normed spaces or finite basic sequences. Formally, we are interested in the following question. Given a space X of n-dimensions, let $\sigma(X)$ be the cardinality of the largest $(1+\epsilon)$-symmetric basic sequence contained in X. Then given some property of n-dimensional spaces, what is the minimal value of σ over all spaces satisfying that property? We shall outline what is known about this quantity for various properties of interest. We shall also consider similar questions about properties of finite sequences, looking in this case for large $(1+\epsilon)$-symmetric block bases. Results about the existence of symmetric sequences, as well as being intrinsically interesting, are useful as steps in the proofs of local versions of Krivine's theorem and the Maurey-Pisier theorem. Indeed, this was the principal motivation of [2,3]. Such results can also be found in [10].

Recall that a sequence of vectors $x_1, \ldots, x_n$ in a normed space X is said to be 1-*symmetric* or simply *symmetric* if, for any permutation $\pi \in S_n$, any choice of signs $\epsilon_1, \ldots, \epsilon_n$ and any sequence of scalars $a_1, \ldots, a_n$,

$$\left\| \sum_1^n \epsilon_i a_i x_{\pi(i)} \right\| = \left\| \sum_1^n a_i x_i \right\|.$$

If under the same conditions one has only

$$\left\|\sum_1^n \epsilon_i a_i x_{\pi(i)}\right\| \leqslant \alpha \left\|\sum_1^n a_i x_i\right\| \tag{1}$$

then the basis $x_1, \ldots, x_n$ is *α-symmetric.* A basis which is $(1+\epsilon)$-symmetric for a fairly small ϵ is often said to be *almost symmetric.*

If (1) holds whenever the permutation π is just the identity permutation, then the basis $x_1, \ldots, x_n$ is said to be *α-unconditional.* A basis is also said to be *almost unconditional* if it is $(1+\epsilon)$-unconditional for a small value of ϵ.

The next definition is not standard, but it is a natural one in this context. Suppose a sequence of scalars $a_1, \ldots, a_n$ is fixed. We shall say that a basis $x_1, \ldots, x_n$ is *α-symmetric at* $a_1, \ldots, a_n$ if (1) holds for the sequence $a_1, \ldots, a_n$. If $\mathbf{a}$ is the vector $\sum_1^n a_i x_i$ we shall also say that $x_1, \ldots, x_n$ is *α-symmetric at* $\mathbf{a}$. If the norm $\|.\|$ being considered is not clear from the context, we shall sometimes say that $x_1, \ldots, x_n$ is *α-symmetric at* $a_1, \ldots, a_n$ or $\mathbf{a}$ *under* $\|.\|$. We shall also speak loosely of a basis being *almost symmetric at* a vector or sequence.

Given a basis $x_1, \ldots, x_n$, a *block basis* is a sequence $y_1, \ldots, y_m$ where each element y_i is a vector of the form $\sum_{j\in A_i} \lambda_j x_j$ and the sets $A_1, \ldots, A_m$ are disjoint. It is more common to require also that if $i_1 < i_2$ and $j_1 \in A_{i_1}$ and $j_2 \in A_{i_2}$ then $j_1 < j_2$, but this definition is more suitable in a local context, and was the definition used by Amir and Milman.

When the arguments of a function are specified, it should be understood that the function depends on these variables only. The results, presented here for spaces with real scalars, apply equally in the complex case. Finally, dimensions are often not written as integers, but all the results are true upon taking integer parts.

§2. Positive Results

The following two theorems may be found in [6,7].

Theorem 1. *Let $\epsilon > 0$, $C > 1$, $1 \leqslant p < \infty$ and let $x_1, \ldots, x_n$ be a sequence of vectors in a normed space, satisfying the condition*

$$\left(\sum_{i=1}^n |a_i|^p\right)^{1/p} \leqslant \left\|\sum_{i=1}^n a_i x_i\right\| \leqslant C\left(\sum_{i=1}^n |a_i|^p\right)^{1/p}$$

for every sequence $a_1, \ldots, a_n$ *of scalars. Then* $x_1, \ldots, x_n$ *has a* $(1+\epsilon)$*-symmetric block basis* $\mathbf{u}_1, \ldots, \mathbf{u}_m$ *with* ± 1*-coefficients and blocks of equal length, of cardinality* $m = \alpha(\epsilon, p, C)\, n/\log n$.

Theorem 2. *Let* $\epsilon > 0$, $1 \leqslant p < 2$ *and let* $x_1, \ldots, x_n$ *be a sequence of unit vectors in a normed space, satisfying the condition*

$$\mathsf{E}\left\| \sum_{i=1}^{n} \epsilon_i x_i \right\| \geqslant n^{1/p} ,$$

where the expectation is taken uniformly over all choices $\epsilon_1, \ldots, \epsilon_n$ *of signs. Then* $x_1, \ldots, x_n$ *has a* $(1+\epsilon)$*-symmetric block basis* $\mathbf{u}_1, \ldots, \mathbf{u}_m$ *with* ± 1*-coefficients and blocks of equal length, of cardinality* $m = \alpha(\epsilon) n^{2/p-1} / \log n$.

These results improve on the bounds of $\alpha(\epsilon, p, C) n^{1/3}$ and $\alpha(\epsilon) n^{(2-p)^2/2p^3}$ obtained by Amir and Milman [2,3], and in fact, as we shall see in the next section, they are close to being best possible. Here we shall give a brief idea of how they are proved, and where the proof differs from that of Amir and Milman. Since the proofs of Theorems 1 and 2 are very similar, we shall concentrate on Theorem 1 in the case $p = 1$.

Let $n = mh$ and let $(\mathbf{R}^n, \|.\|)$ be a normed space such that the standard basis $\mathbf{e}_1, \ldots, \mathbf{e}_n$ satisfies the conditions of Theorem 1 when $p = 1$. Define a random block basis $\mathbf{u}_1, \ldots, \mathbf{u}_m$ as follows. First take a random partition of $[n]$ into m sets $A_1, \ldots, A_m$ of size h, where the set of all such partitions is given the uniform distribution. Then pick a random sequence of signs $(\epsilon_1, \ldots, \epsilon_n) \in \{-1, 1\}^n$, where again these are uniformly distributed. Let the block basis be $\mathbf{u}_1, \ldots, \mathbf{u}_m$, where for each j the vector $\mathbf{u}_j$ is the restriction of $(\epsilon_1, \ldots, \epsilon_n)$ to A_j. The main idea of Amir and Milman was to use concentration of measure to show that if h is large enough, then such a block basis will, with high probability, be almost symmetric. Our proof is also of this kind, but needs a smaller value of h.

Given any sequence $\mathbf{a} = (a_1, \ldots, a_m) \in \mathbf{R}^m$, let $\tilde{\mathbf{a}}$ be the random vector $\sum_{j=1}^{m} a_j \mathbf{u}_j$. By a standard application of a general result of Schechtman [11], or directly from Azuma's inequality, one can show that

$$\mathsf{P}\Big[\big| \|\tilde{\mathbf{a}}\| - \mathsf{E}\,\|\tilde{\mathbf{a}}\| \big| \geqslant \delta \,\|\tilde{\mathbf{a}}\| \Big] \leqslant \exp\Big(-\delta^2 \,\|\tilde{\mathbf{a}}\|_1^2 \Big/ 8C^2 \,\|\tilde{\mathbf{a}}\|_2^2 \Big) . \tag{2}$$

Now, given any vector generated by the random block basis, the ratio of its ℓ_1-norm to its ℓ_2-norm is at least $\sqrt{h}$. Using this estimate in (2), fact that a norm is controlled by its behaviour on a sufficiently fine net of its unit sphere, and the fact that, given the unit sphere of an m-dimensional normed space, it has a δ-net of cardinality at most $(1+2/\delta)^m$, one can show that $\mathbf{u}_1,\ldots,\mathbf{u}_m$ has a high probability of being $(1+\epsilon)$-symmetric provided $h \geqslant \beta(\epsilon,C)n^{1/2}$, and thus whenever $m \leqslant \alpha(\epsilon,C)n^{1/2}$.

In order to improve this estimate to $\alpha(\epsilon,C)n/\log n$, the main idea is to consider various classes of vectors separately. Note that the estimate $\|\tilde{\mathbf{a}}\|_1 / \|\tilde{\mathbf{a}}\|_2 \geqslant \sqrt{h}$ is very weak if for example $\tilde{\mathbf{a}} = \sum_{j=1}^m \mathbf{u}_j$. Eliminating this and similar weaknesses leads to our improved result.

The proof divides naturally into two parts, which we shall not prove.

Lemma 3. *Let $\delta > 0$, let $(\mathbf{R}^m, \|.\|)$ be a normed space and set $N = m^{\delta^{-1}\log(3\delta^{-1})}$. There exist N vectors $\mathbf{a}_1,\ldots,\mathbf{a}_N$ such that if $\|.\|$ is $(1+\delta)$-symmetric at $\mathbf{a}_i$ for every i, then the standard basis of $\mathbf{R}^m$ is $(1+\delta)(1-6\delta)^{-1}$-symmetric.*

Lemma 4. *Let ϵ and C be as in the statement of Theorem 1, let $\delta = \epsilon/11$ and let $h = \beta(\epsilon,C)\log n$. Let $\mathbf{a} = (a_1,\ldots,a_m) \in \mathbf{R}^m$ be a given vector with $\|\mathbf{a}\|_1 = 1$, and for $(\eta,\sigma) \in \{-1,1\}^m \times S_m$ let $\mathbf{a}_{\eta,\sigma}$ denote the vector $(\eta_1 a_{\sigma(1)},\ldots,\eta_m a_{\sigma(m)})$. Then*

$$\mathsf{P}\Big[\exists(\eta,\sigma) \text{ s.t. } \big|\,\|\tilde{\mathbf{a}}_{\eta,\sigma}\| - \mathsf{E}\,\|\tilde{\mathbf{a}}\|\,\big| > (\delta/6C)h\Big] < (1/2)m^{-\delta^{-1}\log(3\delta^{-1})}.$$

The theorem follows easily from these two lemmas. Indeed, suppose they are both true and pick a sequence $\mathbf{a}_1,\ldots,\mathbf{a}_N$ as guaranteed to exist by Lemma 3, taking $\delta = \epsilon/11$. By Lemma 4, the probability that the random block basis fails to be $(1+\delta)$-symmetric at any given vector $\mathbf{a}$ is less than N^{-1}. It follows that it has a positive probability of being $(1+\delta)$-symmetric at each of $\mathbf{a}_1,\ldots,\mathbf{a}_N$. But then, by Lemma 3, there is a positive probability that it is $(1+\epsilon)$-symmetric, since $(1+\delta)(1-6\delta)^{-1} \leqslant (1+\epsilon)$.

The important part of the proof is of course Lemma 4. Since it concerns the deviation of several vectors at once, one might expect the proof to use a result about sub-Gaussian processes. However, it turns out that this is not necessary. For example, if a is the characteristic function of a subset of $[m]$, then Lemma 4 is a trivial consequence of (2): the main difficulty in proving Theorem 1 is in fact the technical problem of coping with vectors with coordinates of widely differing sizes. To prove Lemma 4 for such vectors, one splits them into parts with approximately equal coordinates and applies the triangle inequality rather crudely. Details as well as a proof of Lemma 3 may be found in [7].

It is natural to ask whether a basis equivalent to the unit vector basis of ℓ_∞^n must have a large almost symmetric block basis. The following result is the best that is known in the positive direction.

Theorem 5. *Let $x_1, \ldots, x_n$ be a sequence C-equivalent to the unit vector basis of ℓ_∞^n. Then it has a block basis of cardinality at least $k = n^\alpha$, where $\alpha = \log(1+\epsilon)/2\log C$, which is $(1+\epsilon)$-equivalent to the unit vector basis of ℓ_∞^k.*

This can be proved by a simple technique, essentially due to James. We shall see in the next section that in a sense one cannot do much better than this.

Another very natural question is the following. Suppose X is an arbitrary n-dimensional normed space. How large a $(1+\epsilon)$-symmetric basic sequence must it contain? The following result is due to Alon and Milman [1].

Theorem 6. *Let $0 < \epsilon < 1$, let X be an n-dimensional normed space and let $m = exp(c\sqrt{\epsilon^{-1} log n})$, where c is an absolute constant. Then X contains an m-dimensional subspace Y such that either $d(Y, \ell_2^m) \leqslant 1+\epsilon$ or $d(Y, \ell_\infty^m) \leqslant 1+\epsilon$.*

As remarked by Alon and Milman, this result, as it stands, is best possible (simply consider ℓ_p^n for appropriate $p = p(n)$). However, if one is looking only for a subspace with a $(1+\epsilon)$-symmetric basis, it is not known whether the result can be substantially improved. An upper bound will be given in the next section. The corresponding question about unconditional sequences is also interesting, and Theorem 6 seems to be all that is known in either direction.

Finally, we mention a result of Amir and Milman [2] concerning unconditional block bases of arbitrary bases.

Theorem 7. *Let $\epsilon > 0$ and let $x_1, \ldots, x_n$ be a linearly independent sequence of vectors in a normed space. Then $x_1, \ldots, x_n$ has a $(1+\epsilon)$-unconditional block basis of cardinality at least $\alpha(\epsilon)(\log n)^{1/2}$.*

§3. Negative Results

In this section we shall give an idea of how one constructs sequences and spaces to give upper bounds corresponding to the results in the last section. To begin with Theorem 1, we have the following counterpart.

Theorem 8. *Let $1 \leqslant p < \infty$ and $\epsilon > 0$. For every $n \in \mathbb{N}$ there exists a sequence $x_1, \ldots, x_n$ which is 2-equivalent to the unit vector basis of ℓ_p^n and which has no $(1+\epsilon)$-symmetric block basis of cardinality exceeding m_0, where*

$$m_0 = \begin{cases} \alpha(\epsilon, p) n \log\log n / \log n & 1 < p < \infty \\ \beta(\epsilon) n / \log\log n & p = 1 \end{cases}$$

and $\alpha(\epsilon, p) \to 0$ as $\epsilon \to 0$ or $p \to \infty$, and $\beta(\epsilon) \to 0$ as $\epsilon \to 0$.

The method of proof of Theorem 8 is in a sense the reverse of the method of proof of Theorem 1. In Lemma 3 we found as few vectors as possible with the property that their norms, and the norms or their rearrangements, control the norm on the whole space. The counterpart to Lemma 3 is the next lemma. We shall not give the exact definition of a *standardized* block basis, since it is only needed for technical reasons. The two most important points are that any vector in a standardized block basis is supported on at most $2n/m_0$ points, and that any block basis contains a subset which is a multiple of a standardized block basis.

Lemma 9. *Let $0 < \epsilon < 1/4$, $1 < p < \infty$, $m = n^{3/4}$ and $M = n^{\alpha(p)/\log\log n}$. There exists a sequence of subsets $A_1, \ldots, A_M$ of the unit sphere of ℓ_p^n with the following properties.*

(i) For any $1 \leqslant j \leqslant M$ and any standardized block basis $\mathbf{u}_1, \ldots, \mathbf{u}_m$ there exists a sequence of scalars $a_1, \ldots, a_m$ such that $\sum_1^m a_i \mathbf{u}_i \in A_j$.

(ii) Let $1 \leqslant j \leqslant M$, let $\mathbf{u}_1, \ldots, \mathbf{u}_m$ be a standardized block basis and let $a_1, \ldots, a_m$ be a sequence for which $\sum_1^m a_i \mathbf{u}_i \in A_j$. Then, for any permutation

$\pi \in S_m$ and any sequence $\epsilon_1, \ldots, \epsilon_m$ of signs,

$$\sum_1^m \epsilon_i a_{\pi(i)} \mathbf{u}_i \in A_j \ .$$

(iii) Let $1 \leqslant j < k \leqslant M$, let f be the ℓ_p^n-support functional of a vector in A_j and let $\mathbf{a} \in A_k$. Then $|f(\mathbf{a})| \leqslant (1-4\epsilon)\,\|\mathbf{a}\|_p$.

Lemma 9 is useful because it enables one to construct a norm by constructing its restrictions to the sets $A_1, \ldots, A_M$ separately. Given any $1 \leqslant j \leqslant M$ it turns out to be easy to construct a norm randomly in such a way that its restriction to A_j has two useful properties. First, it is $(1+2\epsilon)$-equivalent to the usual ℓ_p-norm, and second, any given standardized block basis $\mathbf{u}_1, \ldots, \mathbf{u}_m$ has a very small probability of being $(1+2\epsilon)$-symmetric at the vector in A_j which, by Lemma 9 (i), it must generate. (Note that, by Lemma 9 (ii) all the rearrangements of this vector are also in A_j). Let us write S_j for the event that the given block basis *is* symmetric at the vector it generates in A_j.

By far the most important part of Lemma 9 is the third part. This enables us to put together the various restrictions to the sets A_j to give a random norm with the property that the events S_j not only have small probability but are also independent. That is, the restriction of the norm to A_j does not interfere with its restriction to A_k for $k \neq j$. Hence, the probability of a given block basis being $(1+2\epsilon)$-symmetric is at most $\Pi_{j=1}^M \mathbf{P}(S_j)$.

It remains to pick L standardized block bases which form a sort of net, for some L that is not too large. They have the property that if they all fail to be $(1+2\epsilon)$-symmetric, then every standardized block basis (and hence every block basis) fails to be $(1+\epsilon)$-symmetric. This can be done with $L \leqslant (\Pi_{j=1}^M \mathbf{P}(S_j))^{-1}$. Hence, with a non-zero probability, the random norm has the property that no block basis of cardinality m_0 is $(1+\epsilon)$-symmetric.

When $p=1$, the unit sphere of ℓ_p^n is not sufficiently convex to accommodate a large number of classes of vectors on which a norm can be defined independently. A rather delicate construction can be used in a space which is C-equivalent to ℓ_1^n, but even this yields many fewer classes than are obtained when $p > 1$. That is why the upper bound is larger. It seems at least possible that this bound is the correct one, rather than $n/\log n$, but it would probably be hard to prove this.

There is an easy upper bound for Theorem 2. Set $m = 2n^{2/p-1}$, $h = n/m$, let $\mathbf{e}_1, \ldots, \mathbf{e}_m$ be the standard basis of ℓ_1^m and, for $1 \leqslant i \leqslant n$, let $x_i = \mathbf{e}_{\lceil i/h \rceil}$. It is not hard to verify that the conditions of Theorem 2 are satisfied, and since the vectors $x_1, \ldots, x_n$ are in a space of dimension m, there obviously cannot be an almost symmetric block basis of cardinality greater than m.

The following result is more interesting, since the sequence constructed satisfies a lower p-estimate.

Theorem 10. *There exists an absolute constant C such that, for any $1 \leqslant p \leqslant 3/2$ and any $n \in \mathbf{N}$, there is a norm $\|.\|$ on $\mathbf{R}^n$ satisfying the following three conditions:*

(i) the standard basis is normalized;

(ii) for any $\mathbf{a} \in \mathbf{R}^n$, $\|\mathbf{a}\| \geqslant \|\mathbf{a}\|_p$;

(iii) if $k \geqslant Cn^{2/p-1}(\log n)^{4/3}$ then no block basis $\mathbf{u}_1, \ldots, \mathbf{u}_k$ of the standard basis is 2-symmetric.

The construction is a fairly simple random one. For suitable N (a power of n will do) one picks N functionals $f_1, \ldots, f_N$ randomly from the 2^n functionals with ± 1-coordinates, distributed uniformly. Then the random norm on $\mathbf{R}^n$ is given by

$$\|x\| = \max_{1 \leqslant j \leqslant N} |f_j(x)| \vee \|x\|_p \ .$$

The proof that this construction works is quite long, but it is considerably simpler than that of Theorem 8. It fails when $p > 3/2$, because one can, with high probability, find a block basis which is close to the standard basis of ℓ_∞^m, where m is approximately $n^{1/3}$. We shall discuss sequences which satisfy a lower 2-estimate later in the section.

The next example is an explicit construction which can be adapted to give upper bounds for several of the problems mentioned so far. The following result is the most important one connected with the construction. It is proved in detail in [8], as are Theorems 12-14.

Theorem 11. *Let $M > 1$. Then there exists an absolute constant c such that if $n \geqslant f(M)$ is sufficiently large, then there exists an n-dimensional normed space X containing no M-symmetric basic sequence of cardinality exceeding $n^{c/\log\log n}$.*

Theorem 11 gives a negative answer to a question of Milman. He asked (private communication) whether every n-dimensional space contains a 2-symmetric sequence of cardinality proportional to $\sqrt{n}$. It is also relevant to the dependence on ϵ in Dvoretzky's theorem. Bourgain and Lindenstrauss [5] have shown that there exists a function $f(\epsilon, k)$ which is polynomial in ϵ^{-1} for fixed k, such that any space X of more than $f(\epsilon, k)$ dimensions with a 1-symmetric basis has a k-dimensional subspace Y such that $d(Y, \ell_2^k) \leqslant 1 + \epsilon$. Lindenstrauss has informed me that this is also the case if X has a 2-symmetric basis. Hence, if there were some fixed $\gamma > 0$ and $c > 0$ such that any n-dimensional normed space contained a 2-symmetric basic sequence of cardinality at least cn^γ, then the dependence on ϵ in Dvoretzky's theorem would be polynomial for arbitrary n-dimensional spaces. Although Theorem 11 shows that this premise is false, our construction does not give an example of non-polynomial dependence for Dvoretzky's theorem.

The example is constructed as follows. Given any space $X = (\mathbf{R}^m, \|.\|)$ such that the standard basis is 1-unconditional, let $\ell_p^k(X)$ denote the p-direct sum of k copies of X. Then, setting $p = 12$, $k = \log\log n/2\log 12$ and $m = n^{1/k}$, the space X is

$$\ell_\infty^m(\ell_{2p^{k-1}}^m(\ell_{2p^{k-2}}^m(\ldots(\ell_{2p}^m)\ldots))) .$$

This result cannot be improved by choosing a different iterated direct sum of ℓ_p-spaces. It is not hard to deduce from Theorem 2 that any such space contains a $(1+\epsilon)$-symmetric sequence of cardinality $\theta_1(\epsilon)n^{c/\theta_2(\epsilon)+\log\log n}$, for an absolute constant c. Using the same construction, or simple variants of it, the following results can also be obtained.

Theorem 12. *Let $\epsilon > 0$, $C > 1$, let $n \in \mathbf{N}$ and set $\alpha = \log(1+\epsilon)/\log C$ and $k = \log_{12}(1/8\alpha)$. Then there exists a basis C-equivalent to the unit vector basis of ℓ_∞^n with no $(1+\epsilon)$-symmetric block basis of cardinality $n^{8/k}(\log_2 n)^k$.*

Theorem 13. *Let $M \geqslant 1$ and let n be a power of 2. Then there there exists a normalized basis $x_1, \ldots, x_n$ satisfying a lower 2-estimate with no M-symmetric block basis of cardinality $n^{40/\log\log n}$.*

Theorem 14. *There exist absolute constants c, c' such that, for every $q > e^c$ and $M \geqslant 1$, and for n sufficiently large, there exists an n-dimensional space with q-cotype constant at most c', with no subspace of dimension $n^{c/\log q}$ with an M-symmetric basis.*

Theorem 11 can almost certainly be improved. Any improved construction could also be used to improve Theorem 13, and probably adapted to improve Theorems 12 and 14 as well. It seems likely that the result of Alon and Milman (Theorem 6) is the best possible positive result concerning the existence of symmetric sequences in arbitrary spaces. A proof of this would be extremely interesting.

§4. Open Problems

We shall finish the paper by choosing some of the more interesting problems arising from the previous sections.

Problem 1. *Let $C > 1$, let $\epsilon > 0$ and let $x_1, \ldots, x_n$ be a sequence of vectors equivalent to the unit vector basis of ℓ_1^n. Does $x_1, \ldots, x_n$ necessarily have a $(1+\epsilon)$-symmetric block basis of cardinality $\alpha(e, C)n/\log\log n$?*

Problem 2. *Let X be an n-dimensional normed space. How large a 2-symmetric basic sequence must it contain? In particular, does it necessarily contain such a sequence of cardinality substantially larger than $\exp(\sqrt{\log n})$?*

The third problem is the same as Problem 2 with "2-unconditional" replacing "2-symmetric".

Problem 3. *Let X be an n-dimensional normed space. How large a 2-unconditional basic sequence must it contain? In particular, does it necessarily contain such a sequence of cardinality substantially larger than $\exp(\sqrt{\log n})$?*

Problem 4. *Let $x_1, \ldots, x_n$ be any sequence of linearly independent vectors. Then how large a 2-unconditional block basis does it have? In particular, does it have a 2-unconditional block basis of cardinality significantly greater than $\sqrt{\log n}$?*

Problem 5. *Let $x_1, \ldots, x_n$ be any sequence of linearly independent vectors. Then how large a 2-symmetric block basis does it have?*

There does not seem to be strong evidence in either direction for Problems 1 and 3. It does seem quite likely that Problem 2 has a negative answer and Problem 4 a positive one. If Problems 2 and 3 or Problems 4 and 5 could be shown to have different answers, it would be the first non-trivial example of a property of a space or basis which makes it significantly easier to find an unconditional sequence than to find a symmetric one. In connection with Problem 3, it is probably possible to adapt Theorem 10 to give an upper bound of n^α for some fixed $\alpha < 1$, and it is definitely possible to do this for Problem 4. In fact, for each $n \in \mathbb{N}$, there exists a basis of cardinality n with no 2-unconditional block basis of cardinality greater than $Cn^{1/3}(\log n)^{4/3}$, where C is the absolute constant in Theorem 10. This upper bound is of course much larger than the lower bound given by Theorem 7. The best known lower bound for Problem 5 seems to be given by Krivine's theorem, so it must be possible to improve it.

References

[1] N. Alon and V. D. Milman, *Embedding of ℓ_∞^k in finite dimensional Banach spaces*, Israel J. Math. **45**, 265-280.

[2] D. Amir and V. D. Milman, *Unconditional and symmetric sets in n-dimensional normed spaces,* Israel J. Math. **37** (1980), 3-20.

[3] D. Amir and V. D. Milman, *A quantitative finite dimensional Krivine theorem*, Israel J. Math. **50** (1985), 1-12.

[4] K. Azuma, *Weighted sums of certain dependent random variables*, Tôhoku Math. J. **19** (1967), 357-367.

[5] J. Bourgain and J. Lindenstrauss, *Almost Euclidean sections in spaces with a symmetric basis*, GAFA 87-88, Springer Lecture Notes **1376** (1989), 278-288.

[6] W. T. Gowers, *Symmetric block bases in finite-dimensional normed spaces*, Israel J. Math., to appear.

[7] W. T. Gowers, *Symmetric block bases of sequences with large average growth*, Israel J. Math., to appear.

[8] W. T. Gowers, *Symmetric structures in Banach spaces*, Ph. D. thesis, University of Cambridge.

[9] B. Maurey, *Construction de suites symétriques*, C. R. A. S., Paris, **288** (1979), 679-681.

[10] V. D. Milman and G. Schechtman, *Asymptotic theory of finite dimensional normed spaces*, Springer Lecture Notes **1200** (1986), viii + 156 pp.

[11] G. Schechtman, *Lévy type inequality for a class of metric spaces*, Martingale Theory in Harmonic Analysis and Banach Spaces, Springer-Verlag 1981, 211-215.

Some topologies on the space of analytic self-maps of the unit disk

Herbert Hunziker, Hans Jarchow*and Vania Mascioni

Department of Mathematics, University of Zürich

Rämistrasse 74 CH-8001 Zürich (Switzerland)

Abstract

We study some topologies on the space Φ of analytic self-maps of the unit disk induced by composition operators on H^2. Among these is the so-called Hilbert-Schmidt topology, for which we find that the connected components are always arc connected, that they coincide with the sets $\{\psi : C_\psi - C_\varphi \text{ is Hilbert-Schmidt}\}$, and that they are convex subsets of the unit ball of H^∞. These properties are also shared by the other topologies under consideration, and all of them turn out to be intimately related to the classes of composition operators which are order bounded as maps $H^2 \to L^p$.

1 Introduction

Let D be the open unit disk in the complex plane. Consider the class Φ of all analytic functions which map D into D, that is, the unit ball of the Hardy space H^∞ without the constant functions generated by elements of ∂D. Each $\varphi \in \Phi$ is well-known to give rise to a composition operator C_φ on the Hardy space H^2 (see §2 for all definitions), and the usual operator norm $\|\cdot\|$ on $B(H^2)$ induces a metric on Φ defined by

$$d_{\|\cdot\|}(\varphi,\psi) = \|C_\varphi - C_\psi\| \quad .$$

Let us denote the space thus obtained by $(\Phi, \|\cdot\|)$.

*Supported by the Swiss National Science Foundation

In [6,8], MacCluer, Shapiro and Sundberg have studied the connectedness properties of $(\Phi, \|\cdot\|)$. Their research was mainly motivated by a result of Berkson [1], according to which φ is an isolated point in $(\Phi, \|\cdot\|)$ whenever φ is an exposed point of the unit ball of H^∞. In the other direction, it is also known that if φ is isolated in $(\Phi, \|\cdot\|)$, then φ must be an extreme point of Φ [8]. However, the identification of the components of $(\Phi, \|\cdot\|)$ is still incomplete (not even the isolated points have been characterized). Shapiro and Sundberg have conjectured in [8] that the component of φ might coincide with the set $\{\psi : C_\psi - C_\varphi \text{ is compact}\}$.

In this paper we are going to study the set Φ endowed with topologies which are related to the Hilbert-Schmidt norm for operators on H^2. The main instance is the so-called Hilbert-Schmidt topology, which is defined by the "metric"

$$d_2(\varphi, \psi) = \|C_\varphi - C_\psi\|_{HS}$$

(note that the distance between two points may be infinite). Let us agree to call this new topological space $(\Phi, 2)$. We will see that the connected component of φ in $(\Phi, 2)$ is exactly the set $\{\psi : C_\psi - C_\varphi \text{ is Hilbert-Schmidt}\}$ and the latter is always arc connected. Moreover, the isolated points in $(\Phi, 2)$ are exactly the extreme points of Φ. Another feature of $(\Phi, 2)$ is that the components are convex sets.

We thus see that the connectedness question in $(\Phi, 2)$ is completely solved, giving interesting geometric conditions on the ball of H^∞. Actually, $(\Phi, 2)$ is the most prominent example of a class of topological spaces (Φ, p) $(0 \le p \le \infty)$ which correspond in a natural way to the classes of composition operators which can be viewed as order bounded maps $H^2 \to L^p$.

In §2 we fix some notation and in §3 we state some general results on the spaces (Φ, p). Our main results about connected components are contained in §4, and the relationship between the spaces (Φ, p) and $(2, p)$-order boundedness of composition operators is investigated in §5.

2 Notation

Let H^2 be the Hilbert space of all holomorphic functions in the open disk D having square summable Fourier coefficients. If $f \in H^2$ then f has radial limits $f(\zeta) = \lim_{r \to 1} f(r\zeta)$ for almost all $\zeta \in \partial D$, and the scalar product of $f, g \in H^2$ is given by

$$\langle f, g \rangle = \int_{\partial D} f \bar{g} \, d\sigma \quad ,$$

σ being the normalized Lebesgue measure on ∂D.

More generally, if $0 < p < \infty$ then H^p is the (p−)Banach space of all analytic functions f on D such that the integrals $\int_0^{2\pi} |f(re^{i\theta})|^p \, d\theta$, $0 \le r < 1$, are bounded. H^∞ is the Banach space of all uniformly bounded holomorphic functions on D, the norm being given by $\|f\|_{H^\infty} = \text{ess sup}_{\zeta \in \partial D} |f(\zeta)|$.

In the sequel, we shall freely identify functions in H^p with their radial limits. Writing $f \in L^p$, we will always mean that the function f on D has radial limits belonging to $L^p(\partial D, \sigma)$. To simplify the notation, we will write Φ for the unit ball of H^∞ without the constant functions generated by points in ∂D.

It is a well-known consequence of Littlewood's Subordination Principle [5] that each φ in Φ gives rise to a continuous composition operator $C_\varphi : H^p \to H^p$, $f \mapsto f \circ \varphi$ $(0 < p \le \infty)$. It is clear that $\varphi \mapsto C_\varphi$ is an injective map from Φ to $B(H^p)$: in fact, the C_φ are even linearly independent in $B(H^p)$ (cf. [4]).

Recall that an operator T in $B(H^2)$ is Hilbert-Schmidt if

$$\|T\|_{HS} = \Big(\sum_{n \ge 0} \|T\psi_n\|_{H^2}^2\Big)^{1/2} < \infty \quad ,$$

where $(\psi_n)_n$ is the canonical orthonormal basis of H^2 given by $\psi_n(\zeta) = \zeta^n$.

By Shapiro and Taylor [9], a composition operator C_φ on H^2 is Hilbert-Schmidt if and only if $(1 - |\varphi|)^{-1} \in L^1$.

If $\zeta \in D$ is fixed, the composition operator corresponding to the function in Φ taking the constant value ζ may be identified with the evaluation functional δ_ζ at ζ: so $\delta_\zeta(f) = f(\zeta)$. On Φ we can define the $[0,\infty]$-valued "metric" by

$$d_p(\varphi, \psi) = \begin{cases} \exp\left\{\frac{1}{2}\int_{\partial D} \log \|\delta_{\varphi(\zeta)} - \delta_{\psi(\zeta)}\| \, d\sigma(\zeta)\right\} & , \quad p = 0 \\ \left(\int_{\partial D} \|\delta_{\varphi(\zeta)} - \delta_{\psi(\zeta)}\|^p \, d\sigma(\zeta)\right)^{1/p} & , \quad 0 < p < \infty \\ \text{ess sup}_{\zeta \in \partial D} \|\delta_{\varphi(\zeta)} - \delta_{\psi(\zeta)}\| & , \quad p = \infty \end{cases}$$

where all norms are taken in $(H^2)^*$. We will denote by (Φ, p) the space Φ endowed with the topology generated by the open sets

$$U_r^p(\varphi) = \{\psi : d_p(\psi, \varphi) < r\} \quad , \quad \varphi \in \Phi,\ r > 0 .$$

It is easy to verify that $d_p/(1 + d_p)$ is a metric on (Φ, p) if $p \ge 1$. Also, it follows easily from the definition that the identity $(\Phi, p) \to (\Phi, q)$ is continuous if $p \ge q$ (if $p < q$ this does not hold, as we shall see after Proposition 3.5).

Further on, we will show that $(\Phi, 2)$ inherits its topology from the Hilbert-Schmidt norm for operators in H^2.

Given $\varphi, \psi \in \Phi$, let us write $\varphi \sim_p \psi$ if and only if $d_p(\varphi, \psi)$ is finite. In this way, we obtain an equivalence relation together with a corresponding partition of Φ. Let $K_p(\varphi)$ be the equivalence class of φ with respect to $\sim_p$. We will see that the set $K_p(\varphi)$ is convex and coincides with the (arc) connected component of φ in the space (Φ, p), provided $0 \le p < \infty$.

The set of all extreme points of Φ (as a convex subset of the unit ball of H^∞) will be denoted by Φ_{ext}. It is known (see [3]) that $\varphi \in \Phi$ belongs to Φ_{ext} if and only if

$$\int_{\partial D} \log(1 - |\varphi|) = -\infty \quad .$$

We shall use standard notation and terminology from Banach space theory.

3 The spaces (Φ, p)

Let us begin with an easy lemma.

Lemma 3.1 *Let* $\varphi \neq \psi \in \Phi$. *Then*

(i) $\log|\varphi - \psi|$, $\log|1 - \bar{\varphi}\psi| \in L^1$.

(ii) *The following are equivalent:*

(a) $\varphi, \psi \notin \Phi_{\text{ext}}$

(b) $\log(1 - \max\{|\varphi|, |\psi|\}) \in L^1$.

(c) $\log\left(\left|\frac{\varphi-\psi}{1-\bar{\varphi}\psi}\right|^2 \frac{1-|\varphi|^2|\psi|^2}{(1-|\varphi|^2)(1-|\psi|^2)}\right) \in L^1$.

Proof (i). Since $\varphi - \psi \in H^\infty$, we have $\log|\varphi - \psi| \in L^1$ (see [3, p. 53]). On the other hand,

$$\left|\frac{\varphi - \psi}{1 - \bar{\varphi}\psi}\right| \le 1$$

holds, and so we get $\log|1 - \bar{\varphi}\psi| \in L^1$.

(ii). By (a), $\log(1 - |\varphi|), \log(1 - |\psi|) \in L^1$, and so (b) follows easily. The implication (b) $\Rightarrow$ (a) is clear and (b) $\Leftrightarrow$ (c) follows from this, using (i) and the inequalities

$$\frac{1 - |\varphi|^2|\psi|^2}{(1 - |\varphi|^2)(1 - |\psi|^2)} \ge \frac{1}{1 - \max\{|\varphi|, |\psi|\}^2} \ge \frac{1/2}{1 - \max\{|\varphi|, |\psi|\}} \quad .$$

$\square$

Theorem 3.2 *Let $\varphi, \psi \in \Phi$ and $0 \le p \le \infty$. Then $\varphi \sim_p \psi$ if and only if the function*

$$I(\varphi,\psi) = \left|\frac{\varphi-\psi}{1-\bar\varphi\psi}\right|^2 \frac{1-|\varphi|^2|\psi|^2}{(1-|\varphi|^2)(1-|\psi|^2)}$$

belongs to $L^{p/2}$, resp. if and only if $\log I(\varphi,\psi) \in L^1$ when $p = 0$, and in such a case

$$d_p(\varphi,\psi) = \begin{cases} \exp\left\{\frac{1}{2}\int_{\partial D} \log I(\varphi,\psi)\, d\sigma\right\} , & p = 0 \\ \left(\int_{\partial D} I(\varphi,\psi)^{p/2}\, d\sigma\right)^{1/p} , & 0 < p < \infty \\ \operatorname{ess\,sup}_{\partial D} I(\varphi,\psi)^{1/2} , & p = \infty \end{cases}$$

In particular, if $\varphi \sim_p \psi$ and $\varphi \ne \psi$, then neither φ nor ψ is in Φ_{ext}.

Proof First note that, if $\eta, \vartheta \in D$ and $f \in H^2$, then

$$\delta_\eta(f) = f(\eta) = \langle f, k_\eta\rangle \quad ,$$

where $k_\eta(\zeta) = 1/(1-\bar\eta\zeta)$ is the reproducing kernel for H^2. We may thus write

$$\begin{aligned} \|\delta_\eta - \delta_\vartheta\|^2_{(H^2)^*} &= \|k_\eta - k_\vartheta\|^2_{H^2} \\ &= \|k_\eta\|^2_{H^2} + \|k_\vartheta\|^2_{H^2} - 2\Re\langle k_\eta, k_\vartheta\rangle \\ &= \frac{1}{1-|\eta|^2} + \frac{1}{1-|\vartheta|^2} - 2\Re\frac{1}{1-\bar\eta\vartheta} \\ &= \left|\frac{\eta-\vartheta}{1-\bar\eta\vartheta}\right|^2 \frac{1-|\eta|^2|\vartheta|^2}{(1-|\eta|^2)(1-|\vartheta|^2)} \quad . \end{aligned}$$

It is easy to see that, if $\varphi \sim_p \psi$ and $\varphi \ne \psi$, then $\sigma(\{|\varphi| = 1\}) = \sigma(\{|\psi| = 1\}) = 0$. Then the above computation and the definitions give the desired expressions for d_p.

If $\varphi \sim_p \psi$, then $\log I(\varphi,\psi) \in L^1$ and thus by Lemma 3.1(ii) we see that $\varphi, \psi \notin \Phi_{\text{ext}}$.

□

Remarks (1) If $C_\varphi - C_\psi$ is a Hilbert-Schmidt operator on H^2, its Hilbert-Schmidt norm is

$$\begin{aligned} \|C_\varphi - C_\psi\|_{HS} &= \left(\sum_{n=0}^{\infty} \|\varphi^n - \psi^n\|^2_{H^2}\right)^{1/2} \\ &= \left(\int_{\partial D} \sum_{n=0}^{\infty} |\varphi(\zeta)^n - \psi(\zeta)^n|^2\, d\sigma(\zeta)\right)^{1/2} \end{aligned}$$

which is readily seen to be equal to $d_2(\varphi, \psi)$. This explains why we propose to call the topology on Φ induced by d_2 the "Hilbert-Schmidt topology".

(2) If $\varphi, \psi \in \Phi$ we observe that

$$\begin{aligned} \frac{1-|\varphi|^2|\psi|^2}{(1-|\varphi|^2)(1-|\psi|^2)} &= \frac{1}{1-|\varphi|^2} + \frac{1}{1-|\psi|^2} - 1 \\ &\leq \frac{2}{1-\max\{|\varphi|, |\psi|\}^2} \quad , \end{aligned}$$

we may deduce from the last inequality in the proof of Lemma 3.1 that, in general,

$$\left|\frac{\varphi-\psi}{1-\bar{\varphi}\psi}\right|^2 \frac{1}{1-\max\{|\varphi|, |\psi|\}^2} \leq I(\varphi, \psi) \leq \left|\frac{\varphi-\psi}{1-\bar{\varphi}\psi}\right|^2 \frac{2}{1-\max\{|\varphi|, |\psi|\}^2} \; .$$

Proposition 3.3 *(Φ, p) is complete for all $0 \leq p \leq \infty$.*

Proof Let (φ_n) be a Cauchy sequence in (Φ, p). Thus, given $\varepsilon > 0$, we can find an N_ε such that if $m, n \geq N_\varepsilon$ then

$$d_p(\varphi_n, \varphi_m) \leq \varepsilon \quad .$$

Let first $p > 0$. Since $I(\varphi_n, \varphi_m)^{1/2} \geq |\varphi_n - \varphi_m|$, we see that (φ_n) is a Cauchy sequence in H^p and so has, in particular, a pointwise limit φ. Hence, by application of Fatou's Lemma we get

$$d_p(\varphi_n, \varphi) \leq \varepsilon$$

if $n \geq N_\varepsilon$, which shows that (φ_n) converges to φ in (Φ, p).

If $p = 0$, subharmonicity of $\log|\varphi_n - \varphi_m|$ in D together with a standard estimate for Poisson integrals lead us to

$$\log|\varphi_n(\eta) - \varphi_m(\eta)| \leq \frac{2}{1-|\eta|} \int_{\partial D} \log|\varphi_n - \varphi_m| \, d\sigma$$

for all $\eta \in D$. Using again $I(\varphi_n, \varphi_m)^{1/2} \geq |\varphi_n - \varphi_m|$, we see that

$$|\varphi_n(\eta) - \varphi_m(\eta)| \leq d_0(\varphi_n, \varphi_m)^{2/(1-|\eta|)} \quad ,$$

so that (φ_n) converges pointwise. Now we proceed as in the case $p > 0$.

□

Theorem 3.2 enables us to characterize the isolated points in (Φ, p):

Theorem 3.4 *Let $\varphi \in \Phi$ and $0 \leq p \leq \infty$. The following are equivalent:*

(i) $K_p(\varphi) = \{\varphi\}$ *(i.e. no $\psi \neq \varphi$ in Φ satisfies $d_p(\varphi, \psi) < \infty$).*

(ii) $\varphi \in \Phi_{\text{ext}}$.

(iii) φ *is an isolated point in* (Φ, p).

Proof If $\varphi \in \Phi_{\text{ext}}$ then $d_p(\varphi, \psi)$ cannot be finite for any $\psi \neq \varphi$ in Φ, by Theorem 3.2. By definition this implies $K_p(\varphi) = \{\varphi\}$, which in turn readily yields that φ must be isolated in (Φ, p). It remains to prove that (iii) implies (i).

Suppose that $\varphi \notin \Phi_{\text{ext}}$. Following [8], we find an $\omega \in \Phi$ such that $|\omega| = (1 - |\varphi|)^{3/2}$ on ∂D. If we define $\varphi_t \in \Phi$ by $\varphi_t = \varphi + t\omega$, then

$$I(\varphi_s, \varphi_t) \leq \frac{2(s-t)^2}{(1-|s|)(1-|t|)(1-\min\{|s|, |t|\})}$$

for all $-1 < s, t < 1$. This shows, by Theorem 3.2 and dominated convergence, that $t \mapsto \varphi_t$ actually defines a continuous arc in (Φ, p) passing through φ. In particular, φ is not isolated in (Φ, p).

□

By [8], not all extreme points of Φ generate isolated composition operators in $(\Phi, \|\cdot\|)$. In fact, there is even a $\varphi \in \Phi_{\text{ext}}$ such that C_φ is compact: however, all compact composition operators belong to the same (arc) connected component in $(\Phi, \|\cdot\|)$.

In the next proposition we collect some additional information. We denote by 0 the zero-function in Φ.

Proposition 3.5 *Let $0 \leq p \leq \infty$.*

(i) $\varphi \in K_p(0)$ *if and only if*

$$\begin{cases} \varphi \notin \Phi_{\text{ext}} & , \quad p = 0 \\ (1 - |\varphi|)^{-1} \in L^{p/2} & , \quad 0 < p < \infty \\ \|\varphi\|_{H^\infty} < 1 & , \quad p = \infty \end{cases}$$

(ii) *Let $\varphi \sim_p \psi$. Then the following are equivalent:*

(a) $\varphi, \psi \in K_p(0)$

(b) $\varphi\psi \in K_p(0)$.

(iii) *If $\varphi \sim_p \psi$ then* $\frac{\varphi - \psi}{2} \in K_p(0)$.

Proof (i) is a direct consequence of Theorem 3.2.

(ii). If $p = 0$ then the assertion is trivial. Let next $0 < p < \infty$. By Theorem 3.2, $\varphi \sim_p \psi$ if and only if $I(\varphi, \psi) \in L^{p/2}$. By (i), $\varphi \in K_p(0)$ implies $(1 - |\varphi|)^{-1} \in L^{p/2}$, and thus the smaller function $(1 - |\varphi\psi|)^{-1}$ is in $L^{p/2}$, too. Conversely, if $(1 - |\varphi\psi|)^{-1} \in L^{p/2}$, then $\Re[1/(1 - \bar{\varphi}\psi)] \in L^{p/2}$, since

$$\Re\frac{1}{1-\bar{\varphi}\psi} \leq \frac{1}{|1-\bar{\varphi}\psi|} \leq \frac{1}{1-|\varphi\psi|}.$$

Since

$$\frac{1}{1-|\varphi|^2} + \frac{1}{1-|\psi|^2} = I(\varphi,\psi) + 2\Re\frac{1}{1-\bar{\varphi}\psi} \quad ,$$

$I(\varphi, \psi) \in L^{p/2}$ implies that φ and ψ belong to $K_p(0)$.

Finally, let $p = \infty$. Implication (a) $\Rightarrow$ (b) is trivial. To see the converse, note that the hypotheses $\varphi \sim_{(\infty)} \psi$ and $\varphi\psi \in C_{(\infty)}(0)$ (i.e., $\|\varphi\psi\|_{H^\infty} < 1$, by (i)) imply the existence of a constant $K > 0$ such that

$$\left|\frac{\varphi-\psi}{1-\bar{\varphi}\psi}\right|^2 \leq K(1-|\varphi|^2)(1-|\psi|^2)$$

a.e. on ∂D. Writing $|\varphi - \psi|^2 = |1 - \bar{\varphi}\psi|^2 - (1 - |\varphi|^2)(1 - |\psi|^2)$, it is immediate to deduce that

$$\frac{1}{(1-|\varphi|^2)(1-|\psi|^2)} \leq K + \frac{1}{|1-\bar{\varphi}\psi|^2}$$

a.e. on ∂D. Consequently, we must have $\|\varphi\|_{H^\infty}, \|\psi\|_{H^\infty} < 1$, which by (i) concludes this part of the proof.

(iii). The assertion being trivial if $p = 0$ and easy if $p = \infty$, we may immediately pass to $0 < p < \infty$. Using the inequalities

$$\begin{aligned}
\frac{1-|\varphi|^2|\psi|^2}{(1-|\varphi|^2)(1-|\psi|^2)} &= \frac{1}{1-|\varphi|^2} + \frac{1}{1-|\psi|^2} - 1 \\
&\geq \frac{2}{1-\frac{|\varphi|^2+|\psi|^2}{2}} - 1 \\
&\geq \frac{2}{1-\left|\frac{\varphi-\psi}{2}\right|^2} - 1 \quad ,
\end{aligned}$$

we easily get

$$I(\varphi,\psi) \geq \frac{|\varphi-\psi|^2}{2\left(1-\left|\frac{\varphi-\psi}{2}\right|^2\right)} \quad ,$$

whence $\left(1-\left|\frac{\varphi-\psi}{2}\right|^2\right)^{-1} \in L^{p/2}$, which was what we wanted.

□

Remark The description of $K_p(0)$ given in Proposition 3.5(i) allows to conclude that the topologies (Φ, p) are in fact all different.

To see this, it suffices to find a function φ in

$$\bigcap_{q<p} K_q(0) \setminus K_p(0) \quad ,$$

which can be done as follows: take a real function $f \geq 2$ on ∂D such that

$$f \in \bigcap_{q<p/2} L^q \setminus L^{p/2} \quad .$$

Since $\log(1-f^{-1}) \in L^1$, Szegö's Theorem [3, p.53] provides us with a function $\varphi \in \Phi$ such that $|\varphi| = 1 - f^{-1}$ on ∂D. Now, $f = (1-|\varphi|)^{-1}$, and so φ is the function we were looking for, by Proposition 3.5(i).

The relation between two given composition operators $C_{\varphi_0}, C_{\varphi_1}$ and the operators obtained from the functions

$$\varphi_t = (1-t)\varphi_0 + t\varphi_1$$

$(0 \leq t \leq 1)$ will be of fundamental importance for the next section.

Intuition suggests that the operators C_{φ_t} should share some of the properties induced by the "endpoints" C_{φ_0} and C_{φ_1}. That this, however, cannot be completely true, is shown by the functions $\varphi_0(\zeta) = \zeta$ and $\varphi_1(\zeta) = \zeta^2$: here we have $\varphi_t(\zeta) = \zeta(1-t+t\zeta)$. Back to Schwartz [10] goes the observation that the operators C_{φ_t} fail to be compact (see also [8]). Further, if $s \neq t$, we have $\|C_{\varphi_s} - C_{\varphi_t}\| \geq s^{-1/2} + t^{-1/2}$, and so the φ_t do not form a continuous path in $(\Phi, \|\cdot\|)$, cf. [8]. Now, it is also possible to show that, if $0 < s \neq t < 1$, $\varphi_s \sim_p \varphi_t$ if and only if $0 \leq p < 1/2$. In fact, some calculations give

$$I(\varphi_s, \varphi_t)(e^{i\vartheta}) = \frac{(s-t)^2\left[\frac{1}{s(1-s)} + \frac{1}{t(1-t)} - 2(1-\cos\vartheta)\right]}{2(1-\cos\vartheta)\left[(s-t)^2 + 2s(1-s)t(1-t)(1-\cos\vartheta)\right]} ,$$

and this belongs to $L^{p/2}$ only if $p < 1/2$. Consequently, if $p \geq 1/2$ then each φ_t must lie in a different component of (Φ, p). However, if $0 < t < 1$, these operators cannot be isolated in (Φ, p) for any p: in fact, they fail to be extreme points of Φ (see Theorem 3.4).

The following proposition exhibits another elementary but somewhat surprising property of the sets $K_p(0)$: the straight line connecting any $\varphi_1 \in \Phi$ to any $\varphi_0 \in K_p(0)$ consists almost entirely of elements of $K_p(0)$.

Proposition 3.6 *Let $0 \leq p \leq \infty$, $\varphi_0, \varphi_1 \in \Phi$. If $\varphi_0 \in K_p(0)$, then $\varphi_t \in K_p(0)$ for all $0 < t < 1$.*

Proof If $p = 0, \infty$, the proposition is trivial. If $0 < p < \infty$, just note that, for $0 < t < 1$,

$$\begin{aligned} \frac{1}{1-|\varphi_t|} &\leq \frac{1}{(1-t)(1-|\varphi_0|)+t(1-|\varphi_1|)} \\ &\leq \frac{1}{(1-t)(1-|\varphi_0|)} \end{aligned}$$

and that the last function is in $L^{p/2}$, by $\varphi_0 \in K_p(0)$ and Proposition 3.5(i).

□

4 Connected components in (Φ, p)

Let us first prove an elementary inequality for points in D.

Lemma 4.1 *Let $a_0, a_1 \in D$ and $a_t = (1-t)a_0 + ta_1$ for $0 < t < 1$. Then, regardless of $0 \leq s < t \leq 1$,*

$$\begin{aligned} \left|\frac{a_s - a_t}{1-\bar{a}_s a_t}\right|^2 \frac{1-|a_s|^2|a_t|^2}{(1-|a_s|^2)(1-|a_t|^2)} &\leq \\ \leq 2\left|\frac{a_0 - a_1}{1-\bar{a}_0 a_1}\right|^2 \frac{1-|a_0|^2|a_1|^2}{(1-|a_0|^2)(1-|a_1|^2)} &. \end{aligned}$$

Proof Without loss of generality, we may assume $s = 0$. Note first that

$$\begin{aligned} \left|\frac{a_0 - a_t}{1-\bar{a}_0 a_t}\right|^2 &= \frac{t^2|a_0 - a_1|^2}{|(1-t)(1-|a_0|^2)+t(1-\bar{a}_0 a_1)|^2} \\ &\leq \left|\frac{a_0 - a_1}{1-\bar{a}_0 a_1}\right|^2 . \end{aligned} \tag{1}$$

On the other hand, since the function $t \mapsto (1-|a_t|^2)^{-1}$ is convex,

$$\begin{aligned}\frac{1-|a_0|^2|a_t|^2}{(1-|a_0|^2)(1-|a_t|^2)} &= \frac{1}{1-|a_0|^2}+\frac{1}{1-|a_t|^2}-1 \\ &\leq \frac{2-t}{1-|a_0|^2}+\frac{t}{1-|a_1|^2}-1 \\ &\leq 2\frac{1-|a_0|^2|a_1|^2}{(1-|a_0|^2)(1-|a_1|^2)} \quad . \qquad (2)\end{aligned}$$

The assertion follows by combination of (1) and (2).

□

Corollary 4.2 *Let* $0 \leq p < \infty$. *If* $\varphi_0 \sim_p \varphi_1$, *then*

$$t \mapsto (1-t)\varphi_0 + t\varphi_1 \quad , \quad 0 \leq t \leq 1 \, ,$$

provides a continuous arc in (Φ, p) *connecting* φ_0 *and* φ_1.

Proof Define $\varphi_t = (1-t)\varphi_0 + t\varphi_1$. By Lemma 4.1,

$$I(\varphi_s, \varphi_t) \leq 2\, I(\varphi_0, \varphi_1)$$

holds pointwise for all $0 \leq s, t \leq 1$. The corollary follows from Theorem 3.2 and Lebesgue's dominated convergence theorem.

□

Summarizing all the preceding results, we finally arrive at a complete description of the connected components of (Φ, p).

Theorem 4.3 *Let* $0 \leq p < \infty$. *For each* $\varphi \in \Phi$, *the set*

$$K_p(\varphi) = \{\psi : \psi \sim_p \varphi\}$$

is the connected component of (Φ, p) *containing* φ.

$K_p(\varphi)$ is always arc connected in (Φ, p) and convex in H^∞.

$K_p(\varphi)$ reduces to the point φ if and only if φ is an extreme point of the unit ball of H^∞.

The proof follows now directly from Theorem 3.3 and Corollary 4.2 above.

Remark In the case $p = \infty$, the situation seems to be more delicate. From Lemma 4.1 we merely obtain that the functions φ_t belong to the equivalence class of φ_0 and φ_1 with respect to $\sim_\infty$, but we do not know whether $K_\infty(\varphi_0)$ is connected in general (however, it is easy to see that $K_\infty(0)$ *is* connected). Continuity of $t \mapsto \varphi_t$ is here the problem.

5 (r,p)-order boundedness and (Φ,p)

Let $0 < r,p \leq \infty$. We say that an operator $T: H^r \to H^r$ is (r,p)-*order bounded* if there exists a function $h \in L^p(\partial D)$ such that

$$|(Tf)(\zeta)| \leq h(\zeta) \quad \text{a.e. on } \partial D$$

for all $f \in B_{H^r}$.

Thus (r,p)-order boundedness of T means that T can be regarded as a bounded operator $T: H^r \to H^p$, and that $J_pT: H^r \to L^p(\partial D)$ is order bounded in the traditional sense, J_p being the natural embedding $H^p \to L^p(\partial D)$, or majorizing, in the terminology of [7] $(p \geq 1)$.

The following theorem characterizes $(2,p)$-order boundedness of composition operators in terms of the topological spaces (Φ,p):

Theorem 5.1 *Let $0 < p \leq \infty$ and $\varphi, \psi \in \Phi$. Then $C_\varphi - C_\psi$ is $(2,p)$-order bounded if and only if $\varphi \sim_p \psi$.*

Proof Let $f \in B_{H^2}$ and $\zeta \in D$. Then we always have

$$\begin{aligned} |(C_\varphi - C_\psi)(f)(\zeta)| &= |(\delta_{\varphi(\zeta)} - \delta_{\psi(\zeta)})(f)| \\ &\leq \|\delta_{\varphi(\zeta)} - \delta_{\psi(\zeta)}\|_{H^2} \\ &= I(\varphi,\psi)^{1/2}(\zeta) \quad . \end{aligned} \tag{1}$$

Thus, if $\varphi \sim_p \psi$, then $I(\varphi,\psi)^{1/2} \in L^p$ by Theorem 3.2, and so $I(\varphi,\psi)^{1/2}$ acts as an order bound for the functions in $(C_\varphi - C_\psi)(B_{H^2})$, i.e. we get the $(2,p)$-order boundedness of $C_\varphi - C_\psi$.

Conversely, if $C_\varphi - C_\psi$ is $(2,p)$-order bounded, and if $h \in L^p$ is such that

$$|(C_\varphi - C_\psi)(f)(\zeta)| \leq h(\zeta) \quad \text{a.e. on } \partial D$$

for all $f \in B_{H^2}$, then we get $I(\varphi,\psi)^{1/2} \leq h$, by taking the corresponding supremum on the left hand side and using (1). Hence, $I(\varphi,\psi)^{1/2} \in L^p$ and thus $d_p(\varphi,\psi) \leq \|h\|_{L^p}$, i.e., $\varphi \sim_p \psi$.

□

Combining this with Proposition 3.5, we get immediately

Corollary 5.2 *Let $0 < r,p \leq \infty$ and $\varphi \in \Phi$. Then the following are equivalent:*

(i) C_φ is (r,p)-order bounded.

(ii) $\frac{1}{1-|\varphi|} \in L^{p/r}$.

(iii) $(\|\varphi^n\|_{H^1}) \in \ell_{r/p,1}$.

(iv) $(\|\varphi^n\|_{H^p}) \in \ell_{r,p}$.

Proof (i) $\Leftrightarrow$ (ii) : The case $r = 2$ follows immediately from the above theorem and from Proposition 3.5. If $r \neq 2$, just use the fact that

$$\|\delta_\eta\|_{(H^r)^*} = \left(\frac{1}{1-|\eta|^2}\right)^{1/r}$$

for all $\eta \in D$, and argue as above.

(ii) $\Leftrightarrow$ (iii) : It follows from

$$\left(\frac{1}{1-|\varphi|}\right)^{p/r} = \sum_{n=0}^{\infty} \binom{-p/r}{n} (-1)^n |\varphi|^n$$

and from

$$\lim_{n\to\infty} \binom{-p/r}{n} (-1)^n n^{1-p/r} = \Gamma(p/r)$$

that $1/(1-|\varphi|)$ belongs to $L^{p/r}$ if and only if

$$\sum_{n=0}^{\infty} n^{p/r-1} \|\varphi^n\|_{H^1} < \infty \quad ,$$

which means $(\|\varphi^n\|_{H^1}) \in \ell_{r/p,1}$.

(iii) $\Leftrightarrow$ (iv) is trivial.

□

Remarks (1) In particular, we see that C_φ is (r, r)-order bounded if and only if C_φ is Hilbert-Schmidt on H^2. This fact was proved by the first named author in [4]. This work also contains a characterization of composition operators $C_\varphi : H^2 \to H^2$ taking values in H^p for $p > 2$ in terms of Carleson properties of the image measure σ_φ of σ defined by the boundary value function of φ.

(2) Applying to our situation well-known results on majorizing and cone absolutely summing operators (cf. [7, IV.3.8]), we may say that C_φ is $(2, p)$-order bounded if and only if $(J_p C_\varphi)^* : L^{p^*} \to H^2$ is cone absolutely summing ($p \geq 1$). Note also that this is implied by $(J_p C_\varphi)^*$ being "positive p-summing" in the sense of [2].

(3) Let $\varphi, \psi \in \Phi$, $1 \leq r < \infty$ and $1 \leq p < \infty$ be such that $C_\varphi + C_\psi$ is (r,p)-order bounded. Through harmonic functions, it is not difficult to see that $C_\varphi + C_\psi$ can even be regarded as an order bounded operator $L^r \to L^p$. Since composition operators defined on L^r-spaces are trivially positive, C_φ and C_ψ must both be (r,p)-order bounded. Taking $r = p = 2$ and observing that $C_\varphi + C_\psi$ is $(2,2)$-order bounded if and only if $C_\varphi + C_\psi$ is Hilbert-Schmidt on H^2, we obtain that $C_\varphi + C_\psi$ is Hilbert-Schmidt on H^2 if and only if both C_φ and C_ψ are Hilbert-Schmidt. Generalizations of this will be discussed elsewhere.

Our final theorem exhibits a somewhat unexpected "ideal" property of the components of (Φ, p):

Theorem 5.3 *Let $0 \leq p \leq \infty$ and let $\chi \in \Phi$ be arbitrary. Then*

$$\varphi \sim_p \psi \quad \Longrightarrow \quad \begin{cases} \chi \circ \varphi \sim_p \chi \circ \psi \\ \varphi \circ \chi \sim_p \psi \circ \chi \end{cases} .$$

In particular, if $\varphi \in K_p(0)$, then $\varphi \circ \chi$ and $\chi \circ \varphi$ belong to $K_p(0)$, for no matter which $\chi \in \Phi$.

Proof The case $p = 0$ follows easily from Proposition 3.5(i). Let $0 < p \leq \infty$, $\varphi \sim_p \psi$ and $\chi \in \Phi$ be arbitrary. By Theorem 5.1, $C_\varphi - C_\psi$ is $(2,p)$-order bounded. It is a direct consequence of the definition that $(C_\varphi - C_\psi) \circ C_\chi = C_{\chi \circ \varphi} - C_{\chi \circ \psi}$ is $(2,p)$-order bounded, too, whence $\chi \circ \varphi \sim_p \chi \circ \psi$.

On the other hand we know that, for all $f \in B_{H^2}$,

$$|f(\varphi(\zeta)) - f(\psi(\zeta))| \leq I(\varphi, \psi)^{1/2}(\zeta) \quad \text{a.e. on } \bar{D},$$

and that $I(\varphi, \psi)^{1/2} \in L^p$. As a consequence, we get that

$$|f(\varphi(\chi(\zeta))) - f(\psi(\chi(\zeta)))| \leq (I(\varphi, \psi)^{1/2} \circ \chi)(\zeta) \quad \text{a.e. on } \partial D .$$

Now, via appropriate spaces of harmonic functions, C_χ induces a continuous operator on L^p, and thus $I(\varphi, \psi)^{1/2} \circ \chi \in L^p$ appears as an order bound for the functions in $(C_{\varphi \circ \chi} - C_{\psi \circ \chi})(B_{H^2})$. Hence, $\varphi \circ \chi \sim_p \psi \circ \chi$.

□

Remarks (1) To conclude, let us briefly look at what happens if our composition function χ is a Möbius transform. Given $\eta \in D$, write

$$\tau_\eta(\zeta) = \frac{\eta - \zeta}{1 - \bar{\eta}\zeta}$$

and consider the map

$$i_\eta : (\Phi, p) \to (\Phi, p) \ , \ \varphi \mapsto \tau_\eta \circ \varphi \quad .$$

A computation leads to

$$\frac{1}{1-|\varphi|^2} - \frac{1}{1-|\eta|^2} \leq I(\varphi, \tau_\eta \circ \varphi) \leq \frac{5-|\eta|^2}{1-|\eta|^2} \frac{1}{1-|\varphi|^2} ,$$

and so $\varphi \sim_p \tau_\eta \circ \varphi$ holds if and only if $\varphi \in K_p(0)$. Since

$$\begin{aligned} d_p(\tau_\eta \circ \varphi, \tau_\eta \circ \psi) &\leq \|C_{\tau_\eta}\|_{H^2} \, d_p(\varphi, \psi) \\ &\leq \left(\frac{1+|\eta|}{1-|\eta|}\right)^{1/2} d_p(\varphi, \psi) \quad , \end{aligned}$$

we get that the map i_η is a homeomorphism of (Φ, p) onto itself. Moreover, i_η induces a permutation of the components of (Φ, p), with "fixed component" $K_p(0)$.

(2) Concerning composition with $\psi_n(\zeta) = \zeta^n$ $(n \geq 1)$, we can prove that (for $0 \leq p \leq \infty$)

(i) $\varphi \sim_p \psi_n \circ \varphi \Leftrightarrow \varphi \in K_p(0)$

(ii) $\varphi \circ \psi_n \sim_p \psi \circ \psi_n \Leftrightarrow \varphi \sim_p \psi$

(iii) $\varphi \in K_p(0) \Leftrightarrow \varphi \circ \psi_n \in K_p(0) \Leftrightarrow \psi_n \circ \varphi \in K_p(0)$.

Details will appear elsewhere.

References

[1] E. BERKSON, Composition operators isolated in the uniform operator topology, *Proc. Amer. Math. Soc.* **81** (1981) 230–232.

[2] O. BLASCO, Positive p-summing operators on L^p spaces, *Proc. Amer. Math. Soc.* **100** (1987) 275–280.

[3] K. HOFFMAN, *Banach spaces of analytic functions*, Prentice-Hall, 1962.

[4] H. HUNZIKER, Kompositionsoperatoren auf Hardyräumen, Thesis, Univ. of Zürich, 1989.

[5] J. E. LITTLEWOOD, On inequalities in the theory of functions, *Proc. London Math. Soc.* **23** (1925) 481–519.

[6] B. D. MACCLUER, Components in the space of composition operators, Preprint.

[7] H. H. SCHAEFER, *Banach lattices and positive operators*, Springer Verlag, Berlin 1974.

[8] J. H. SHAPIRO AND C. SUNDBERG, Isolation amongst the composition operators, Preprint.

[9] J. H. SHAPIRO AND P. D. TAYLOR, Compact, nuclear, and Hilbert-Schmidt composition operators on H^p, *Indiana Univ. Math. J.* **23** (1973) 471–496.

[10] H. J. SCHWARTZ, Composition operators on H^p, Thesis, Univ. of Toledo, 1969.

Minimal and strongly minimal Orlicz sequence spaces

N.J. KALTON*
DEPARTMENT OF MATHEMATICS, UNIVERSITY OF MISSOURI
COLUMBIA, MO. 65211

(*): This research was supported by NSF-grant DMS 8901636.

1. Introduction.

The structure theory of Orlicz sequence spaces was initiated in work of Lindberg [6] and Lindenstrauss and Tzafriri ([7], [8] and [9]; see also [10]) in the early seventies. Recently this study has been continued by Hernandez and Rodriguez-Salinas ([2], [3]). In their work, Lindenstrauss and Tzafriri introduced the class of minimal Orlicz sequence spaces and conjectured that these spaces are prime. The only separable prime spaces known are the ℓ_p and c_0, but Lindenstrauss and Tzafriri gave other examples of minimal Orlicz spaces. The purpose of this note is to show that this conjecture is false in general, but that a smaller non-trivial class of strongly minimal spaces is introduced which still has the potential to contain new prime spaces. Our results are achieved by introducing separate necessary and sufficient conditions for a reflexive Orlicz sequence space ℓ_G to be complemented in another such space ℓ_F.

We now proceed to a more detailed discussion of the basic definitions and our results. We refer to [10] for the basic facts about Orlicz sequence spaces, but review here some key definitions and ideas. It will be convenient to allow Orlicz functions to be possibly non-convex, even though we do not wish to discuss non-locally convex examples. Thus we will for the purposes of this note consider an Orlicz function to be a continuous function $F : [0,\infty) \to [0,\infty)$ satisfying $F(0) = 0$, such that $F(x) > 0$ if $x > 0$ and satisfying, for a suitable constant C, $F(tx) \leq CtF(x)$ whenever $0 \leq t \leq 1$ and $0 < x < \infty$. We say that F satisfies the Δ_2-condition if $F(2x) \leq KF(x)$ for a suitable constant K and all x. Two Orlicz functions F and G are equivalent if $\log F(x)/G(x)$ is bounded on $(0,\infty)$ and equivalent near zero if $\log F(x)/G(x)$ is bounded on $(0,1)$. Any Orlicz function satisfying our definition and the Δ_2-condition is then equivalent to a convex Orlicz function.

If F is an Orlicz function satisfying the Δ_2-condition then we define the Orlicz function space $L_F = L_F(0,\infty)$ to be the space of all measurable real functions f on $(0,\infty)$ satisfying

$$\int_0^\infty F(|f(t)|)\,dt < \infty$$

and this is isomorphic to a Banach space if we equip it with the quasi-norm whose unit ball B is given by $B = \{f : \int F(|f|)\,dt \leq 1\}$. Similarly the Orlicz sequence space ℓ_F consists of all sequences $(x_n)_{n=1}^\infty$ such that $\sum_{n=1}^\infty F(|x_n|) < \infty$; its unit ball consists of all sequences for which $\sum F(|x_n|) \leq 1$. ℓ_F and ℓ_G coincide as Banach spaces if and only F and G are equivalent near zero. $L_F(0,\infty)$ is reflexive if and only if there exists $\alpha > 0$ and a constant C so that if $0 \le t \le 1$ and $0 < x < \infty$, $F(tx) \le Ct^{1+\alpha}F(x)$. ℓ_F is reflexive if F is equivalent near zero to a function G for which $L_G(0,\infty)$ is reflexive.

If F is convex and L_F is reflexive we define F^* by $F^*(t) = \sup_{0<s<\infty}(st - F(s))$. Then F^* is also a convex Orlicz function satisfying the Δ_2–condition, and L_F^* can be naturally identified with L_{F^*}. Similarly, ℓ_F^* can be naturally identified with ℓ_{F^*}.

Let us now assume that $F(x) = \exp(\phi(\log x))$ for $x > 0$ where ϕ is a uniformly Lipschitz function; this is, in particular the case when F is convex and satisfies the Δ_2–condition. Then we define E_F to be the closure of the set of functions $F_t(x) = F(tx)/F(t)$, for $0 < t \le 1$, in $C[0,1]$; E_F is a compact set in $C[0,1]$. We let C_F be the closed convex hull of E_F. It is shown in [6] (see [10]) that ℓ_G is isomorphic to a closed subspace of ℓ_F if and only if G is equivalent near zero to some $G_1 \in C_F$. Lindenstrauss and Tzafriri [8] showed that if $G \in E_F$ then ℓ_G is isomorphic to a complemented subspace of ℓ_F. They asked if the converse was true, i.e. if ℓ_G is isomorphic to a complemented subspace of ℓ_F, is G equivalent near zero to some $G_1 \in E_F$? The author gave a counter-example to this with $G(x) = x^p$ in [5].

However, these considerations lead Lindenstrauss and Tzafriri to introduce the class of minimal Orlicz functions. We define an Orlicz function satisfying the above conditions to be minimal if $F \in E_G$ whenever $G \in E_F$. It is easy to see that if F is minimal then it is equivalent to a convex and minimal Orlicz function. In fact if $\tilde{F}$ is equivalent to F and is convex then there exists $G \in E_{\tilde{F}}$ which is minimal; further some $F_1 \in E_G$ is equivalent to F by the minimality of F so that F_1 is both minimal and convex and equivalent to F. We shall refer to any Orlicz sequence space ℓ_F where F is equivalent near zero to a minimal Orlicz function as a minimal Orlicz sequence space. It is shown in [8] that if ℓ_F is a minimal Orlicz sequence space and $G \in E_F$ then $\ell_F \approx \ell_G$ and this suggests the conjecture that each such space is prime. In [8] non-trivial examples of minimal Orlicz sequence spaces were constructed. More recently, Hernandez and Rodriguez-Salinas ([3]) gave an explicit example, which actually was introduced for different purposes in [4]. This example, which is reflexive, is given by

$$(*) \qquad F(t) = t^p \exp\left(\sum_{n=0}^{\infty}(1 - \cos(2\pi(\log t)/2^n)\right)$$

where $1 < p < \infty$.

Let us define a minimal Orlicz sequence space ℓ_F to be *strongly minimal* if, whenever ℓ_G is an Orlicz sequence space which is isomorphic to a complemented subspace of ℓ_F then G is equivalent to function in E_F. We shall show that there is a non-trivial minimal reflexive Orlicz sequence space which is not strongly minimal and in fact contains a complemented copy of ℓ_p for some p; this space cannot be prime so that the Lindenstrauss-Tzafriri conjecture is false. However, we show that the space ℓ_F with F given by (*) is strongly minimal. This does not show that the space is prime; however it does suggest the possibility in view of the following:

THEOREM 1.1. *Let ℓ_F be a strongly minimal reflexive Orlicz sequence space. Let X be a complemented subspace of ℓ_F with an unconditional basis. Then X contains a complemented subspace isomorphic to ℓ_F. In particular if $X \approx X \times X$ then $X \approx \ell_F$.*

PROOF: It is easy to show that some subsequence of the unconditional basis is equivalent to the canonical basis of an Orlicz sequence space ℓ_G. But then G is equivalent near zero to a function in E_F and so (see [8]) $\ell_G \approx \ell_F$. The last assertion is a well-known form of the Pelczynski decomposition technique.∎

In view of Theorem 1.1 we can relate our example to three open problems (this observation is due to Peter Casazza). If ℓ_F is not a new prime space then it either has a complemented subspace X which fails to have an unconditional basis, or ℓ_F fails to have the Schroeder-Bernstein Property [1].

This research was initiated during a visit to Spain in May 1989. The author would like to thank Francisco Hernandez for some interesting discussions on this topic and the Universities of Madrid and Zaragoza for their hospitality.

2. Complemented subspaces of Orlicz sequence spaces.

We recall that a basic sequence $(x_n)_{n=1}^{\infty}$ in a Banach space X dominates a basic sequence $(y_n)_{n=1}^{\infty}$ in a Banach space Y provided there is a constant M so that for all $a_1, \ldots, a_n$ and $n \in \mathbf{N}$,

$$\|\sum_{i=1}^{n} a_i y_i\| \leq M \|\sum_{i=1}^{n} a_i x_i\|.$$

If (x_n) is a basis of X then (x_n^*) denotes the biorthogonal functionals in X^*. The following lemma is very well-known and we only sketch the proof.

LEMMA 2.1. *Let X and Y be reflexive Banach spaces with symmetric bases $(x_n)_{n=1}^{\infty}$ and $(y_n)_{n=1}^{\infty}$, respectively. In order that X be isomorphic to a complemented subspace of Y it is necessary and sufficient that there is an increasing sequence of positive integers $(p_n)_{n=0}^{\infty}$ with $p_0 = 0$ and block basic sequences $u_n = \sum_{i=p_{n-1}+1}^{p_n} \alpha_i y_i$, $\phi_n = \sum_{i=p_{n-1}+1}^{p_n} \beta_i y_i^*$ so that (x_n) dominates (u_n), (x_n^*) dominates (ϕ_n) and $\inf |\phi_n(u_n)| > 0$.*

PROOF: One direction follows easily from standard "gliding hump" techniques. For the other assume that $(u_n), (\phi_n)$ satisfy the given conditions. We may assume without loss of generality that $\phi_n(u_n) = 1$ for all n. Define $A : Y \to X$ by $Ay = \sum \phi_n(y) x_n$ and $B : X \to Y$ by $Bx = \sum x_n^*(x) u_n$. It suffices to show that both A and B are well-defined and bounded. For $y \in Y$, $x^* \in X^*$, and $m \in \mathbf{N}$,

$$\begin{aligned} |\sum_{k=1}^{m} x^*(x_k)\phi_k(y)| &\leq \|y\| \|\sum_{k=1}^{m} x^*(x_k)\phi_k\| \\ &\leq M\|y\|\|x^*\| \end{aligned}$$

for a suitable constant M independent of m, y, x^*. This shows that A is indeed bounded. A very similar argument shows that B is bounded. ∎

LEMMA 2.2. *Let F be a convex Orlicz function such that $L_F(0,\infty)$ is reflexive (so that F and F^* are both continuous.) Then for $0 \le x < \infty$ we have $x \le F^{-1}(x)(F^*)^{-1}(x) \le 2x$.*

PROOF: The right-hand side inequality is obvious from $uv \le F(u) + F^*(v)$ upon substituting $F(u) = F^*(v) = x$. For the left-hand inequality, suppose $x = F(u)$ and note that $F(u)/u \le F'(u)$ where F' denotes the left-hand derivative of F. Then $F^*(F'(u)) = uF'(u) - F(u)$ and $(F^*)'(F'(u)) \le u$. Thus $(F^*)'(x/u) \le u$ and hence $F^*(x/u) \le x$. Thus $x \le u(F^*)^{-1}(x) = F^{-1}(x)(F^*)^{-1}(x)$.∎

THEOREM 2.3. *Let F, G be convex Orlicz functions such that $L_F(0,\infty)$ and $L_G(0,\infty)$ are reflexive. Then for ℓ_G to be isomorphic to a complemented subspace of ℓ_F it is necessary and sufficient that there is a constant C and a sequence μ_n of probability measures each with compact support in $(0,1]$ such that:*

$$(1) \qquad \int F(F^{-1}(t)x)\frac{d\mu_n(t)}{t} \le CG(x) \qquad 2^{-n} \le x \le 1$$

$$(2) \qquad \int F^*((F^*)^{-1}(t)x)\frac{d\mu_n(t)}{t} \le CG^*(x) \qquad 2^{-n} \le x \le 1.$$

PROOF: Suppose (1) and (2) hold. Since $\int t^{-1}d\mu_n < \infty$ for each n, we may find an increasing sequence of positive integers $(p_n)_{n=0}^{\infty}$ with $p_0 = 0$ and a sequence of nonnegative measurable functions $(f_n)_{n=0}^{\infty}$ with support f_n contained in $[p_{n-1}, p_n]$, so that f_n is nonincreasing on $[p_{n-1}, p_n]$ and such that $\lambda(f_n > t) = \int_{(t,1]} s^{-1}d\mu_n(s)$.

Let $u_n(t) = F^{-1}(f_n(t))$ and let $v_n(t) = (F^*)^{-1}(f_n(t))$. Then clearly

$$\int_0^\infty F(u_n(t))dt = \int_0^\infty F^*(v_n(t))dt = \int_0^\infty f_n(t)dt = 1.$$

Thus $\|u_n\|_{L_F} = \|v_n\|_{L_{F^*}} = 1$.

Suppose $a_1, \ldots, a_n \in \mathbf{R}$ with $\sum_{i=1}^n G(|a_i|) \le C^{-1}$. Let $J = \{i : 1 \le i \le n,\ |a_i| \ge 2^{-i}\}$. Then

$$\begin{aligned}\int_0^\infty F(\sum_{i\in J} a_i u_i(t))dt &= \sum_{i\in J}\int_{p_{i-1}}^{p_i} F(|a_i|F^{-1}(f_i(t))dt \\ &= \sum_{i\in J}\int F(|a_i|F^{-1}(t))\frac{d\mu_i(t)}{t} \\ &\le C\sum_{i\in J} G(|a_i|) \le 1.\end{aligned}$$

Thus $\|\sum_{i\in J} a_i u_i\|_{L_F} \le 1$ while $\|\sum_{i\notin J} a_i u_i\|_{L_F} \le \sum_{i=1}^n 2^{-i} \le 1$. It follows that the unit vector basis of ℓ_G dominates (u_n) and a similar argument shows that the unit vector basis of ℓ_{G^*} dominates (v_n).

Now let P be the natural averaging projection of L_F onto ℓ_F, (which we identify as the subspace of L_F of functions constant on each interval $(n-1, n]$ for $n \in \mathbf{N}$) i.e.

$$Pf = (\int_{n-1}^n f(t)dt)_{n=1}^\infty.$$

We also use P for the same projection on L_{F^*}. Let $\tilde{u}_n = Pu_n$ and $\tilde{v}_n = Pv_n$. Since P is bounded on both L_F and L_{F^*}, (u_n) dominates the block basic sequence $(\tilde{u}_n)$ and (v_n) dominates the sequence $(\tilde{v}_n)$. Now since u_n and v_n are each nonincreasing on $[p_{n-1}, p_n]$ we have

$$\begin{aligned}
\sum_{i=p_{n-1}}^{p_n} \tilde{u}_{n,i}\tilde{v}_{n,i} &= \sum_{i=p_{n-1}+1}^{p_n} \int_{i-1}^{i} u_n(t)dt \int_{i-1}^{i} v_n(t)dt \\
&\geq \sum_{i=p_{n-1}+1}^{p_n} u_n(i)v_n(i) \\
&\geq \int_{p_{n-1}+1}^{p_n} u_n(t)v_n(t)dt \\
&= \int_{p_{n-1}+1}^{p_n} F^{-1}(f_n(t))(F^*)^{-1}(f_n(t))dt \\
&\geq \int_{p_{n-1}+1}^{p_n} f_n(t)dt \\
&= 1 - \int_{p_{n-1}}^{p_{n-1}+1} f_n(t)dt.
\end{aligned}$$

Now we split into two cases. If $\liminf \int_{p_{n-1}}^{p_{n-1}+1} f_n dt < \frac{1}{2}$ we can pass to a subsequence and apply Lemma 1 to obtain the conclusion. In the other case we can suppose $\int_{p_{n-1}}^{p_{n-1}+1} f_n dt \geq \frac{1}{2}$ for all n. In this case $f_n(p_{n-1} + \frac{1}{4}) \geq \frac{1}{4}$, and hence $\tilde{u}_{n,p_{n-1}+1} \geq \frac{1}{4}F^{-1}(\frac{1}{4})$. Thus $(\tilde{u}_n)$ dominates the unit vector basis of ℓ_F; similarly $(\tilde{v}_n)$ dominates the unit vector basis of ℓ_{F^*}. As $(\tilde{u}_n)$ and $(\tilde{v}_n)$ are dominated respectively by the unit vector bases of ℓ_G and ℓ_{G^*} we conclude that G is equivalent to F in this case.

Conversely, let us suppose that ℓ_G is equivalent to a complemented subspace of ℓ_F. We denote by (e_n) and (e_n^*) the canonical bases in ℓ_F and ℓ_F^*. We may suppose that there exist normalized block basic sequences (u_n) in ℓ_F and (v_n) in ℓ_{F^*}, equivalent respectively to the unit vector bases of ℓ_G and ℓ_{G^*}, of the form $u_n = \sum_{p_{n-1}+1}^{p_n} \alpha_i e_i$ and $v_n = \sum_{p_{n-1}+1}^{p_n} \beta_i e_i^*$ where $0 = p_0 < p_1 < \ldots < p_n < \ldots$ and so that for some $\delta > 0$,

$$|\sum_{i=p_{n-1}+1}^{p_n} \alpha_i\beta_i| \geq \delta.$$

Note also that

$$\sum_{i=p_{n-1}+1}^{p_n} |\alpha_i\beta_i| \leq \sum_{i=p_{n-1}}^{p_n} (F(|\alpha_i|) + F^*(|\beta_i|)) \leq 2.$$

Let $K = 4/\delta$ and $A_n = \{i : p_{n-1}+1 \le i \le p_n,\ \max(F(|\alpha_i|), F^*(|\beta_i|)) \le K|\alpha_i\beta_i|\}$. Let B_n be the complement of A_n relative to $\{i : p_{n-1}+1 \le i \le p_n\}$. Then

$$\begin{aligned}\sum_{i\in B_n} |\alpha_i\beta_i| &\le \frac{1}{K}\sum_{i\in B_n} F(|\alpha_i|) + F^*(|\beta_i|)\\ &\le \frac{2}{K} = \frac{\delta}{2}.\end{aligned}$$

Thus if $\sigma_n = \sum_{i\in A_n} |\alpha_i\beta_i|$ then $\sigma_n \ge \frac{\delta}{2}$.

Now for $i \in A_n$ we have

$$|\alpha_i| \le F^{-1}(K|\alpha_i||\beta_i|) \le \frac{2K|\alpha_i\beta_i|}{(F^*)^{-1}(K|\alpha_i\beta_i|)}$$

so that $|\beta_i| \ge (2K)^{-1}(F^*)^{-1}(K|\alpha_i\beta_i|)$. This and a similar inequality for α_i imply the existence of a constant $c_1 > 0$ depending only on F, F^* and δ so that, for $i \in A_n$,

$$\begin{aligned}|\alpha_i| &\ge c_1 F^{-1}(|\alpha_i\beta_i|)\\ |\beta_i| &\ge c_1 (F^*)^{-1}(|\alpha_i\beta_i|).\end{aligned}$$

Now let $\tilde{u}_n = \sum_{i\in A_n} \alpha_i e_i$. Plainly the canonical basis of ℓ_G dominates $(\tilde{u}_n)$. Thus there exists a constant C_1 so that for any $\gamma > 0$ there exists $n_0 = n_0(\gamma)$ so that if $n \ge n_0$ and $\gamma \le x \le 1$,

$$\sum_{i\in A_n} F(x|\alpha_i|) \le C_1 G(x).$$

Thus for $n \ge n_0$ and $\gamma \le x \le 1$,

$$\sum_{i\in A_n} F(xc_1F^{-1}(|\alpha_i\beta_i|)) \le C_1G(x),$$

and by utilizing the Δ_2-condition, we obtain for a suitable constant C_2 depending only F, F^* and δ,

$$\sum_{i\in A_n} F(xF^{-1}(|\alpha_i\beta_i|) \le C_2 G(x).$$

A similar argument can be applied to $\tilde{v}_n = \sum_{i\in A_n} \beta_i e_i^*$ in ℓ_{F^*}. By passing to a subsequence we can for a suitable constant C_3 require that for $2^{-n} \le x \le 1$,

$$\sum_{i\in A_n} F(xF^{-1}(|\alpha_i\beta_i|) \le C_3 G(x)$$

$$\sum_{i\in A_n} F^*(x(F^*)^{-1}(|\alpha_i\beta_i|)) \le C_3 G^*(x).$$

Now let μ_n be the probability measure supported on a finite subset of $(0,1]$ given by

$$\mu_n = \frac{1}{\sigma_n}\sum_{i\in A_n} |\alpha_i\beta_i| \epsilon_{|\alpha_i\beta_i|}$$

where ϵ_a denotes the Dirac measure at a. Then

$$\int F(xF^{-1}(t))\frac{d\mu_n(t)}{t} = \frac{1}{\sigma_n}\sum_{i\in A_n} F(xF^{-1}(|\alpha_i\beta_i|))$$
$$\leq 2C_3\delta^{-1}G(x)$$

as long as $2^{-n} \leq x \leq 1$. We also obtain the similar inequality for F^* and G^* and hence the theorem is proved.∎

Let us now introduce some notation. If ℓ_F is a reflexive Orlicz sequence space and $0 < \lambda < \infty$ we shall say that G is λ−represented in F if there is a constant C and a sequence ν_n of probability measures with compact support in $(0,1]$ such that for $2^{-n} \leq x \leq 1$,

$$\left(\int \max\left(\frac{F(tx)}{G(x)F(t)}, \frac{G(x)F(t)}{F(tx)}\right)^{\lambda} d\nu_n(t)\right)^{\frac{1}{\lambda}} \leq C.$$

THEOREM 2.4. *Let ℓ_F be a reflexive Orlicz sequence space. Then there exist constants $0 < \lambda_0 < 1 < \lambda_1 < \infty$ with the property that for any Orlicz sequence space ℓ_G to be isomorphic to a complemented subspace of ℓ_F it is necessary that G is λ_0−represented in F and sufficient that G is λ_1−represented in F.*

PROOF: Clearly it suffices to consider the case when F is convex. Since ℓ_F is reflexive, we may assume that L_F is reflexive so that there is a constant $0 < \alpha < 1$ and a constant $c > 0$ so that whenever $\xi \geq 1$ and $0 \leq x < \infty$, we have $F(\xi x) \geq c\xi^{1+\alpha}F(x)$ and $F^*(\xi x) \geq c\xi^{1+\alpha}F^*(x)$.

For ξ, x we define $A(\xi,x) = \{t : F(t)G(x) \geq \xi F(tx)\}$. Also let $B(\xi,x) = \{t : F^*(tx) \geq \xi F^*(t)G^*(x)\}$. We make two claims:

CLAIM 1: If $0 < y \leq 1$, and $\xi \geq 2$ then

$$F(A(\xi, G^{-1}(y))) \subset F^*(B(c\xi^{\alpha}2^{-(1+\alpha)}, (G^*)^{-1}(y))).$$

CLAIM 2: If $0 < y \leq 1$, and $\xi \geq 2$, then

$$F^*(B(\xi, (G^*)^{-1}(y))) \subset F(A(c\xi^{\alpha}2^{-(1+\alpha)}, G^{-1}(y))).$$

PROOF OF CLAIM 1: Suppose $F^{-1}(s) \in A(\xi, G^{-1}(y))$. Then $sy \geq \xi F(F^{-1}(s)G^{-1}(y))$ so that $F^{-1}(s)G^{-1}(y) \leq F^{-1}(sy/\xi)$. By Lemma 2.2, this implies

$$\frac{1}{2}\xi(F^*)^{-1}(sy/\xi) \leq (F^*)^{-1}(s)(G^*)^{-1}(y).$$

Thus applying F^* to both sides we obtain

$$\frac{c}{2^{1+\alpha}}\xi^{\alpha}sy \leq F^*((F^*)^{-1}(s)(G^*)^{-1}(y)).$$

Now if we substitute $\tau = (F^*)^{-1}(s)$ and $z = (G^*)^{-1}(y)$,

$$\frac{c}{2^{1+\alpha}}\xi^{\alpha}F^*(\tau)G^*(z) \leq F^*(\tau z).$$

This implies that $\tau = (F^*)^{-1}(s) \in B(c\xi^\alpha 2^{-(1+\alpha)}, (G^*)^{-1}(y))$ as required.

PROOF OF CLAIM 2: Suppose $(F^*)^{-1}(s) \in B(\xi, (G^*)^{-1}y)$. Then

$$\xi sy \leq F^*((F^*)^{-1}(s)(G^*)^{-1}(y))$$

so that $(F^*)^{-1}(\xi sy) \leq (F^*)^{-1}(s)(G^*)^{-1}(y)$. Hence $\frac{1}{2}\xi F^{-1}(s)G^{-1}(y) \leq F^{-1}(\xi sy)$ and so $F(\frac{1}{2}\xi F^{-1}(s)G^{-1}(y)) \leq \xi sy$. Now, by the assumptions on F,

$$\frac{c}{2^{1+\alpha}}\xi^\alpha F(F^{-1}(s)G^{-1}(y)) \leq sy$$

and $F^{-1}(s) \in A(c\xi^\alpha/2^{1+\alpha}, G^{-1}(y))$ as required.

We proceed to the proof of the theorem. We pick λ_0 so that $0 < \lambda_0 < \alpha$. Suppose first that ℓ_G is isomorphic to a complemented subspace of ℓ_F. Then we may choose measures μ_n as in Theorem 2.3. Let $\nu_n = \mu_n \circ F$. Then we have

$$\int \frac{F(tx)}{F(t)G(x)} d\nu_n(t) \leq C \tag{3}$$

for $2^{-n} \leq x \leq 1$.

Next suppose $2^{-n} \leq (G^*)^{-1}(G(x)) = y \leq 1$. Then for $\xi \geq 2$,

$$\begin{aligned}
\nu_n(A(\xi, x)) &= \mu_n(F(A(\xi, x))) \\
&\leq \mu_n(F^*B(\frac{c}{2^{1+\alpha}}\xi^\alpha, y)) \\
&\leq \frac{2^{1+\alpha}}{c}\xi^{-\alpha}\int \frac{F^*((F^*)^{-1}(t)y)}{G^*(y)}\frac{d\mu_n(t)}{t} \\
&\leq Cc^{-1}2^{1+\alpha}\xi^{-\alpha}.
\end{aligned}$$

Now if $\lambda_0 < \alpha$ this leads to an estimate

$$\left(\int \left(\frac{F(t)G(x)}{F(tx)}\right)^{\lambda_0} d\nu_n(t)\right)^{1/\lambda_0} \leq C_1$$

for $2^{-n} \leq y \leq 1$ where C_1 depends only on C, α and λ_0. Passing to a suitable subsequence and combining with (3) gives the result in one direction.

For the converse direction, pick λ_1 so that $\lambda_1\alpha > 1$. This time we suppose ν_n are given as in the definition of λ-representability. Let $\mu_n = \nu_n \circ F^{-1}$. Then since $\lambda_1 > 1$ it is easy to see that (1) of Theorem 2.3 is satisfied.

Now suppose x is such that $2^{-n} \leq G^{-1}(G^*(x)) = y \leq 1$. Suppose $\xi \geq 2$; then for a suitable constant C independent of n,

$$\begin{aligned}
\mu_n(F^*B(\xi, x)) &\leq \mu_n(F(A(\frac{c}{2^{1+\alpha}}\xi^\alpha, y)) \\
&= \nu_n(A(\frac{c}{2^{1+\alpha}}\xi^\alpha, y)) \\
&\leq C\xi^{-\lambda_1\alpha}.
\end{aligned}$$

This leads to an estimate, for $2^{-n} \leq y \leq 1$,

$$\int \frac{F^*(tx)}{F^*(t)G^*(x)} d\mu_n \circ F^*(t) \leq C_1$$

for a suitable C_1. Changing variables and passing to a suitable subsequence gives (2) of Theorem 2.3.∎

3. Examples of minimal Orlicz sequence spaces.

We will now describe a method of construction of minimal Orlicz functions suggested by the work of Hernandez and Rodriguez-Salinas [3]. First we fix $p > 1$. Identify the unit circle $\mathbf{T}$ with $\mathbf{R}/2\pi\mathbf{Z}$. We shall suppose that $(f_n)_{n=0}^{\infty}$ is a sequence of C^1-functions on $\mathbf{T}$ (i.e. $2\pi-$periodic functions on $\mathbf{R}$) satisfying $f_n(0) = 0$ and such that the series $\sum_{n=0}^{\infty} 2^{-n} L_n$ converges where $L_n = \|f_n'\|_\infty$. For convenience we let $R_n = \sum_{k=n+1}^{\infty} 2^{-k} L_k$ so that $\lim_{n\to\infty} R_n = 0$. We then define the C^1-function ϕ on $\mathbf{R}$ by

$$\phi(u) = \sum_{n=0}^{\infty} f_n(\frac{2\pi u}{2^n}).$$

We define for $t > 0$, $F(t) = t^p \exp(\phi(-\log t))$.

We also introduce the functions g_n on $\mathbf{T}$ defined by $g_n(\theta) = \sum_{k=0}^{n} f_k(2^{n-k}\theta)$. Let us then note that for any u, v we have

$$|\phi(u+v) - \phi(v) - g_n(\frac{2\pi(u+v)}{2^n}) + g_n(\frac{2\pi v}{2^n})| \leq \sum_{k=n+1}^{\infty} |f_k(\frac{2\pi(u+v)}{2^k}) - f_k(\frac{2\pi v}{2^k})|$$

and hence

$$|\phi(u+v) - \phi(v) - g_n(\frac{2\pi(u+v)}{2^n}) + g_n(\frac{2\pi v}{2^n})| \leq 2\pi|u|R_n. \tag{4}$$

PROPOSITION 3.1. *ℓ_F is a minimal reflexive Orlicz sequence space.*

PROOF: First, for any $\epsilon > 0$ there exists $N = N(\epsilon)$ so that

$$2\pi R_N = 2\pi \sum_{n=N+1}^{\infty} 2^{-n} L_n < \epsilon$$

and hence ϕ satisfies an estimate $|\phi(u+v) - \phi(v)| < M(\epsilon) + \epsilon|u|$ where $M(\epsilon) = 2\|g_N\|_\infty$. From this it follows without difficulty that F is equivalent an Orlicz function G with the property that $x^{-\alpha}G(x)$ is increasing for some $\alpha > 1$. Thus ℓ_F is a reflexive Orlicz space.

To demonstrate minimality, suppose $G \in E_F$. Then there is a sequence $\sigma_k \geq 0$ such that if $G(x) = x^p \exp(\psi(-\log x))$ then $\lim_{k\to\infty}(\phi(u+\sigma_k) - \phi(\sigma_k)) = \psi(u)$, uniformly on compact subsets of $[0,\infty)$. By passing to a subsequence we may suppose that $2\pi\sigma_k/2^n$ converges in $\mathbf{T}=\mathbf{R}/2\pi\mathbf{Z}$ for each fixed n to some a_n where $0\le a_n<2\pi$. Thus

$$|\psi(u)-g_n(\frac{2\pi u}{2^n}+a_n)+g_n(a_n)|\le 2\pi R_n|u|.$$

If we let $\tau_n=2^n(1-(2\pi)^{-1}a_n)$ then

$$|\psi(u+\tau_n)-\psi(\tau_n)-g_n(\frac{2\pi u}{2^n})|\le 4\pi R_n|u|$$

and hence

$$|\psi(u+\tau_n)-\psi(\tau_n)-\phi(u)|\le 6\pi R_n|u|.$$

This implies that $F\in E_G$ and so F is minimal.∎

We now impose an additional constraint.

PROPOSITION 3.2. *Suppose for some increasing sequence of integers N_n we have $\sup 2^{N_n}R_{N_n}=A<\infty$. If $F(x)$ is equivalent to x^r for some r then $\|g_{N_n}\|_\infty$ is bounded. Furthermore $G(x)=x^p\exp(\psi(-\log x))$ is λ−represented in F if and only if there is a constant C so that for every n there is a probability measure μ_n on $\mathbf{T}$ such that*

$$\int \exp(\lambda|g_{N_n}(\theta+\theta_0)-g_{N_n}(\theta)-h_n(\theta_0)|)d\mu_n(\theta)\le C,\qquad 0\le\theta_0<2\pi \tag{5}$$

where $h_n(\theta)=\psi(2^{N_n}\theta/2\pi)$.

PROOF: Notice first that $|\phi(u)-g_{N_n}(2\pi u/2^{N_n})|\le 2\pi A$ for $|u|\le 2^{N_n}$. In particular $|\phi(2^{N_n})|\le 2\pi A$. Thus if F is equivalent to x^r then $r=p$. But then ϕ is bounded and so there is also a uniform bound on g_{N_n}.

For the second part, we observe that if G is λ−represented in F then there is a constant C_0 and a sequence of compactly supported probability measures ν_n on $[0,\infty)$ such that

$$\int \exp(\lambda|\phi(u+v)-\phi(u)-\psi(v)|)d\nu_n(u)\le C_0$$

for $0\le v\le 2^{N_n}$. If $0\le v\le 2^{N_n}$, (4) gives us the estimate

$$|\phi(u+v)-\phi(u)-g_{N_n}(\frac{2\pi(u+v)}{2^{N_n}})+g_{N_n}(\frac{2\pi u}{2^{N_n}})|\le 2\pi A.$$

If we define μ_n on $\mathbf{T}$ by $\int f(\theta)d\mu_n(\theta)=\int f(2\pi t/2^{N_n})d\nu_n(t)$ (5) will follow with $C=C_0e^{2\pi\lambda A}$.

Conversely we assume (5) we quickly get that for $0\le\theta_0<2\pi$,

$$\int \exp(\lambda|\phi(\frac{2^{N_n}(\theta+\theta_0)}{2\pi})-\phi(\frac{2^{N_n}\theta}{2\pi})-\psi(\frac{2^{N_n}\theta_0}{2\pi})|)d\mu_n\le Ce^{2\pi\lambda A}.$$

It then easily follows by a change of variables that G is λ−represented in F.∎

THEOREM 3.3. *Suppose $1 < p < \infty$. There exists a minimal reflexive Orlicz sequence space ℓ_F which is not isomorphic to ℓ_p but which contains ℓ_p as a complemented subspace. In particular, ℓ_F is not prime.*

PROOF: We pick a sequence of C^1-functions $(h_n)_{n=0}^{\infty}$ on $\mathbf{T}$ with $h_n(0) = h_n(2\pi) = 0$ and such that if $M_n = \|h_n\|_\infty$ then we have both $M_n > 2^n + \sum_{k=1}^{n-1} M_k$ and

$$\int_0^{2\pi} \exp(4^n|h_n(\theta)|)\frac{d\theta}{2\pi} \le 2$$

for all n.

Let $B_n = \|h_n'\|_\infty$. We pick a strictly increasing sequence of integers N_k such that $B_k 2^{N_{k-1}-N_k} \le 1$ for all k, where $N_0 = 0$. Then define f_n by $f_n = h_{N_k}$ if $n = N_k$ and $f_n = 0$ if $n \notin (N_k)_{k=0}^{\infty}$. Thus

$$\begin{aligned} 2^{N_k} R_{N_k} &= \sum_{j=k+1}^{\infty} B_j 2^{N_k - N_j} \\ &\le \sum_{j=k+1}^{\infty} 2^{N_k - N_{j-1}} \\ &\le 2 \end{aligned}$$

so that we can apply Proposition 3.2. Observe first that F cannot be equivalent to any x^r since $\|g_{N_k}\|_\infty \ge 2^k$.

To complete the proof we show that x^p is $\lambda-$represented in F for every λ. Thus Theorem 2.4 will give the result. To show this we estimate by convexity of the exponential function,

$$\int_0^{2\pi} \exp(\lambda|g_{N_k}(\theta+\theta_0) - g_{N_k}(\theta)|)\frac{d\theta}{2\pi} \le \int_0^{2\pi} \exp(2\lambda|g_{N_k}(\theta)|)\frac{d\theta}{2\pi}.$$

However writing $g_{N_k} = 2^{-(k+1)}(0) + \sum_{j=0}^{k} 2^{-(j+1)}(2^{j+1} f_{N_j})$ and again using convexity the integral is estimated by

$$2^{-(k+1)} + \sum_{j=0}^{k} 2^{-(j+1)} \int \exp(\lambda 2^{j+2}|f_{N_j}(\theta)|)\frac{d\theta}{2\pi} =$$

$$= 2^{-(k+1)} + \sum_{j=0}^{k} 2^{-(j+1)} \int \exp(\lambda 2^{j+2}|h_j(\theta)|)\frac{d\theta}{2\pi}$$

which is bounded, independent of k, for every λ. We can now apply Proposition 3.2. ∎

We now turn to the construction of strongly minimal Orlicz sequence spaces. We recall that ℓ_F is strongly minimal if whenever ℓ_G is isomorphic to a complemented subspace of ℓ_F then G is equivalent to F. Our example is of the above form with $f_n(x) = \alpha(1-\cos x)$ for every n, so that $\phi(u) = \alpha(\sum_{n=0}^{\infty}(1-\cos(2\pi u/2^n)))$. This example was first discussed in a function space context by Johnson, Maurey, Schechtman and Tzafriri [4] and later its minimality was proved by Hernandez and Rodriguez-Salinas

[3], who also observe that for small enough α, F is actually convex, (it is, of course, always equivalent to a convex function).

THEOREM 3.4. *Suppose $1 < p < \infty$ and $|\alpha| > 0$. Then if*

$$F(x) = x^p \exp \alpha(\sum_{n=0}^{\infty}(1 - \cos\left(\frac{2\pi \log x}{2^n}\right)))$$

then F is not equivalent to any x^r and ℓ_F is a strongly minimal Orlicz sequence space.

PROOF: In this case we can apply Proposition 3.2 (take $N_n = n$.) Thus our proof reduces to analyzing the functions $g_n(\theta) = \alpha \sum_{k=0}^{n}(1-\cos(2^k\theta))$ on $\mathbf{T}$. The fact that F is non-equivalent to any x^r is proved in [4], or may be proved from Proposition 3.2 by estimating $\|g_n\|_2$. We therefore have only to establish that if $G(x) = x^p \exp(\psi(-\log x))$ is λ-represented for some $\lambda > 0$ in F then G is equivalent to some $G_1 \in E_F$. This in turn will be achieved by establishing the following result of possibly independent interest.

THEOREM 3.5. *For any K there is a constant $C = C(K)$ (independent of n) so that if $g_n(\theta) = \sum_{k=0}^{n}(1 - \cos 2^k\theta)$, μ is a probability measure on $\mathbf{T}$ and h is a function on $\mathbf{T}$ such that for every $0 \leq \theta_0 < 2\pi$ we have*

$$\int |g_n(\theta + \theta_0) - g_n(\theta) - h(\theta_0)|^2 d\mu(\theta) \leq K^2$$

then there exists σ, $0 \leq \sigma < 2\pi$ such that

$$|h(\theta) - (g_n(\theta + \sigma) - g_n(\sigma))| \leq C$$

for $0 \leq \theta < 2\pi$.

PROOF: We first introduce the angular distance on $\mathbf{T}$, $\delta(\theta_1, \theta_2) = d(\theta_1 - \theta_2, 2\pi\mathbf{Z})$. We will be interested in ways of measuring the spread of μ. For $0 \leq k \leq n$, we define:

$$\alpha_k = \int\int \sin^2(2^{k-1}(\theta_1 - \theta_2))d\mu(\theta_1)d\mu(\theta_2)$$

$$\beta_k = \int\int \delta(2^k\theta_1, 2^k\theta_2)d\mu(\theta_1)d\mu(\theta_2).$$

In order to estimate these we choose τ_k for $0 \leq k \leq n$ to minimize

$$\int \sin^2(2^{k-1}(\theta - \tau_k))d\mu(\theta).$$

It is clear that such a minimizer exists and further if $1 \leq k \leq n$, we can choose τ_k from amongst at least 2^k possibilities so that if $A_k = \{\theta : \delta(2^{k-1}\theta, 2^{k-1}\tau_k) > \pi/2\}$ and $a_k = \mu A_k$ then $a_k \leq \frac{1}{2}$. We further introduce $B_k = \{\theta : \delta(2^k\theta, 2^k\tau_k) > \pi/4\}$ and $b_k = \mu B_k$ for $0 \leq k \leq n$. Let $C_k = (\mathbf{T} \setminus A_k) \cap (\mathbf{T} \setminus B_k)$ and $D_k = (\mathbf{T} \setminus B_k) \cap A_k$. Then $\mu C_k \geq 1 - a_k - b_k$ and $\mu D_k \geq a_k - b_k$.

We next relate a_k, b_k to α_k. Clearly by integrating over B_k we have

$$\sin^2(\frac{\pi}{8})b_k \leq \int \sin^2(2^{k-1}(\theta - \tau_k))d\mu(\theta) \leq \alpha_k.$$

Thus

$$b_k \leq 10\alpha_k.$$

Thus as long as $\alpha_k \leq 1/40$ we have $b_k \leq 1/4$. Hence in this case we have $\mu C_k \geq 1/4$. Now if $k \geq 1$ and $\theta_1 \in C_k$ and $\theta_2 \in D_k$ we have $\delta(2^{k-1}\theta_1, 2^{k-1}\tau_k) \leq \pi/8$ but $\delta(2^{k-1}\theta_1, 2^{k-1}\tau_k) \geq 3\pi/8$. Thus $\delta(2^{k-1}\theta_1, 2^{k-1}\theta_2) \geq \pi/4$ and $\sin^2(2^{k-2}\theta_1, 2^{k-2}\theta_2) \geq \sin^2(\pi/8) \geq 1/10$. Integrating over $C_k \times D_k \cup D_k \times C_k$ gives

$$\mu(C_k)\mu(D_k) \leq 5\alpha_{k-1}.$$

Thus if $\alpha_k \leq 1/40$ we obtain $\mu D_k \leq 20\alpha_{k-1}$ and hence $a_k \leq b_k + 20\alpha_{k-1} \leq 20\alpha_{k-1} + 10\alpha_k$. Since $a_k \leq \frac{1}{2}$ we can say in general that for $k \geq 1$,

$$a_k \leq 20(\alpha_{k-1} + \alpha_k).$$

Now suppose $\theta_1, \theta_2 \in C_k$. Then, if $k \geq 1$, we clearly have $\delta(2^{k-1}\theta_1, 2^{k-1}\theta_2) \leq \frac{1}{2}\delta(2^k\theta_1, 2^k\theta_2)$. Hence by integration we obtain

$$\begin{aligned}\beta_{k-1} &\leq \frac{1}{2}\beta_k + \pi(1 - \mu(C_k)^2) \\ &\leq \frac{1}{2}\beta_k + 2\pi(1 - \mu(C_k)) \\ &\leq \frac{1}{2}\beta_k + 2\pi(a_k + b_k).\end{aligned}$$

By induction, we obtain

$$\beta_l \leq 2^{l-n}\beta_n + 2\pi \sum_{j=l+1}^{n} 2^{l+1-j}(a_j + b_j)$$

and hence

$$\begin{aligned}\sum_{l=0}^{n} \beta_l &\leq 2(\beta_n + 2\pi(\sum_{j=1}^{n}(a_j + b_j))) \\ &\leq 2\pi(1 + \sum_{j=1}^{n}(20\alpha_{j-1} + 30\alpha_j)) \\ &\leq 2\pi(1 + 50\sum_{j=0}^{n}\alpha_j).\end{aligned}$$

Now returning to the original statement of the theorem we observe that if the hypotheses on μ hold then:

$$\left(\int\int |g_n(\theta_1+\theta_0)-g_n(\theta_2+\theta_0)|^2 d\mu(\theta_1)d\mu(\theta_2)\right)^{\frac{1}{2}} \leq 2K$$

for every $0 \leq \theta_0 < 2\pi$. Hence integrating again and using Fubini's theorem,

$$\int\int\left(\int_0^{2\pi} |g_n(\theta_1+\theta_0)-g_n(\theta_2+\theta_0)|^2 \frac{d\theta_0}{2\pi}\right) d\mu(\theta_1)d\mu(\theta_2) \leq 4K^2.$$

However,

$$g_n(\theta_1+\theta_0)-g_n(\theta_2+\theta_0) = 2\sum_{k=0}^{n} \sin(2^{k-1}(\theta_1-\theta_2))\sin(2^k\theta_0+2^{k-1}(\theta_1+\theta_2))$$

so that the integral can be rewritten as

$$\int\int\sum_{k=0}^{n}\sin^2(2^{k-1}(\theta_1-\theta_2))\,d\mu(\theta_1)d\mu(\theta_2) \leq 2K^2$$

or

$$\sum_{k=0}^{n}\alpha_k \leq 2K^2.$$

This in turn yields

$$\sum_{k=0}^{n}\beta_k \leq 2\pi(1+100K^2).$$

It follows that we can pick $\sigma \in \mathbf{T}$ so that

$$\int\sum_{k=0}^{n}\delta(2^k\theta, 2^k\sigma)d\mu(\theta) \leq 2\pi(1+100K^2).$$

For this choice of σ notice that, for $0 \leq \theta_0 < 2\pi$,

$$\int\left|\sum_{k=0}^{n}(\cos 2^k(\theta+\theta_0)-\cos 2^k(\sigma+\theta_0))\right| d\mu(\theta) \leq 2\pi(1+100K^2)$$

and so, for $0 \leq \theta_0 < 2\pi$,

$$\int |g_n(\theta+\theta_0)-g_n(\sigma+\theta_0)|d\mu(\theta) \leq 2\pi(1+100K^2),$$

It now follows easily that

$$|h(\theta_0)-(g_n(\sigma+\theta_0)-g_n(\sigma))| \leq 4\pi(1+100K^2)+K$$

and the theorem is proved.∎

PROOF OF THEOREM 3.4, CONTINUED: This is almost immediate. If $G(x) = x^p \exp(\psi(-\log x))$ is λ-represented in F for some λ then setting $h_n(\theta) = \psi(2^n(\frac{\theta}{2\pi}))$ for $0 \leq \theta < 2\pi$ we certainly obtain that there exists a probability measure μ_n on $\mathbf{T}$ so that

$$\int |g_n(\theta+\theta_0) - g_n(\theta) - h_n(\theta_0)|^2 d\mu_n(\theta) \leq K^2$$

where K is independent of n. Thus there exists σ_n with

$$|h_n(\theta) - (g_n(\theta+\sigma_n) - g_n(\sigma_n))| \leq C$$

for $0 \leq \theta < 2\pi$ where C is independent of n. But then if $\alpha_n = 2^n(\sigma_n/2\pi)$ we have that

$$|\psi(u) - (\phi(u+\alpha_n) - \phi(u))| \leq C'$$

for $0 \leq u \leq 2^n$ where C' is independent of n. It quickly follows that G is equivalent to a function in E_F.∎

References.

1. P.G. Casazza, The Schroeder-Bernstein Property for Banach spaces, Cont. Math. 85 (1989) 61-77.
2. F.L. Hernandez and B. Rodriguez-Salinas, On ℓ^p-complemented copies in Orlicz spaces, Israel J. Math. 62 (1988) 37-55.
3. F.L. Hernandez and B. Rodriguez-Salinas, On ℓ^p-complemented copies in Orlicz spaces II, Israel J. Math. 68 (1989) 27-55.
4. W.B. Johnson, B. Maurey, G. Schechtman and L. Tzafriri, *Symmetric structures in Banach spaces,* Mem. Amer. Math. Soc. 217, 1979.
5. N.J. Kalton, Orlicz sequence spaces without local convexity, Math. Proc. Cambridge Philos. Soc. 81 (1977) 253-277.
6. K.J. Lindberg, On subspaces of Orlicz sequence spaces, Studia Math. 45 (1973) 119-146.
7. J. Lindenstrauss and L. Tzafriri, On Orlicz sequence spaces, Israel J. Math. 10 (1971) 379-390.
8. J. Lindenstrauss and L. Tzafriri, On Orlicz sequence spaces II, Israel J. Math. 11 (1972) 355-379.
9. J. Lindenstrauss and L. Tzafriri, On Orlicz sequence spaces III, Israel J. Math. 14 (1973) 368-389.
10. J. Lindenstrauss and L. Tzafriri, *Classical Banach spaces I: sequence spaces,* Springer, Berlin-Heidelberg-New York 1977.

Type and Cotype in Musielak-Orlicz spaces

by

A. Kamińska
A. Mickiewicz University
and
Oakland University

and

B. Turett
Oakland University

Abstract. The type and cotype of a Musielak-Orlicz space over a nonatomic measure space are characterized in terms of the Young function that generates the space.

The notions of type and cotype of Banach spaces were introduced by B. Maurey and G. Pisier [13,14] in the mid-1970's. Since then, these notions have found frequent use in the geometry of Banach spaces [6,9]. In particular, the type and cotype of Orlicz spaces, Lorentz spaces, and general rearrangement invariant Banach function spaces have been investigated [1,2,9,10]. In this paper, the type and cotype of Musielak-Orlicz spaces, a class of function spaces which are, in general, not rearrangement invariant, are characterized in terms of the Young function that generates the space. As a corollary, a characterization of the types and cotypes of Orlicz spaces is obtained.

Recall that a Banach space X is of *type* p for some $1 < p \leq 2$ if, whenever (x_n) is a sequence in X satisfying $(\|x_n\|) \in \ell^p$, then $\sum_{n=1}^{\infty} x_n r_n$ converges almost everywhere on $[0, 1]$. (Here, and throughout this paper, r_n will denote the n-th Rademacher function.) Similarly, a Banach space X is of *cotype* q for some $2 \leq q < \infty$ if, whenever (x_n) is a sequence in X satisfying $\sum_{n=1}^{\infty} x_n r_n$ converges almost everywhere, then $(\|x_n\|) \in \ell^q$. In certain settings, type and cotype are dual notions. Pisier [16] has shown that a B-convex Banach space is of type p if and only if its dual space is of cotype q where $\frac{1}{p} + \frac{1}{q} = 1$.

Let us agree on some notation. (T, Σ, μ) will denote a measure space and $\Phi : \mathbf{R}^+ \times T \to \mathbf{R}_e^+$ will denote a Young function (with parameter); i.e., Φ is a non-negative, extended real-valued function such that, for almost all t, $\Phi(0,t) = 0$, $\Phi(u,t) > 0$ for $u > 0$, and $\Phi(u,t)$ is convex with respect to u, and such that, for all $u \geq 0$, $\Phi(u,t)$ is Σ-measurable with respect to t. The Musielak-Orlicz space L^Φ is then defined as the set of equivalence classes of measurable functions $f : T \to \mathbf{R}$ such that $\int_T \Phi(k|f(t)|, t)\, d\mu(t) < \infty$ for some $k > 0$. Under the norm given by $\|f\|_\Phi = \inf\{r > 0 : \int_T \Phi(|f(t)|/r, t)\, d\mu(t) \leq 1\}$, L^Φ is a Banach space. If Φ does not depend on the parameter t, i.e., if $\Phi(u,t) = \Phi(u)$ for all $t \in T$, then L^Φ is an Orlicz space. More information about Musielak-Orlicz spaces can be found in [7], [15].

In this paper, several notions restricting the growth of the Young function Φ will be considered. The most classical notion is that of a Δ_2-condition. A Young function Φ

satisfies a Δ_2-*condition* if there exists a positive constant K and a nonnegative integrable function h such that $\Phi(2u,t) \leq K\Phi(u,t) + h(t)$ for all $u \geq 0$ and almost all $t \in T$. Growth conditions intimately connected to the Δ_2-condition will now be defined.

Definition 1. A Young function Φ satisfies *condition* Δ^q $(q \geq 1)$ if there exists $K > 0$ and a nonnegative integrable function h such that $\Phi(\lambda u, t) \leq K\lambda^q(\Phi(u,t) + h(t))$ for $\lambda \geq 1$, $u \geq 0$, and almost all $t \in T$.

A Young function Φ satisfies *condition* Δ^{*p} $(p \geq 1)$ if there exists $K > 0$ and a nonnegative integrable h such that $\Phi(\lambda u,t) \geq K\lambda^p\,(\Phi(u,t) - h(t))$ for $\lambda \geq 1$, $u \geq 0$, and almost all t.

Recall that, if Φ_1 and Φ_2 are Young functions, Φ_1 is said to be equivalent to Φ_2, denoted by $\Phi_1 \sim \Phi_2$, if there exist positive constants K_1 and K_2 and nonnegative integrable functions h_1 and h_2 satisfying $\Phi_1(K_1u,t) \leq \Phi_2(u,t) + h_1(t)$ and $\Phi_2(K_2u,t) \leq \Phi_1(u,t) + h_2(t)$ for all $u \geq 0$ and almost all $t \in T$. It is easy to check that conditions Δ^q and Δ^{*p} are preserved under equivalence and that a Young function satisfying condition Δ^q also satisfies a Δ_2-condition.

The following lemmas collect some equivalent formulations of the conditions Δ^q and Δ^{*p}. In order to state the next lemma, a generalization of the notion of a Young function with parameter is needed. An *Orlicz function (with parameter)* is a nonnegative, extended real-valued function $\Phi : \mathbf{R}^+ \times T \to \mathbf{R}_e^+$ such that, for almost all t, $\Phi(0,t) = 0$, $\Phi(u,t)$ is nondecreasing and continuous in u, and $\Phi(u,t)$ tends to ∞ as u tends to ∞, and such that, for all $u \geq 0$, $\Phi(u,t)$ is Σ-measurable with respect to t. The notion of equivalence will be used between a Young function and an Orlicz function.

Some of the ideas in the next lemma can be found in [12].

Lemma 2. Let Φ be a Young function and $q \geq 1$. The following assertions are equivalent:

(a) Φ satisfies condition Δ^q.

(b) There exist $K > 0$ and a nonnegative measurable function g such that $\int \Phi(g(t),t)dt < \infty$ and, for almost all $t \in T$, $\Phi(\lambda u,t) \leq K\lambda^q \Phi(u,t)$ for all $\lambda \geq 1$ and $u \geq g(t)$.

(c) There exists an Orlicz function $\tilde{\Phi}$ equivalent to Φ such that, for almost all $t \in T$, $\tilde{\Phi}(\lambda u,t) \leq \lambda^q \tilde{\Phi}(u,t)$ for $u \geq 0$ and $\lambda \geq 1$.

(d) There exists an Orlicz function $\tilde{\Phi}$ equivalent to Φ such that, for almost all $t \in T$, $\tilde{\Phi}(u^{1/q},t)$ is concave in u.

Proof. The proof that (a) implies (b) is straightforward. Indeed, with h as given in the definition of condition Δ^q, define g to be the (measurable) function satisfying $\Phi(g(t),t) = h(t)$.

Assume that (b) is satisfied and define

$$\varphi(v,t) = \begin{cases} \dfrac{\Phi(g(t),t)}{g(t)^q} & \text{if } 0 \leq v \leq g(t) \\ \inf_{g(t)\leq u \leq v} \dfrac{\Phi(u,t)}{u^q} & \text{if } v > g(t). \end{cases}$$

Then φ is nonincreasing and continuous and, if $v > g(t)$, $\dfrac{1}{K}\dfrac{\Phi(v,t)}{v^q} \leq \varphi(v,t) \leq \dfrac{\Phi(v,t)}{v^q}$. Define $\tilde{\Phi}(u,t) = \int_0^u \varphi(v,t)v^{q-1}dv$. $\tilde{\Phi}$ is an Orlicz function (but perhaps not a Young function).

To check that $\tilde{\Phi}$ is equivalent to Φ, note that if $u \geq g(t)$, $\tilde{\Phi}(u,t) \geq \varphi(u,t)\,u^q/q \geq \Phi(u,t)/Kq \geq \Phi(u/Kq,t)$. Thus, for all $u \geq 0$, $\Phi(u/Kq,t) \leq \tilde{\Phi}(u,t) + \Phi(g(t),t)$. Since $\Phi(g(\cdot),\cdot) \in L^1$, this is half of what is needed. For the other inequality, note

$$\begin{aligned} \tilde{\Phi}(u,t) &\leq \int_0^{g(t)} \frac{\Phi(g(t),t)}{g(t)^q} v^{q-1}dv + \int_{g(t)}^u \frac{\Phi(v,t)}{v}\,dv \\ &\leq \frac{\Phi(g(t),t)}{q} + \frac{\Phi(u,t)}{u}(u - g(t)) \\ &\leq \Phi(u,t) + \Phi(g(t),t). \end{aligned}$$

Again, since $\Phi(g(\cdot),\cdot) \in L^1$, $\tilde{\Phi} \sim \Phi$.

Next, an easy change of variables and the nonincreasing nature of φ yields that $\tilde{\Phi}(\lambda u,t) \leq \lambda^q \tilde{\Phi}(u,t)$ for all $\lambda \geq 1, u \geq 0$, and almost all t. This shows that (b) implies (c).

Since condition Δ^q is preserved under equivalence, (c) implies (a) is clear. To show (b) implies (d), consider the $\tilde{\Phi}$ constructed above. Then $\tilde{\Phi}(u^{1/q},t) = \frac{1}{q}\int_0^u \varphi(w^{1/q},t)\,dw$ and, since the integrand is a decreasing function, $\tilde{\Phi}(u^{1/q},t)$ is concave in u.

Finally, to show that (d) implies (c), note that concavity implies that, for $\lambda \geq 1$ and $u \geq 0$, $\dfrac{\Phi((\lambda u)^{1/q},\ t)}{\lambda u} \leq \dfrac{\Phi(u^{1/q},t)}{u}$. Thus $\Phi(\lambda^{1/q}u^{1/q},\ t) \leq \lambda\Phi(u^{1/q},\ t)$ for $\lambda \geq 1$ and $u \geq 0$. This completes the proof of Lemma 2.

There are analagous characterizations of condition Δ^{*p}.

Lemma 3. Let Φ be a Young function and $p \geq 1$. The following assertions are equivalent:

(a) Φ satisfies condition Δ^{*p}.

(b) There exist $K > 0$ and a nonnegative measurable function g such that $\int \Phi(g(t),t)\,dt < \infty$ and, for almost all $t \in T$, $\Phi(\lambda u,t) \geq K\lambda^q \Phi(u,t)$ for all $\lambda \geq 1$ and $u \geq g(t)$.

(c) There exists a Young function $\bar{\Phi}$ equivalent to Φ such that, for almost all $t \in T$, $\bar{\Phi}(\lambda u,t) \geq \lambda^p \bar{\Phi}(u,t)$ for $u \geq 0$ and $\lambda \geq 1$.

(d) There exists a Young function $\bar{\Phi}$ equivalent to Φ such that, for almost all $t \in T$, $\bar{\Phi}(u^{1/p},t)$ is convex in u.

The proof of this lemma is analogous to the proof of Lemma 2 where, in (b) implies (c),

$$\varphi(v,t) = \begin{cases} \dfrac{\Phi(g(t),t)}{g(t)^{p+1}}\, v & \text{if } 0 \leq v \leq g(t) \\ \sup_{g(t)\leq u\leq v} \dfrac{\Phi(u,t)}{u^p} & \text{if } v > g(t). \end{cases}$$

In this case, note that $\bar{\Phi}$ is a Young function.

The connections between the Δ_2-condition, condition Δ^q, and condition Δ^{*p} are given below. The conjugate function of a Young function Φ is defined by

$\Phi^*(v,t) = \sup\{uv - \Phi(u,t) : u \geq 0\}$. It is easy to check that, for a Young function Φ, $(\Phi^*)^* = \Phi$ and that, if two Orlicz functions are equivalent, so are their conjugates.

Proposition 4. The following assertions are equivalent:

(a) Φ satisfies a Δ_2-condition.

(b) Φ satisfies condition Δ^q for some $q \geq 1$.

(c) Φ^* satisfies condition Δ^{*p} for some $p > 1$.

In fact, if $p, q > 1$ satisfy $\frac{1}{p} + \frac{1}{q} = 1$, Φ satisfies condition Δ^q if and only if Φ^* satisfies condition Δ^{*p}.

Proof. It is clear that (b) implies (a). That condition (a) implies condition (b) follows from the proofs of Theorem 4.1 in [8, p. 24] and Lemma 3.1 in [11, pp. 21-22] and the fact that if Φ satisfies a Δ_2-condition, there exists an equivalent $\bar{\Phi}$ satisfying $\bar{\Phi}(2u,t) \leq K\bar{\Phi}(u,t)$ for some $K > 0$, $u \geq g(t)$ where $\bar{\Phi}(g(\cdot),\cdot)$ is integrable, and almost all $t \in T$. The equivalence of conditions (b) and (c) follow from conjugate duality. Indeed, assume condition (b) is satisfied; so, by virtue of Lemma 2, there is no loss of generality in assuming that $\Phi(\lambda u,t) \leq K\lambda^q\Phi(u,t)$ for all $\lambda \geq 1$, $u \geq 0$, and almost all t. Taking the conjugate of each side yields $\Phi^*(v/\lambda,t) \geq K\lambda^q\Phi^*(v/K\lambda^q,t)$ for all $\lambda \geq 1$, $v \geq 0$, and almost all t. With $x = v/K\lambda^q$ and $\mu = K\lambda^{q-1}$, it follows that $\Phi^*(\mu x,t) \geq K^{1/(q-1)}\mu^p\Phi^*(x,t)$ which completes the proof since the condition Δ^{*p} is preserved under equivalence.

With these preliminaries on growth conditions behind us, their connections to the type and cotype of Musielak-Orlicz spaces can be stated.

Theorem 5. Let (T,Σ,μ) be nonatomic and let Φ be a Young function with parameter. Let $1 < p \leq 2 \leq q < \infty$.

a. The Musielak-Orlicz space L^Φ is of cotype q if and only if Φ satisfies condition Δ^q.

b. The Musielak-Orlicz space L^Φ is of type p if and only if Φ satisfies a Δ_2-condition and condition Δ^{*p}.

The sufficiency in (a) is proven in a manner similar to the proof of Theorem 2 in [2] by Z. G. Gorgadze and V.I. Tarieladze; the extra ingredients needed are contained in Lemma 2.

In order to prove the necessity in (a), a few technical lemmas will be used.

Lemma 6. Let (T, Σ, μ) be a nonatomic measure space. For natural numbers m, n, and j, let $f_{n,j}: T \to \mathbf{R}^+$ be measurable functions satisfying $f_{n,j+1} \leq f_{n,j}$ and $\int_T \sup_{n \geq m} f_{n,j}\, d\mu = \infty$. Then there exists a sequence (A_n) of pairwise disjoint sets in Σ and a strictly increasing sequence $(\ell_j)_{j \geq 0}$ of nonnegative integers with $\ell_0 = 0$ such that $\sum_{n=\ell_{j-1}+1}^{\ell_j} \int_{A_n} f_{nj}\, d\mu = 1$ for each $j \in \mathbf{N}$.

Proof. By the decreasing nature on $(f_{n,j})_j$, either $\sup_{n \geq 1} f_{n,j}$ is almost everywhere finitely-valued for large j or, for each j, $\sup_{n \geq 1} f_{n,j}$ takes on the value $+\infty$ on a set of positive measure. In the first case, assume, without loss of generality, that $\sup_{n \geq 1} f_{n,j}$ is everywhere finitely-valued for all $j \in \mathbf{N}$. The nonatomic nature of the measure spaces gives rise to $C_1' \in \Sigma$ such that $\int_{C_1'} \sup_{n \geq 1} f_{n,1}\, d\mu = 2$. Since $(\max_{1 \leq n \leq \ell} f_{n,1})$ increases to $\sup_{n \geq 1} f_{n,1}$, there exists $\ell_1 \in \mathbf{N}$ such that $1 \leq \int_{C_1'} \max_{1 \leq n \leq \ell_1} f_{n,1}\, d\mu \leq 2$. Choose a measurable set $C_1 \subset C_1'$ with $\int_{C_1} \max_{1 \leq n \leq \ell} f_{n,1}\, d\mu = 1$. For $n = 1, \ldots, \ell_1$, define $A_n = \{t \in C_1 : \max_{1 \leq k \leq \ell_1} f_{k,1}(t) = f_{n,1}(t) > f_i(t) \text{ for } i = 1, \ldots, n-1\}$. Then $\max_{1 \leq n \leq \ell_1} f_{n,1} = \sum_{n=1}^{\ell_1} f_{n,1}\, \chi_{A_n}$ and $\sum_{n=1}^{\ell_1} \int_{A_n} f_{n,1}\, d\mu = 1$.

Since $\int_{C_1} \sup_{n \geq \ell_1+1} f_{n,2}\, d\mu \leq \int_{C_1} \sup_{n \geq 1} f_{n,1}\, d\mu = 1$, it follows that $\int_{T \setminus C_1} \sup_{n \geq \ell_1+1} f_{n,2}\, d\mu = \infty$. Repeating the above argument yields $\ell_2 > 0$, and sets $A_{\ell_1+1}, \ldots A_{\ell_2}$ with the desired property. The remainder of the proof, in this case, follows similarly.

In the second case, assume that, for every natural number j, D_j is the set of positive measure on which $\sup_{n\geq 1} f_{n,j}$ takes on the value $+\infty$. Then the sequence (D_j) decreases to a set D. Depending on whether μD is positive or zero, the nonatomic nature of μ or disjunctification yields a sequence (B_k) of disjoint sets on which $\sup_{n\geq 1} f_{n,k}(t) = \infty$ for $t \in B_k$. An argument similar to the first case now completes the proof.

The next lemma provides an equivalent formulation of condition Δ^q. The definition says that the (nonnegative) function $H_K(t) \equiv \sup_{\substack{u\geq 0\\ \lambda\geq 1}} \left\{\frac{1}{K\lambda^q}\,\Phi(\lambda u,t) - \Phi(u,t)\right\}$ should be integrable for some $K > 0$. Here it is noted that one need only consider λ that are large powers of 2.

Lemma 7. The Young function Φ satisfies condition Δ^q if and only if Φ satisfies a Δ_2-condition and there exists $K > 0$ and a natural number m such that $\int \sup_{n\geq m} \sup_{u\geq 0} \left\{\frac{1}{K\cdot 2^{nq}}\,\Phi(2^n u,t) \;-\; \Phi(u,t)\right\} d\mu < \infty$.

Proof. In order to conserve space, we shall denote the above integrand by $h_{K,m}(t)$. Since the necessity is clear, assume Φ satisfies a Δ_2-condition and $\int h_{K,m}\,d\mu < \infty$. Let $K_1 > 2$ and $0 \leq h_1 \in L^1$ satisfy $\Phi(2u,t) \leq K_1\Phi(u,t) + h_1(t)$ for $u \geq 0$ and $t \in T$ and let $\ell > \max\{K_1^m, KK_1\}$. Then

$$
\begin{aligned}
H_\ell(t) = & \sup_{1\leq\lambda\leq 2^m} \sup_{u\geq 0} \left\{\frac{1}{\ell\lambda^q}\,\Phi(\lambda u,t) \;-\Phi(u,t)\right\} \\
& + \sup_{\lambda\geq\lambda_m} \sup_{u\geq 0} \left\{\frac{1}{\ell\lambda^q}\,\Phi(\lambda\mu,t) \;-\; \Phi(u,t)\right\}.
\end{aligned}
$$

Consider the first term:

$$\begin{aligned}\sup_{1\le\lambda\le 2^m}\ \sup_{u\ge 0}\left\{\frac{1}{\ell\lambda^q}\Phi(\lambda u,t)\ -\ \Phi(u,t)\right\} &\le \sup_{u\ge 0}\left\{\frac{1}{\ell}\Phi(2^m u,t)-\Phi(u,t)\right\}\\ &\le \sup_{u\ge 0}\left\{\frac{K_1^m}{\ell}\ \Phi(u,t)\ +\ \frac{K_1^m-1}{\ell(K_1-1)}\ h_1\ -\ \Phi(u,t)\right\}\\ &\le \frac{K_1^m-1}{\ell(K_1-1)}\ h_1\\ &\le h_1.\end{aligned}$$

In order to bound the second term, let $\lambda\ge 2^m$ and choose $n\in\mathbf{N}$ such that $2^n\le\lambda\le 2^{n+1}$. Then

$$\begin{aligned}\sup_{u\ge 0}\left\{\frac{1}{\ell\lambda^q}\ \Phi(\lambda u,t)-\Phi(u,t)\right\} &\le \sup_{u\ge 0}\left\{\frac{1}{\ell\cdot 2^{nq}}\ \Phi(2^{n+1}u,t)-\Phi(u,t)\right\}\\ &\le \sup_{u\ge 0}\left\{\frac{1}{\ell\cdot 2^{nq}}\ \left(K_1\Phi(2^n u,t)\ +\ h(t)\right)-\Phi(u,t)\right\}\\ &\le \sup_{u\ge 0}\left\{\frac{1}{K\cdot 2^{nq}}\ \Phi(2^n u,t)-\Phi(u,t)\right\}+\frac{1}{\ell\cdot 2^{nq}}h_1(t)\\ &\le h_{K,m}(t)+h_1(t).\end{aligned}$$

Combining the above bounds yields that $H_\ell(t)\le h_{K,m}(t)+2h_1(t)$ for $t\in T$. Therefore $H_\ell\in L^1$ and the proof is complete.

We are now prepared to prove the remainder of Theorem 5.

Proof of the necessity in Theorem 5a. Assume that L^Φ has finite cotype q but that Φ fails to satisfy condition Δ^q. The finite cotype of L^Φ implies that Φ satisfies a Δ_2-condition since otherwise L^Φ would contain an isometric copy of ℓ^∞ [4]. Thus, by Lemma 7 and the notation defined in its proof, for all $K>0$ and all natural numbers m, $\int h_{K,m}\,d\mu\ =\ \infty$. In particular, for all natural numbers m and j, $\int\sup_{n\ge m}\ \sup_{u\ge 0}\left\{\frac{1}{2^j 2^{nq}}\ \Phi(2^n u,t)\ -\ \Phi(u,t)\right\}d\mu\ =\ \infty$. Let

$\{u_i : i \in \mathbb{N}\}$ denote the set of nonnegative rational numbers with $u_1 = 0$; taking the supremum over the nonnegative reals or over $\{u_i\}$ in the preceding integral yields the same result. Now define $A_{nij} = \{t \in T : \frac{1}{2^{j+nq}} \Phi(2^n u_i, t) \geq \Phi(u_i, t)\}$ and $x_{n,j}(t) = \sup_i u_i \chi_{A_{nij}}(t)$. Then $\int \sup_{n \geq m} \frac{1}{2^{nq}} \Phi(2^n x_{nj}(t), t)\, d\mu = \infty$ for all m, j in $\mathbb{N}$. By the definition of A_{nij}, $A_{n,i,j+1} \subset A_{n,ij}$ and hence $x_{n,\, j+1} \leq x_{nj}$. Finally, defining $f_{nj}(t) = \frac{1}{2^{nq}} \Phi(2^n x_{nj}(t), t)$ yields a double sequence (f_{nj}) of measurable functions on T such that $f_{n,j+1} \leq f_{nj}$ and $\int \sup_{n \geq m} f_{nj}\, d\mu = \infty$ for all natural numbers m and j. If we knew that the f_{nj}'s assumed only real values (almost everywhere and for sufficiently large n and j), Lemma 6 would apply.

So, assume momentarily that there exist increasing sequences (n_k) and (j_k) of natural numbers and a set $C_k \in \Sigma$ with $\mu C_k > 0$ such that $f_{n_k, j_k}(t) = \infty$ for all $t \in C_k$. The (C_k) may be chosen to be pairwise disjoint. Then, for all $t \in C_k$, $\lim_{\ell \to \infty} \Phi(2^{n_k} \max_{1 \leq i \leq \ell} u_i \chi_{A_{n_k,\, i,\, j_k}}(t),\ t) = \infty$. Choose $\ell_k \in \mathbb{N}$ and a measurable subset D_k of C_k such that $\int_{D_k} \frac{1}{2^{n_k q}} \Phi(2^{n_k} \max_{1 \leq i \leq \ell} u_i \chi_{A_{n_k,i,j_k}}(t), t)\, d\mu = 1$. Define $g_k = \max_{1 \leq i \leq \ell} u_i \chi_{A_{n_k,i,j_k}} \chi_{D_k}$; then, by the definition of A_{n_k,i,j_k}, $1 = \int \frac{1}{2^{n_k q}} \Phi(2^{n_k} g_k(t), t)\, d\mu \leq 2^{j_k} \int \Phi(g_k(t), t)\, d\mu$. Thus $\int \Phi(g_k(t), t)\, d\mu = 2^{-j_k}$ and $\int_{D_k} \Phi(2^{n_k} g_k(t), t)\, d\mu = 2^{n_k q}$. Let $T_1, \cdots, T_{[2^{n_k q}]}$ be pairwise disjoint subsets of D_k such that $\int_{T_i} \Phi(2^{n_k} g_k(t), t)\, d\mu = 1$ for $i = 1, \cdots, [2^{n_k q}]$. Then $\|g_k \chi_{T_i}\|_\Phi = 2^{-n_k}$ for $i = 1, \cdots, [2^{n_k q}]$ and

$$\sum_{k=1}^{\infty} \sum_{i=1}^{[2^{n_k q}]} \|g_k \chi_{T_i}\|_\Phi^q = \sum_{k=1}^{\infty} \frac{[2^{n_k q}]}{2^{n_k q}}$$

$$\geq \sum_{k=1}^{\infty} \left(1 - \frac{1}{2^{n_k q}}\right)$$

$$= \infty.$$

Moreover

$$\begin{aligned}\int \Phi\left(\sum_{k=1}^{\infty}\sum_{i=1}^{[2^{n_k q}]} g_k\chi_{T_i}(t),t\right)d\mu &= \sum_{k=1}^{\infty}\int \Phi\left(\sum_{i=1}^{[2^{n_k q}]} g_k\chi_{T_i}(t),t\right)d\mu \\ &\le \sum_{k=1}^{\infty}\int \Phi(g_k(t),t)\,d\mu \\ &\le \sum_{k=1}^{\infty} 2^{-n_k} \\ &< \infty.\end{aligned}$$

Since Φ is an even function, when the functions $g_k\chi_{T_i}$ are ordered in the order that they are summed above, the computations yield a sequence (z_n) in L^Φ such that $\sum_n z_n r_n$ converges almost everywhere but $\sum \|z_n\|^q = \infty$. This contradicts that L^Φ has cotype q.

We may therefore assume that f_{nj} is real-valued for each n, j in $\mathbf{N}$. Lemma 6 then implies that there exists a sequence (A_k) of pairwise disjoint sets in Σ and a sequence $(\ell_n)_{n=0}^{\infty}$ with $\ell_0 = 0$ such that $\sum_{k=\ell_{i-1}+1}^{\ell_i} \int_{A_k} \frac{1}{2^{n_k q}}\Phi(2^{n_k}\, x_{ki}(t),t)\,d\mu = 1$. For each fixed i, consider the set $S_i = \{k \in [\ell_{i-1}+1,\ \ell_i] \ :\ \int_{A_k} \Phi(2^{n_k}x_{ki}(t),t)\,d\mu < 1\}$. Then $\sum_{k\in S_i}\int_{A_k}\frac{1}{2^{n_k q}}\Phi(2^{n_k}x_{ki}(t),t)\,d\mu < \frac{1}{2}$. Therefore, by eliminating the k in S_i, there exists a (new) sequence (A_k) and a (new) sequence $(\ell_n)_{n=0}^{\infty}$ with $\ell_0 = 0$ such that, for each $i \in \mathbf{N}$,

$$\frac{1}{2} \le \sum_{k=\ell_{i-1}+1}^{\ell_i} \int_{A_k} \frac{1}{2^{n_k q}}\,\Phi(2^{n_k}x_{ki}(t),t)\,d\mu \le 1$$

and, for each $k = \ell_{i-1}+1,\ \cdots,\ \ell_i$,

$$\int_{A_k} \Phi(2^{n_k}x_{ki}(t),t)\,d\mu \ge 1.$$

Let $m_k = [\![\int_{A_k} \Phi(2^{n_k}x_{ki}(t),t)\,d\mu]\!]$ and choose pairwise disjoint measurable subsets $T_1^k, \cdots, T_{m_k}^k$ of A_k such that $\int_{T_j^k} \Phi(2^{n_k}x_{ki}(t),t)\,d\mu = 1$ for $j = 1, \cdots, m_k$.

Hence, for $j = 1, \cdots, m_k$, $\|x_{ki}\chi_{T_j^k}\|_\Phi = 2^{-n_k}$. Then

$$\sum_{i=1}^{\infty} \sum_{k=\ell_{i-1}+1}^{\ell_i} \sum_{j=1}^{m_k} \|x_{ki}\, \chi_{T_j^k}\|_\Phi = \sum_{i=1}^{\infty} \sum_{k=\ell_{i-1}+1}^{\ell_i} 2^{-n_k q}\, m_k$$

$$\geq \sum_{i=1}^{\infty} \sum_{k=\ell_{i-1}+1}^{\ell_i} 2^{-n_k q} \int_{A_k} \Phi(2^{n_k} x_{ki}(t), t)\, d\mu$$

$$\geq \sum_{i=1}^{\infty} \frac{1}{2}$$

$$= \infty.$$

On the other hand,

$$\int \Phi(\sum_{i=1}^{\infty} \sum_{k=\ell_{i-1}+1}^{\ell_i} \sum_{j=1}^{m_k} x_{ki}\, \chi_{T_j^k}(t), t)\, d\mu$$

$$= \sum_{i=1}^{\infty} \sum_{k=\ell_{i-1}+1}^{\ell_i} \int \Phi(\sum_{j=1}^{m_k} x_{ki}\, \chi_{T_j^k}(t), t)\, d\mu$$

$$\leq \sum_{i=1}^{\infty} \sum_{k=\ell_{i-1}+1}^{\ell_i} \int_{A_k} \Phi(x_{ki}(t), t)\, d\mu$$

$$\leq \sum_{i=1}^{\infty} \sum_{k=\ell_{i-1}+1}^{\ell_i} 2^{-(i+n_k q)} \int_{A_k} \Phi(2^{n_k} x_{ki}(t), t)\, d\mu$$

$$= \sum_{i=1}^{\infty} 2^{-i} \left(\sum_{k=\ell_{i-1}+1}^{\ell_i} 2^{-n_k q} \int_{A_k} \Phi(2^{n_k} x_{ki}(t), t)\, d\mu \right)$$

$$\leq \sum_{i=1}^{\infty} 2^{-i}$$

$$< \infty.$$

As in the preceding case, this contradicts that L^Φ has finite cotype q and the proof of Theorem 5a is complete.

Part (b) of Theorem 5 follows from part (a) via duality and the following facts: the result of Pisier mentioned in the introduction; a result of Pisier [17] which states that a Banach space has type $p > 1$ if and only if it does not contain ℓ_n^1's uniformly, i.e., if and only if the Banach space is B-convex; the result that, when Φ satisfies a Δ_2-condition, $(L^\Phi)^* = L^{\Phi^*}$ (see [7] or [15]); and the fact that, in a Musielak-Orlicz space, reflexivity and B-convexity are equivalent [5,6] and can be characterized by both the Young function Φ and its conjugate function satisfying a Δ_2-condition.

As an immediate corollary of Theorem 5, Proposition 4, and the result of Maurey and Pisier [13] that a Banach space has cotype $q < \infty$ if and only if it does not contain ℓ_n^∞'s uniformly, we obtain:

Corollary 8. A Musielak-Orlicz space L^Φ does not contain ℓ_n^∞'s uniformly if and only if Φ satisfies a Δ_2-condition.

An analagous characterization of when L^Φ contains ℓ_n^1's uniformly is already known (see [5],[6]).

It should be noted that the proof of Theorem 5, although cumbersome in the setting of Musielak-Orlicz spaces, simplifies significantly if one only considers Orlicz spaces. In fact, in the Orlicz space setting, we have the following:

Corollary 9. Let L^Φ be an Orlicz space over a nonatomic measure space and $1 < p \leq 2 \leq q < \infty$. If $\mu T < \infty$, then:

a. L^Φ has cotype q if and only if there exists $K > 0$ and $u_0 \geq 0$ such that $\Phi(\lambda u) \leq K\lambda^q \Phi(u)$ for all $\lambda \geq 1$, and $u \geq u_0$.

b. L^Φ has type p if and only if Φ satisfies a Δ_2-condition and there exists $K > 0$ and $u_0 \geq 0$ such that $\Phi(\lambda u) \geq K\lambda^p \Phi(u)$ for all $\lambda \geq 1$, and $u \geq u_0$.

If $\mu T = \infty$, the the above inequalities have to hold for all $u \geq 0$.

The above characterization of cotype of an Orlicz space has been done previously in [9] and [10] and the characterization of type was done in [9] under the assumption of a Δ_2-condition. In fact, in [10], Maleev and Troyanski define a modulus on *arbitrary* Banach lattices and employ this modulus to characterize the cotype of Banach lattices. In particular, they compute the cotype of L^Φ when T is $(0, \infty)$, $[0, 1]$, or $[1, \infty)$.

Additionally, in the Orlicz space case, these results can be stated in terms of indices. If $\mu T < \infty$, define lower and upper indices

$$\alpha_\Phi = \sup \{p : \inf_{\lambda,u\geq 1} \frac{\Phi(\lambda u)}{\lambda^p \Phi(u)} > 0\}$$

$$\beta_\Phi = \inf \{q : \sup_{\lambda,u\geq 1} \frac{\Phi(\lambda u)}{\lambda^q \Phi(u)} < \infty\};$$

If $\mu T = \infty$, the supremum and infimum are taken over all positive u and all $\lambda \geq 1$. For information on indices and their applications, see [9] or [11]. The following are essentially known and rephrases our results in the Orlicz space case in terms of indices.

Corollary 10. Let (T, Σ, μ) be a nonatomic measure space and let Φ be a Young function (without parameter).

a. The Orlicz space L^Φ has finite cotype if and only if $\beta_\Phi < \infty$.

b. If $q > \max\{\beta_\Phi, 2\}$, the Orlicz space L^Φ has cotype q.

c. If the Orlicz space L^Φ is of cotype q, then $q \geq \max\{\beta_\Phi, 2\}$.

d. The Orlicz space L^Φ has type larger than 1 if and only if $1 < \alpha_\Phi \leq \beta_\Phi < \infty$.

e. If $\beta_\Phi < \infty$ and $1 \leq p < \min\{\alpha_\Phi, 2\}$, then the Orlicz space L^Φ is of type p.

f. If the Orlicz space L^Φ is of type $p > 1$, then $\beta_\Phi < \infty$ and $p \leq \min\{\alpha_\Phi, 2\}$.

We note that the main result can be aplied to the Nakano space; i.e., the Musielak-Orlicz space with $\Phi(u,t) = u^{p(t)}$ and the function $p : T \to [1,\infty)$ measurable. In fact, the Nakano space has cotype q exactly when $p(t) \leq q$ for almost all $t \in T$ and it has type p exactly when $p(t) \geq p$ for almost all $t \in T$.

We conclude with some examples:

1. For any $p \in [1,2]$ and $q \in [2,\infty)$, there exists an Orlicz spaces of type p and cotype q. Indeed, it suffices to take $L^\Phi = L^p \cap L^q$ where $\Phi(u) = \max(u^p, u^q)$.

2. An Orlicz space need not have its upper index as a cotype. Let $q \geq 2$ and define

$$\Phi(u) = \begin{cases} 0 & \text{if } u = 0 \\ \frac{u^q}{|ln\ u|} & \text{if } 0 < u \leq \frac{1}{e} \\ \left(\frac{1}{q}+1\right) u^q - \frac{1}{q}e^{-q} & \text{if } u > \frac{1}{e}. \end{cases}$$

Then $\beta_\Phi = q$, but $\sup_{\substack{\lambda \geq 1 \\ u \geq 0}} \frac{\Phi(\lambda u)}{\lambda^q \Phi(u)} = \infty$. Thus L^Φ has cotype $q + \epsilon$ for every ϵ, but does not have cotype q.

In closing, we want to thank Professor Troyanski for informing us of his and Professor Maleev's results.

REFERENCES

1. J. Creekmore, Type and cotype in Lorentz L_{pq} spaces, *Proc. Acad. Amsterdam*, A 84 (1981), 145-152.

2. Z. G. Gorgadze and V.I. Tarieladze, On geometry of Orlicz spaces, *Probability Theory on Vector Spaces*, Lecture Notes in Mathematics, Volume 828, Springer-Verlag, 47-51.

3. J. Hoffmann-Jørgensen, *Probability in B-spaces*, Lecture Note Series, No. 48, Aarhus Univ., 1977.

4. H. Hudzik, On some equivalent conditions in Musielak-Orlicz spaces, *Commentationes Math.* 24 (1984), 57-64.

5. H. Hudzik and A. Kamińska, On uniformly convexifiable and B-convex Musielak-Orlicz spaces. *Commentationes Math.* 25 (1985), 59-75.

6. H. Hudzik, A. Kamińska, and W. Kurc, Uniformly non-$\ell^1(n)$ Musielak-Orlicz spaces, *Bull. Acad. Polon. Sci.* 35 (1987), 441-448.

7. A. Kozek, Orlicz spaces of functions with values in Banach spaces, *Commentationes Math.* 19 (1976), 259-288.

8. M. A. Krasnasel'skii and Ya. B. Rutickii, *Convex Functions and Orlicz Spaces*, P. Noordhoff, Ltd., Groningen, 1961.

9. J. Lindenstrauss and L. Tzafriri, *Classical Banach Spaces II*, Springer-Verlag, 1979.

10. R.P. Maleev and S. L. Troyanski, On cotypes of Banach lattices, *Constructive Function Theory '81*, Sofia, 1983, 429-441.

11. L. Maligranda, Indices and Interpolation, *Dissertationes Mathematicae (Rozprawy Mathematyczne)*, No. 234, Warsaw, 1985.

12. W. Matuszewska and W. Orlicz, On certain properties of φ-functions, *Bull. Acad. Polon. Sci.* 8 (1960), 439-443.

13. B. Maurey and G. Pisier, Caractérisation d'une classe d'espace de Banach par des propriétés de séries aléatoires vectorielles, *C. R. Acad. Sci. Paris* 277, Series A(1973), 687-690.

14. B. Maurey and G. Pisier, Séries de variables aléatoires vectorielles indépendantes et propriétés géométriques des espaces de Banach, *Studia Math.* 58 (1976), 45-90.

15. J. Musielak, *Orlicz Spaces and Modular Spaces*, Lecture Notes in Mathematics, Volume 1038, Springer-Verlag, 1983.

16. G. Pisier, Holomorphic semi-groups and the geometry of Banach spaces, *Annals of Math.* 115 (1982), 375-392.

17. G. Pisier, Sur les espaces de Banach qui ne contiennent pas uniformément de ℓ_n^1, *C. R. Acad. Sci. Paris* 277, Series A(1973), 991-994.

ON THE COMPLEX GROTHENDIECK CONSTANT IN THE n-DIMENSIONAL CASE

Hermann König (Kiel)

Abstract. We derive some new bounds for the complex Grothendieck constants $K_G^{\mathbb{C}}(n)$ for n-dimensional Hilbert spaces. One has e.g. $1.152 \le K_G^{\mathbb{C}}(2) \le 1.216$. For $n \longrightarrow \infty$ the estimates yield Haagerup's bound $K_G^{\mathbb{C}} \le 1.405$.

1. Introduction

Grothendieck's inequality is the following well-known [LP].

Theorem 1. *For any $n \ge 2$ there is $K_G(n) > 0$ with the following property: Given any finite matrix $(a_{jk})_{j,k=1}^m$ with*

$$\left| \sum_{j,k=1}^m a_{jk}\, t_j\, s_k \right| \le \sup_{1\le j,k\le m} |t_j|\,|s_k| ,$$

and given any vectors $(x_j)_{j=1}^m$, $(y_k)_{k=1}^m$ in an n-dimensional Hilbert space H_n, one has

$$\left| \sum_{j,k=1}^m a_{jk} < x_j, y_k > \right| \le K_G(n) \sup_{1\le j,k\le m} \|x_j\|\,\|y_k\| .$$

We use superscripts $\mathbb{R}$ and $\mathbb{C}$ to indicate the different values in the real and complex cases; the complex constants are <u>smaller</u> than the real ones. Clearly, the sequence $(K_G(n))_{n\ge 2}$ is increasing in either case. The best known upper estimates for the Grothendieck constant $K_G := \lim_{n\to\infty} K_G(n)$ are due to Krivine and Haagerup,

$$K_G^{\mathbb{R}} \le \frac{\pi}{2\ln(1+\sqrt{2})} \cong 1.782 \qquad \text{[Kr]}$$

$$K_G^{\mathbb{C}} \le \frac{8}{\pi(k_0+1)} \cong 1.405 \qquad \text{[Ha]} .$$

Here k_0 is the solution of a certain equation involving elliptic integrals, see below. The upper estimate for $K_G^{\mathbb{R}}$ is conjectured to be the exact value by Krivine, whereas the bound for $K_G^{\mathbb{C}}$ is not considered to be precise by Haagerup.

We remark that Pisier [Pi] first proved that $K_G^{\mathbb{C}} < K_G^{\mathbb{R}}$ by showing $K_G^{\mathbb{C}} < \pi/2$; the inequality $\pi/2 \le K_G^{\mathbb{R}}$ is fairly easy. The real constants $K_G^{\mathbb{R}}(n)$ have been estimated by Krivine [Kr] too, using spherical function techniques. In particular, $K_G^{\mathbb{R}}(2) = \sqrt{2}$ and $K_G^{\mathbb{R}}(4) \le \pi/2$. Modifying and combining techniques of [Kr] and [Ha], we give bounds for the constants $K_G^{\mathbb{C}}(n)$. The constants $K_G(n)$ occur in inequalities related to absolutely summing operators; they are the smallest constants such that the following statements hold, cf. [LT], [LP], [TJ, ch. 4]:

a) Any continuous linear operator $T : l_1 \longrightarrow l_2^n$ is 1-summing with

$$\pi_1(T) \le K_G(n) \| T \| .$$

b) Any continuous linear operator $T : l_\infty \longrightarrow l_1$ is 2-summing. If the 2-summing norm is calculated only on n vectors, one has

$$\pi_2^{(n)}(T) \le K_G(n) \| T \| .$$

c) For any operator between Banach lattices, $T : X \longrightarrow Y$, and any sequence $x_1, \cdots x_n \in X$, one has

$$\| (\sum_{i=1}^{n} | T x_i |^2)^{1/2} \| \le K_G(n) \| T \| \| (\sum_{i=1}^{n} | x_i |^2)^{1/2} \|$$

To state the estimate for $K_G^{\mathbb{C}}(n)$, we define $\varphi : [-1,1] \longrightarrow [-1,1]$ by

$$\varphi(s) = s \int_0^{\pi/2} \frac{\cos^2 \alpha}{\sqrt{1-s^2 \sin^2 \alpha}} \, d\alpha \quad , \quad s \in [-1,1] .$$

By Haagerup [Ha], $\varphi^{-1} : [-1,1] \longrightarrow [-1,1]$ exists and admits a convergent power series expansion in $[-1,1]$

$$\varphi^{-1}(t) = \sum_{k=0}^{\infty} \beta_{2k+1} t^{2k+1} \quad , \quad t \in [-1,1]$$

with $\beta_1 = 4/\pi$ and $\beta_{2k+1} \le 0$ for $k \ge 1$. Formal calculations yield

$$\beta_3 = -\frac{8}{\pi^3} , \ \beta_5 = 0 , \ \beta_7 = -\frac{16}{\pi^7} , \ \beta_9 = -\frac{80}{\pi^9} , \ \beta_{11} = -\frac{480}{\pi^{11}} , \ \beta_{13} = -\frac{3136}{\pi^{13}} .$$

Theorem 2. *Let* $A_{2k+1}(n) = \frac{n!\,(k+1)!}{(k+n)!}$ *for* $k \in \mathbb{N}_0$, $n \geq 2$. *There is a unique solution* s_n *of the equation*

$$2 \sum_{k=0}^{\infty} A_{2k+1}(n)\ \beta_{2k+1}\ \varphi(s_n)^{2k+1} = 1 + s_n \ ,\ 0 < s_n < 1 \ .$$

For this value s_n, $K_G^{\mathbb{C}}(n) \leq \frac{1}{\varphi(s_n)}$. *The sequence* $\left(\frac{1}{\varphi(s_n)}\right)_{n\geq 2}$ *is increasing.*

As $n \longrightarrow \infty$, $A_{2k+1}(n) \longrightarrow \delta_{k,0}$. Thus the limiting equation is

$$2\beta_1\ \varphi(s_0) = 1 + s_0 \ , \quad \text{i.e.} \quad \varphi(s_0) = \frac{\pi}{8}\ (1 + s_0) \ .$$

This is Haagerup's estimate $K_G^{\mathbb{C}} \leq \frac{1}{\varphi(s_0)} = \frac{8}{\pi(1+s_0)} \cong 1.4049$. He conjectures that this might be slightly improved to

$$K_G^{\mathbb{C}} \leq |\,\varphi(i)\,|^{-1} = \left(\int_0^{\pi/2} \frac{\cos^2\alpha}{\sqrt{1+\sin^2\alpha}}\, d\alpha\right)^{-1} \cong 1.4046$$

(extending φ to the upper half-plane). This is motivated by the fact that $K_G^{\mathbb{R}} \leq \frac{\pi}{2\ln(1+\sqrt{2})} = |\varphi(i)|^{-1}$ for the corresponding function $\varphi(s) = \frac{2}{\pi} \sin^{-1} s$ in the real case. A corresponding conjecture for a slight improvement of the bound for $K_G^{\mathbb{C}}(n)$ of the theorem is the

Conjecture. *Let* $B_{2k+1}(n) = (-1)^k \frac{(n-3)\cdots(n-(2k+1))}{(n+1)\cdots(n+(2k-1))}$ *for* $k \in \mathbb{N}_0$, $n \geq 2$. *Then there is a unique solution* $0 < c_n < 1$ *of*

$$\sum_{k=0}^{\infty} B_{2k+1}(n)\ \beta_{2k+1}\ c_n^{2k+1} = 1$$

and $K_G^{\mathbb{C}}(n) \leq \frac{1}{c_n}$. *The sequence* $\left(\frac{1}{c_n}\right)_{n\geq 2}$ *is increasing.*

This would be a formal analogue of the results of Krivine [Kr] for $K_G^{\mathbb{R}}(n)$ and would give $K_G^{\mathbb{C}}(3) \leq 4/\pi$, in particular. We make a few comments in section 3 on the origin of the formula in the conjecture.

The equations in the theorem and the conjecture can be solved numerically by using tables for the complete elliptic integrals E and K and noting that

$$\varphi(s) = \frac{1}{s}\ (E(s) - (1 - s^2)\ K(s)) \ .$$

Lower numerical bounds for $K_G^{\mathbb{C}}(n)$ can be given by using the (unpublished) method of Davie who showed that

$$K_G^{\mathbb{R}} \geq 1.677 \quad \text{and} \quad K_G^{\mathbb{C}} \geq 1.338.$$

Some remarks on this are found in section 4. The following table gives numerical values for the bounds on $K_G^{\mathbb{C}}(n)$, rounded to 4 decimal digits, for some values of n. The upper and lower bounds differ by about 5 %.

n	$K_G^{\mathbb{C}}(n) \geq$	$K_G^{\mathbb{C}}(n) \leq$	Conj. $K_G^{\mathbb{C}}(n) \leq$
2	1.1526	1.2157	1.2146
3	1.2108	1.2744	1.2732 $(4/\pi)$
4	1.2413	1.3048	1.3036
5	1.2600	1.3236	1.3224
10	1.2984	1.3628	1.3618
20	1.3181	1.3834	1.3827
50	1.3300	1.3962	1.3957
∞	1.3381	1.4049	1.4046

2. The addition formula for complex spherical harmonics

The proof of the theorem relies in an essential way on the addition formula for sperical functions. To define complex spherical harmonics in n variables, fix $n \in \mathbb{N}$ and choose $k \in \mathbb{N}_0$. Let $m = [k/2]$ and $l = k - m$, i.e. $l = m$ if k is even and $l = m + 1$ if k is odd. Let H_k be a complex polynomial in the 2n variables $z_1, \cdots z_n$, $\bar{z}_1, \cdots \bar{z}_n$ which is harmonic and homogeneous of degrees (l,m), i.e. which satisfies

$$\text{(a)} \quad \sum_{i=1}^{n} \frac{\partial^2}{\partial z_i \, \partial \bar{z}_i} H_k(z, \bar{z}) = 0$$

$$\text{(b)} \quad H_k(\lambda z, \mu \bar{z}) = \lambda^l \mu^m H_k(z, \bar{z}) \quad , \quad \lambda, \mu \in \mathbb{C}$$

where $z = (z_1, \cdots z_n)$, $\bar{z} = (\bar{z}_1, \cdots \bar{z}_n)$. On the complex n-dimensional sphere $S^{n-1}(\mathbb{C})$, define $Y_k : S^{n-1}(\mathbb{C}) \longrightarrow \mathbb{C}$ by

$$Y_k(\zeta) := H_k(\zeta, \bar{\zeta}) \quad , \quad \zeta \in S^{n-1}(\mathbb{C}) \; .$$

This "restriction" of H_k to $S^{n-1}(\mathbb{C})$ is called a spherical function of order k in n dimensions. The set of all such spherical functions is denoted by $\mathscr{H}_k = \mathscr{H}_{n,k}$; we fix n and often omit this index. Let $N_k := \dim_{\mathbb{C}} \mathscr{H}_k$. (One has [DGS]

$$N_k = \binom{n+k-1}{n-1}\binom{n+l-1}{n-1} - \binom{n+k-2}{n-1}\binom{n+l-2}{n-1} \quad).$$

For $\alpha, \beta > -1$, denote by $P_k^{(\alpha,\beta)}$ the Jacobi polynomial of degree k on $\mathbb{R}$, normalized by $P_k^{(\alpha,\beta)}(1) = 1$. Thus the $(P_k^{(\alpha,\beta)})_{k \in \mathbb{N}_0}$ are (the) orthogonal polynomials on $[-1,1]$ with respect to the weight function $(1-t)^\alpha \ (1+t)^\beta$. For $t \in \mathbb{R}$, let

$$q_k(t) := \begin{cases} P_{[k/2]}^{(n-2,0)}(2t^2-1) & k \text{ even} \\ t\, P_{[k/2]}^{(n-2,1)}(2t^2-1) & k \text{ odd}. \end{cases}$$

The $(q_k)_{k \in \mathbb{N}_0}$ are orthogonal on $[-1,1]$ with respect to $|t| \ (1-t^2)^{n-2}$. Further, for $z \in \mathbb{C}$, put

$$Q_k(z) := \begin{cases} q_k(|z|) & k \text{ even} \\ \frac{z}{|z|}\, q_k(|z|) & k \text{ odd}. \end{cases}$$

For the purpose of surface integration, identify $S^{n-1}(\mathbb{C}) = S^{2n-1}(\mathbb{R})$ and denote by $d\omega = d\omega_n$ the (rotation invariant) surface measure on $S^{n-1}(\mathbb{C})$, normalized by $\int_{S^{n-1}(\mathbb{C})} d\omega_n = 1$. Consider $\mathscr{H}_k = \mathscr{H}_{n,k}$ as a subspace of $L_2 = L_2(S^{n-1}(\mathbb{C}), d\omega_n)$ and denote by $(Y_{kj})_{j=1}^{N_k}$ an arbitrary but fixed orthonormal basis of $\mathscr{H}_k$ in L_2. The following result is stated by Delsarte-Goethals-Seidel [DGS] who refer to Koornwinder [Ko].

Addition theorem for spherical functions. *For all* $k \in \mathbb{N}_0$ *and* $x, y \in S^{n-1}(\mathbb{C})$

$$N_k\, Q_k(<x,y>) = \sum_{j=1}^{N_k} Y_{kj}(x)\, \overline{Y_{kj}(y)}.$$

In the real case, a corresponding formula holds, if the polynomial q_k is taken to be $P_k^{(\frac{n-3}{2}, \frac{n-3}{2})}$, cf. Müller [Mü].

Corollary. *For all* $k \in \mathbb{N}_0$ *and* $x, y \in S^{n-1}(\mathbb{C})$

$$N_k \int_{S^{n-1}} Q_k(<x,y>)\, \overline{Q_k(<z,y>)}\, d\omega_n(y) = Q_k(<x,z>).$$

Proof of the corollary. The left side equals by the addition theorem

$$\frac{1}{N_k} \sum_{i,j=1}^{N_k} Y_{ki}(x) \overline{Y_{kj}(z)} \int_{S^{n-1}} \overline{Y_{ki}(y)} \; Y_{kj}(y) \; d\omega_n(y)$$

$$= \frac{1}{N_k} \sum_{j=1}^{N_k} Y_{kj}(x) \overline{Y_{kj}(z)} = Q_k(<x,z>) \; . \qquad \blacksquare$$

Thus $N_k Q_k(<\cdot,\cdot>) = \sum_{j=1}^{N_k} Y_{kj} \otimes \overline{Y_{kj}}$ is the kernel of the orthogonal projection in $L_2(S^{n-1}, d\omega)$ onto $\mathcal{H}_{n,k}$ (as an integral operator).

The addition theorem may be useful for other purposes in the geometry of Banach spaces. No simple, explicit proof is given in [DGS], [Ko] (in particular, not for the case of "k odd" which we need). We thus supply a sketch of a fairly simple proof obtained by modifying the arguments of Müller [Mü] in the real case.

Sketch of proof.

(i) The function $F_k : S^{n-1}(\mathbb{C}) \times S^{n-1}(\mathbb{C}) \longrightarrow \mathbb{C}$ given by

$$F_k(x,y) = \sum_{j=1}^{N_k} Y_{kj}(x) \overline{Y_{kj}(y)}$$

is unitarily invariant: For any unitary $(n \times n)$-matrix A, $F_k(Ax,Ay) = F_k(x,y)$. To see this, note that $x \longmapsto Y_{kj}(Ax)$ is a spherical function in $\mathcal{H}_k$, too (for all j). Basis expansion gives $Y_{kj}(Ax) = \sum_{r=1}^{N_k} c_{jr}^k \, Y_{kr}(x)$. The matrix $C^k = (c_{jr}^k)_{j,r=1}^{N_k}$ is unitary for all $k \in \mathbb{N}_0$ since by the orthogonality of the Y_{kj} and invariance under A

$$\delta_{ij} = \int_{S^{n-1}} Y_{ki}(x) \overline{Y_{kj}(x)} \; d\omega(x)$$

$$= \int_{S^{n-1}} Y_{ki}(Ax) \overline{Y_{kj}(Ax)} \; d\omega(x) = \sum_{r=1}^{N_k} c_{ir}^k \overline{c_{jr}^k} \; .$$

Using this, direct calculation yields $F_k(Ax,Ay) = F_k(x,y)$.

(ii) Let $x,y \in S^{n-1}(\mathbb{C})$. Choose $A \in U(n)$ unitary with $Ay = e_n = (0,\cdots 0,1)$. Let $\tilde{\zeta} := Ax = \eta \, e_n + \sqrt{1-|\eta|^2} \; \eta^{n-1}$ with $\eta \in \mathbb{C}$, $\eta^{n-1} \in S^{n-2}(\mathbb{C})$. Then

$$< x,y > \; = \; < Ax, Ay > \; = \; <\tilde{\zeta}, e_n> \; = \eta \ , \ F_k(x,y) = F_k(\tilde{\zeta}, e_n) \ .$$

Since $G := \{B \in U(u) \mid Be_n = e_n\} \cong U(n-1)$, for any two $\eta^{n-1}, \tilde{\eta}^{n-1} \in S^{n-2}(\mathbb{C})$ there is $B \in G$ with $B\eta^{n-1} = \tilde{\eta}^{n-1}$. Thus $F(\tilde{\zeta}, e_n)$ is independent of the special choice of $\eta^{n-1} \in S^{n-2}(\mathbb{C})$. Hence with $\zeta = \eta e_n + \sqrt{1-|\eta|^2}\, e_{n-1}$

$$F_k(x,y) = F_k(\zeta, e_n) = \Phi_k(\eta) = \Phi_k(< x,y >)$$

for some suitable function $\Phi_k : \{z \in \mathbb{C} \mid |z| \le 1\} \longrightarrow \mathbb{C}$.

(iii) We claim that $\quad \Phi_k(\eta) = \begin{cases} \Phi_k(|\eta|) & k \text{ even} \\ \operatorname{sgn}\eta\, \Phi_k(|\eta|) & k \text{ odd} \end{cases}$

with $\operatorname{sgn}\eta = \eta / |\eta|$: By definition $F_k(\cdot, e_n) \in \mathscr{H}_k$. Thus with $m = [k/2], l = k-m$ there is a (l,m)-homogeneous polynomial H_k with $F_k(\zeta, e_n) = H_k(\zeta, \overline{\zeta}), \zeta$ as in (ii). Thus for all $\Theta \in [0, 2\pi)$

$$\begin{aligned} \Phi_k(e^{i\Theta}\eta) &= F_k(e^{i\Theta}\zeta, e_n) = H_k(e^{i\Theta}\zeta, e^{-i\Theta}\overline{\zeta}) \\ &= (e^{i\Theta})^l \, (e^{-i\Theta})^m \, H_k(\zeta, \overline{\zeta}) = (e^{i\Theta})^{\varepsilon} \, \Phi_k(\eta) \end{aligned}$$

with $\varepsilon = 0$ if k is even and $\varepsilon = 1$ if k is odd.

(iv) We claim that Φ_k is a polynomial of degree k of the form

$$\Phi_k(t) = \begin{cases} \varphi_m(t^2) & k \text{ even} \\ t\,\varphi_m(t^2) & k \text{ odd} \end{cases}$$

with some polynomial φ_m of degree $m = [k/2]$: By the (l,m)-homogeneity of the function H_k of (iii)

$$H_k(u_1, \cdots u_n, \overline{u}_1, \cdots, \overline{u}_n) = \sum_{i=0}^{l} \sum_{j=0}^{m} A_{l-i,m-j}(u_1, \cdots u_{n-1}, \overline{u}_1, \cdots \overline{u}_{n-1})\, u_n^i \, \overline{u}_n^j$$

where $A_{l-i,m-j}$ is $(l-i, m-j)$-homogenous in the first pair of $(n-1)$ variables. Let $\tilde{\zeta} = \eta e_n + \sqrt{1-|\eta|^2}\; \eta^{n-1}$ as above. Since $F_k(\tilde{\zeta}, e_n) = H_k(\tilde{\zeta}, \overline{\tilde{\zeta}})$ is independent of $\eta^{n-1} \in S^{n-2}(\mathbb{C})$, the coefficient functions $A_{l-i,m-j}$ only do depend on

$$|u_1|^2 + \cdots + |u_{n-1}|^2 = u_1 \overline{u_1} + \cdots + u_{n-1} \overline{u_{n-1}} \ .$$

Being $(l-i, m-j)$-homogeneous polynomials, they are of the form

$$A_{l-i,m-j}(u_1,\cdots u_{n-1},\overline{u}_1,\cdots\overline{u}_{n-1}) = c_{ij}^k\ (|u_1|^2+\cdots+|u_{n-1}|^2)^{\frac{k-(i+j)}{2}}$$

with $\frac{k-(i+j)}{2}\in\mathbb{N}_0$. Consider $\zeta = te_n + \sqrt{1-t^2}\ e_{n-1}$ for $t\in[-1,1]$. Then

$$\Phi_k(t) = F_k(\zeta,e_n) = H_k(\zeta,\overline{\zeta})$$

$$= \sum_{i=0}^{l}\sum_{j=0}^{m} c_{ij}^k\ (1-t^2)^{\frac{k-(i+j)}{2}}\ t^{i+j}\ ,$$

where $c_{ij}^k \neq 0$ only if $\frac{k-(i+j)}{2}\in\mathbb{N}_0$. If k is even (odd), thus (i+j) has to be even (odd). Hence Φ_k has the form as claimed.

(v) If $f:[0,1]\longrightarrow\mathbb{R}$ is an integrable function and $e\in S^{n-1}(\mathbb{C})$, introduction of (2n-1) polar coordinates in $S^{n-1}(\mathbb{C}) = S^{2n-1}(\mathbb{R})$ shows that (independently of e)

$$\int_{S^{n-1}(\mathbb{C})} f(|<x,e>|)\,d\omega_n(x) = 2(n-1)\int_0^1 f(t)\ t(1-t^2)^{n-2}\,dt\ .$$

(vi) We found for $x,y\in S^{n-1}(\mathbb{C})$ and $m = [k/2]$ that

$$F_k(x,y) = \Phi_k(<x,y>) = \begin{cases} \varphi_m(|<x,y>|^2) & k \text{ even} \\ <x,y>\varphi_m(|<x,y>|^2) & k \text{ odd}\ . \end{cases}$$

As in the real case [Mü], Green's formula and the harmonic property of the H_k's yield that spherical functions of different orders k and j are $L_2(S^{n-1},d\omega)$-orthogonal. Thus for odd k and j, with $m = [k/2]$ and $l = [j/2]$ and $e\in S^{n-1}(\mathbb{C})$

$$0 = \int_{S^{n-1}(\mathbb{C})} \Phi_k(<x,e>)\ \overline{\Phi_j(<x,e>)}\ d\omega(x)$$

$$= \int_{S^{n-1}(\mathbb{C})} |<x,e>|^2\ \varphi_m(|<x,e>|^2)\ \varphi_l(|<x,e>|^2)\ d\omega(x)$$

$$= 2(n-1)\int_0^1 t^2\ \varphi_m(t^2)\ \varphi_l(t^2)\ t(1-t^2)^{n-2}\,dt$$

$$= \frac{n-1}{2^n}\int_{-1}^1 \varphi_m(\tfrac{1+u}{2})\ \varphi_l(\tfrac{1+u}{2})\ (1-u)^{n-2}\ (1+u)\,du$$

using (v) and substituting $u = 2t^2-1$. This shows that, for k odd and $m = [k/2]$, $\varphi_m(\frac{1+u}{2})$ is a multiple of $P_m^{(n-2,1)}(u)$, hence with some

$M_k \in \mathbb{R}$, $\varphi_m(t^2) = M_k \; P_m^{(n-2,1)}(2t^2-1)$. We thus get

$$\sum_{i=1}^{N_k} Y_{ki}(x) \; \overline{Y_{ki}(y)} = F_k(<x,y>) = M_k <x,y> \; P_m^{(n-2,1)}(2|<x,y>|^2-1)$$

$$= M_k \; Q_k(<x,y>) \; .$$

Take $x = y$ to find $M_k = N_k$ from the fact that $P_m^{(n-2,1)}(1) = 1$ and the orthonormality of (Y_{ki}). The case of k even is similar; the missing factor $<x,e>$ leads to $P_m^{(n-2,0)}$ instead of $P_m^{(n-2,1)}$. ∎

3. The upper bound for $K_G^{\mathbb{C}}(n)$

We now prove that the upper bound for $K_G^{\mathbb{C}}(n)$ given in section 1 is valid.

Proof of theorem 2.

(i) Let $A = (a_{ij})_{i,j=1}^{m}$ be a (complex) matrix with

$$\sup \left\{ \left| \sum_{i,j=1}^{m} a_{ij} \; t_i \; s_j \right| \; \Big| \; |t_i| \leq 1, \; |s_j| \leq 1 \right\} \leq 1 \; . \tag{1}$$

Let $N \in \mathbb{N}$. Averaging over the sphere $S^{N-1}(\mathbb{C})$ yields for $x, y \in S^{N-1}(\mathbb{C})$

$$\int_{S^{N-1}} \operatorname{sgn} <x,u> \; \operatorname{sgn} <\overline{u},y> \; d\omega_N(u) = \Phi(<x,y>) \tag{2}$$

where $\Phi(z) := \operatorname{sgn} z \; \varphi(|z|)$, $\varphi(s) := s \int_0^{\pi/2} \frac{\cos^2 \alpha}{\sqrt{1-s^2 \sin^2 \alpha}} \, d\alpha$.

Given any elements $x_1, \cdots x_m, y_1, \cdots y_m \in S^{N-1}(\mathbb{C})$ and replacing t_i, s_j in (1) by $\operatorname{sgn} <x_i,u>$, $\operatorname{sgn} <\overline{u},y_j>$, we find by averaging over $S^{N-1}(\mathbb{C})$

$$\left| \sum_{i,j=1}^{m} a_{ij} \; \Phi(<x_i,y_j>) \right| \leq 1 \tag{3}$$

Haagerup [Ha] shows this by Gaussian averaging. The spherical average in (2), however, is as easily deduced from real (2N)-dimensional averaging - where the corresponding function φ is $\varphi(s) = \frac{2}{\pi} \sin^{-1} s$ [LP] - and the relation

$$\operatorname{sgn} z = \frac{1}{4} \int_0^{2\pi} \operatorname{sgn} \operatorname{Re} (e^{-i\vartheta} z) \; e^{i\vartheta} \; d\vartheta$$

given by [Ha]. Using this, calculation shows that the left side of (2) equals

$$\frac{1}{8\pi}\int_0^{2\pi}\int_0^{2\pi} e^{i(\vartheta-\eta)} \sin^{-1} \mathrm{Re}\,(e^{i(\eta-\vartheta)} <x,y>)\, d\vartheta\, d\eta$$

which is evaluated to be $\Phi(<x,y>)$ as in [Ha].

For $n \in \mathbb{N}$ consider the Hilbert space l_2-sum $H := l_2(L_2(S^{n-1}(\mathbb{C})), d\omega_n)$. We will define nonlinear functions $f, g : S^{n-1}(\mathbb{C}) \longrightarrow H$ (depending on n) such that

$$\mathrm{Im}\, f, \mathrm{Im}\, g \subset S(H) = \{h \in H \mid \|h\| = 1\} \tag{4}$$

$$\exists_{0<c_n<1}\ \forall_{x,y \in S^{n-1}(\mathbb{C})} \quad <f(x), g(y)>_H = \Phi^{-1}(c_n <x,y>_{\mathbb{C}^n}) . \tag{5}$$

($\Phi : \mathbb{D} \longrightarrow \mathbb{D}$ is invertible by [Ha] if $\mathbb{D} = \{z \in \mathbb{C} \mid |z| \le 1\}$.)

Now let A be as in (1) and take any elements $x_1, \cdots x_m, y_1, \cdots y_m \in S^{n-1}(\mathbb{C})$. By (4), the elements $\tilde{x}_i := f(x_i), \tilde{y}_j := g(y_j)$ belong to the unit sphere of some N-dimensional subspace H_N of H with $N \le 2m$. Thus (3) and (5) can be applied, yielding

$$c_n \left|\sum_{i,j=1}^m a_{ij} <x_i, y_j>\right| = \left|\sum_{i,j=1}^m a_{ij}\, \Phi(<\tilde{x}_i, \tilde{y}_j>_{H_N})\right| \le 1 .$$

This implies $K_G^{\mathbb{C}}(n) \le \frac{1}{c_n}$ (since x_i, y_j may be taken to be normalized by homogeneity).

(iii) It remains to construct f, g with (4),(5) and to estimate c_n. Let q_k be the polynomials given in section 2. They were orthogonal polynomials on [-1,1] with respect to the weight function $w(t) = |t|\ (1-t^2)^{n-2}$ and satisfy $|q_k(t)| \le 1$ for $|t| \le 1$.

Take any $0 < c_n < 1$. Since $\varphi^{-1} \in L_2([-1,1], w(t)\,dt)$, we have an $L_2(w)$-convergent series expansion

$$\varphi^{-1}(c_n t) = \sum_{\substack{k=1\\ k \text{ odd}}}^{\infty} \alpha_k(c_n)\, q_k(t) \quad , \ t \in [-1,1] . \tag{6}$$

Only the q_k of odd order occur because φ^{-1} is an odd function. We shall see in (iv) and use now that (6) is an absolutely convergent series, i.e.

$\sum_{\substack{k=1\\k \text{ odd}}}^{\infty} |\alpha_k(c_n)| < \infty$. Let Q_k be as in section 2. Then $Q_k(<x,\cdot>) \in L_2(S^{n-1})$ for any $x \in S^{n-1}$ and we can define $f, g : S^{n-1} \longrightarrow H = l_2(L_2(S^{n-1}))$ by

$$f(x) := (\operatorname{sgn} \alpha_k(c_n) \sqrt{|a_k(c_n)|\, N_k}\ Q_k(<x,\cdot>))_{k \text{ odd}}$$

$$g(y) := (\sqrt{|\alpha_k(c_n)|\, N_k}\ Q_k(<y,\cdot))_{k \text{ odd}} .$$

By the corollary to the addition theorem for spherical functions

$$<f(x), g(y)>_H = \sum_{\substack{k=1\\k \text{ odd}}}^{\infty} \alpha_k(c_n)\, N_k \int_{S^{n-1}} Q_k(<x,z>)\, \overline{Q_k(<y,z>)}\, d\omega(z)$$

$$= \sum_{k=1}^{\infty} \alpha_k(c_n)\, Q_k(<x,y>) = \Phi^{-1}(c_n <x,y>) .$$

The last equality follows from (6) and the fact that for $z \in \mathbb{D}$

$$\Phi^{-1}(z) = \operatorname{sgn} z\ \varphi^{-1}(z) , \quad Q_k(z) = \operatorname{sgn} z\ q_k(z) \qquad (k \text{ odd}) .$$

Thus (5) is verified. Concerning (4), i.e. $\|f(x)\|_H = \|g(x)\|_H = 1$, we get a condition on c_n.

$$<f(x), f(x)>_H = <g(x), g(x)>_H$$

$$= \sum_{\substack{k=1\\k \text{ odd}}}^{\infty} |\alpha_k(c_n)|\ Q_k(<x,x>) = \sum_{\substack{k=1\\k \text{ odd}}}^{\infty} |\alpha_k(c_n)| .$$

The absolute convergence of (6) ensures that, in fact, $f(x), g(x) \in H$. To satisfy (4), we have to require $\sum_{\substack{k=1\\k \text{ odd}}}^{\infty} |\alpha_k(c_n)| = 1$, i.e. we have to find $0 < c_n \le 1$ maximal with this property. Then $K_G^{\mathbb{C}}(n) \le \frac{1}{c_n}$.

(iv) By Haagerup [Ha], for any $0 < c_n \le 1$, there is an absolutely convergent expansion

$$\varphi^{-1}(c_n t) = \sum_{\substack{j=1\\j \text{ odd}}}^{\infty} \beta_j\, c_n^j\, t^j , \quad t \in [-1,1] \tag{7}$$

with $\beta_1 = 4/\pi$ and $\beta_j \le 0$ for $j \ge 3$ and odd. For j odd, write

$$t^j = \sum_{\substack{k=1\\k \text{ odd}}}^{j} \gamma_{jk}\, q_k(t) , \quad t \in [-1,1] . \tag{8}$$

(The γ_{jk} are non-zero only for odd k, since t^j is an odd function.)
We claim that $\gamma_{jk} > 0$ for all odd $k \le j$: Since the q_k's are orthogonal with respect to w, one has $\gamma_{jk} = (t^j, q_k)_w / (q_k, q_k)_w$ and thus

$$\operatorname{sgn} \gamma_{jk} = \operatorname{sgn} (t^j, q_k)_w = \operatorname{sgn} \int_0^1 t^j q_k(t) \, t(1-t^2)^{n-2} \, dt \, .$$

Let $l := [k/2]$, $m := [j/2]$. Hence $l \le m$ and $\frac{j+1}{2} = m+1$. The definition of q_k yields after substitution of $s = 2t^2 - 1$

$$\operatorname{sgn} \gamma_{jk} = \operatorname{sgn} \int_{-1}^1 P_l^{(n-2,1)}(s) \, (1-s)^{n-2} \, (1+s)^{m+1} \, ds \, .$$

By Rodriguez' formula [Sz]

$$P_l^{(n-2,1)}(s) = (-1)^l c_{nl} \, (1-s)^{-(n-2)} \, (1+s)^{-1} \, D^l[(1-s)^{l+n-2} \, (1+s)^{l+1}]$$

with $c_{nl} > 0$. Hence

$$\operatorname{sgn} \gamma_{jk} = \operatorname{sgn} (-1)^l \int_{-1}^1 D^l[(1-s)^{l+n-2} \, (1+s)^{l+1}] \, (1+s)^m \, ds$$

$$= \operatorname{sgn} \int_{-1}^1 (1-s)^{l+n-2} \, (1+s)^{l+1} \, D^l[(1+s)^m] \, ds$$

after m-fold integration. Since $l \le m$, the integrand is strictly positiv and thus $\gamma_{jk} > 0$ for all odd $k \le j$.

For $t = 1$, (8) yields $1 = \sum_{\substack{k=1 \\ \text{odd}}}^{j} \gamma_{jk} = \sum_{k=1}^{j} |\gamma_{jk}|$. Inserting (8) into (7) thus gives an absolutely convergent expansion of $\varphi^{-1}(c_n t)$ in terms of the q_k's,

$$\varphi^{-1}(c_n t) = \sum_{\substack{j=1 \\ j \text{ odd}}}^{\infty} \beta_j c_n^j \Big(\sum_{\substack{k=1 \\ k \text{ odd}}}^{j} \gamma_{jk} \, q_k(t) \Big)$$

$$= \sum_{\substack{k=1 \\ k \text{ odd}}}^{\infty} \Big(\sum_{\substack{j=k \\ j \text{ odd}}}^{\infty} \beta_j c_n^j \gamma_{jk} \Big) q_k(t) = \sum_{\substack{k=1 \\ k \text{ odd}}}^{\infty} \alpha_k(c_n) \, q_k(t)$$

which is (6) with $\alpha_k(c_n) = \sum_{\substack{j=k \\ j \text{ odd}}}^{\infty} \beta_j \gamma_{jk} c_n^j$. Since $\gamma_{jk} > 0$ and $\beta_j \le 0$ for $j > 3$ odd, we conclude that $\alpha_k(c_n) < 0$ for all $k \ge 3$ odd and $0 < c_n \le 1$. On the other hand $\alpha_1(c_n) > 0$ since for $t = 1$

$$\alpha_1(c_n) = \varphi^{-1}(c_n) - \sum_{\substack{k=3 \\ k \text{ odd}}}^{\infty} \alpha_k(c_n)$$

with $\varphi^{-1}(c_n) > 0$ on $(0,1]$. In particular, $\alpha_1(1) = 1 - \sum_{k=3}^{\infty} \alpha_k(1) > 1$. The sign-character of $(\alpha_k(c_n))$ is thus the same as the one of (β_k). The condition on c_n, $\sum_{\substack{k=1 \\ k \text{ odd}}}^{\infty} |\alpha_k(c_n)| = 1$, is equivalent to

$$1 = 2\alpha_1(c_n) - \sum_{\substack{k=1 \\ k \text{ odd}}}^{\infty} \alpha_k(c_n) = 2\alpha_1(c_n) - \varphi^{-1}(c_n) \ . \tag{9}$$

(v) To evaluate $\alpha_1(c_n)$, put $A_j(n) := \gamma_{j1}$ (which depends on n, too). Note that $q_1(t) = t$. For odd j and $l = [j/2]$, calculation shows

$$A_j(n) = (t^j, t)_w / (t,t)_w = \int_0^1 t^{j+2}(1-t^2)^{n-2}\, dt \Big/ \int_0^1 t^3(1-t^2)^{n-2}\, dt$$

$$= \frac{n!\ (l+1)!}{(n+l)!} \ , \quad A_1(n) = 1 \ .$$

For this value of $A_j(n)$, $\alpha_1(c_n) = \sum_{\substack{j=1 \\ j \text{ odd}}}^{\infty} A_j(n)\, \beta_j\, c_n^j$. Since $\varphi : [0,1] \longrightarrow [0,1]$ is bijective, we can put $s_n = \varphi^{-1}(c_n) \in (0,1]$ for $c_n \in (0,1]$ and find that (9) is equivalent to the equation in theorem 2, namely

$$2\alpha_1(\varphi(s_n)) - s_n = 2 \sum_{\substack{j=1 \\ j \text{ odd}}}^{\infty} A_j(n)\, \beta_j\, \varphi(s_n)^j - s_n = 1 \ . \tag{10}$$

We claim that (10) has exactly one solution $s_n \in (0,1]$. To prove this, let $f(s) := 2\alpha_1(\varphi(s)) - s$. Note that $f(0) = 0$ and $f(1) = 2\alpha_1(1) - 1 > 1$ since $\alpha_1(1) > 1$ as seen in (iv). Thus it suffices to prove that f is strictly increasing on $(0,1]$. We show that $f' > 0$ on $[0,1]$. First of all, $\varphi'(s) \geq \pi/4$ for $s \in [0,1]$ as easily seen from the series expansion of the function φ, cf. [Ha]. Thus, using $\beta_1 = 4/\pi$, $\beta_j \leq 0$ for $j \geq 3$, $A_1(n) = 1$ and $\varphi(s) \leq 1$,

$$f'(s) = 2\varphi'(s) \cdot \sum_{\substack{j=1 \\ j \text{ odd}}}^{\infty} A_j(n)\, \beta_j\, j\, \varphi(s)^{j-1} - 1$$

$$\geq \frac{\pi}{2} \left(\frac{4}{\pi} - \sum_{\substack{j=3 \\ j \text{ odd}}}^{\infty} A_j(n)\, j\, |\beta_j| \right) - 1 \ . \tag{11}$$

Now $A_j(n+1) = \frac{n+1}{n+1+[j/2]} A_j(n) \leq A_j(n)$ and thus for all $n \geq 2$

$$A_j(n) \cdot j \le A_j(2) \cdot j = \frac{2}{[j/2]+2}\, j = \frac{4j}{j+3} \le 4 .$$

We have $\sum_{\substack{j=3 \\ j \text{ odd}}}^{\infty} |\beta_j| = \frac{4}{\pi} - 1$ since

$$1 = \varphi^{-1}(1) = \sum_{\substack{j=1 \\ j \text{ odd}}}^{\infty} \beta_j = \frac{4}{\pi} - \sum_{\substack{j=3 \\ j \text{ odd}}}^{\infty} |\beta_j| .$$

Using $|\beta_3| = \frac{8}{\pi^3}$, we get

$$\sum_{\substack{j=3 \\ j \text{ odd}}}^{\infty} A_j(n)\, j\, |\beta_j| \le 2\,|\beta_3| + 4 \sum_{\substack{j=5 \\ j \text{ odd}}}^{\infty} |\beta_j|$$

$$= 4 \sum_{\substack{j=3 \\ j \text{ odd}}}^{\infty} |\beta_j| - 2\,|\beta_3| = 4\left(\frac{4}{\pi} - 1\right) - \frac{16}{\pi^3} .$$

This and (11) yield

$$f'(s) \ge 2\pi + \frac{8}{\pi^2} - 7 \ge 0.1 > 0 .$$

Hence (10) has exactly one solution $s_n \in (0,1)$. Since $A_1(n) = 1$ and $A_j(n+1) \le A_j(n)$, we have with $f_n(s) := \sum_{\substack{j=1 \\ j \text{ odd}}}^{\infty} A_j(n)\, \beta_j\, \varphi(s)^j - s$ that

$f_n(s) \le f_{n+1}(s)$ and thus $1 = f_{n+1}(s_{n+1}) = f_n(s_n) \le f_{n+1}(s_n)$. Since f_{n+1} is an increasing function, $(s_n)_{n\ge2}$ is decreasing and $\left(\frac{1}{\varphi(s_n)}\right)_{n\ge2}$ is increasing. This ends the proof of theorem 2. ■

We end this section with a few comments on the conjecture in section 1. As mentioned there, Haagerup [Ha] conjectures that $K_G^{\mathbb{C}} \le 1/|\varphi(i)|$. Now $|\varphi(i)| = c$ iff $\varphi^{-1}(ic) = i$ iff

$$\sum_{\substack{k=1 \\ k \text{ odd}}}^{\infty} (-1)^{[k/2]}\, \beta_k\, c^k = 1 . \tag{12}$$

This would replace his proven bound $K_G^{\mathbb{C}} \le 1/\varphi(s_0)$ where $\varphi(s_0) = c$ is determined from

$$\sum_{\substack{k=1 \\ k \text{ odd}}}^{\infty} |\beta_k|\, c^k = 1 . \tag{13}$$

Since $\beta_5 = 0$, the first term where the series (12) and (13) differ is $|\beta_9| c^9$ which is already fairly small; thus there would be only a small improvement of the $K_G^{\mathbb{C}}$-bound.

The analogues of the coefficients $\beta_k c^k$ in the n-dimensional case are the coefficients $\alpha_k(c)$, with condition (13) replaced by $\sum_{\substack{k=1 \\ k \text{ odd}}}^{\infty} |\alpha_k(c)| = 1$. Thus (12) ought to be replaced by

$$\sum_{\substack{k=1 \\ k \text{ odd}}}^{\infty} (-1)^{[k/2]} \alpha_k(c) = 1 , \tag{14}$$

and then $K_G^{\mathbb{C}}(n) \le 1/\varphi(c)$. The previous proof showed that

$$\alpha_k(c) = \sum_{\substack{j=k \\ j \text{ odd}}}^{\infty} \beta_j \, \gamma_{jk}(n) \, c^j , \quad k \text{ odd}$$

where the coefficients $\gamma_{jk} = \gamma_{jk}(n)$ of the expansion of t^j in terms of the polynomials q_k can be calculated explicitely,

$$\gamma_{jk}(n) = (n+2m) \frac{l!}{(l-m)!\, m!} \; \frac{(l+1)!\,(m+n-1)!}{(l+m+n)!} \; \frac{(m+n-2)!}{(m+1)!(n-2)!}$$

with $m = [k/2]$, $l = [j/2]$. Equation (14) then reads

$$1 = \sum_{\substack{k=1 \\ k \text{ odd}}}^{\infty} (-1)^{[k/2]} \alpha_k(c) = \sum_{\substack{j=1 \\ j \text{ odd}}}^{\infty} B_j(n) \, \beta_j \, c^j$$

where $\quad B_j(n) := \sum_{\substack{k=1 \\ k \text{ odd}}}^{\infty} (-1)^{[k/2]} \gamma_{jk}(n) = (-1)^{[j/2]} \dfrac{(n-3)(n-5)\cdots(n-j)}{(n+1)(n+3)\cdots(n+j-2)}$.

If this conjecture were true, it would be in complete formal analogy with the results of Krivine [Kr] in the real case of $K_G^{\mathbb{R}}(n)$. The equations for $n = 3,5,7$ would be

$$\beta_1 c = 1 \quad , \quad \beta_1 c + |\beta_3|/3 \, c^3 = 1 \quad , \quad \beta_1 c + |\beta_3|/2 \, c^3 = 1$$

noting that $\beta_5 = 0$. This would give bounds $K_G^{\mathbb{C}}(3) \le 4/\pi$,

$$K_G^{\mathbb{C}}(5) \le \left(\sqrt[3]{\frac{\sqrt{41}+3}{16}} - \sqrt[3]{\frac{\sqrt{41}-3}{16}} \right) \pi , \; K_G^{\mathbb{C}}(7) \le \left(\sqrt[3]{\frac{\sqrt{91/3}+3}{24}} - \sqrt[3]{\frac{\sqrt{91/3}-3}{24}} \right) \pi .$$

The difference to the proven bounds is about 10^{-3} for small values of n.

4. Some remarks on the lower bounds for $K_G^{\mathbb{C}}(n)$

A slight variation of A. M. Davie's very clever method for the lower bound $K_G^{\mathbb{C}} \geq 1.338$ [Da] yields lower bounds for $K_G^{\mathbb{C}}(n)$. Unfortunately, Davie's arguments are still unpublished. I would like to thank G. Jameson for sending me some notes on this lower bound. A few remarks on the lower bounds for $K_G^{\mathbb{C}}(n)$ might be in order, nevertheless. For $1 \leq p \leq \infty$, let $L_p^n = L_p(S^{n-1}, d\omega_n)$ and P_n denote the orthogonal projection (in L_2^n) onto the n-dimensional subspace $E_n = \text{Span } \{<\cdot, x> \mid x \in \mathbb{K}^n\}$ of (restrictions of) linear functions. P_n is an integral operator with kernel $n<x,y>$ on S^{n-1}, it may act between any L_p^n-spaces. It is easily checked that $\| P_n : L_1^n \longrightarrow L_2^n \| \leq \sqrt{n}$. Let $0 \leq \rho \leq 1$ and $T_n = T_n(\rho) = P_n - \rho \, \text{Id}$. Denoting the nuclear and 1-summing norms, respectively, by ν and π_1, respectively, we have $\sqrt{n} \leq \nu(P_n : L_\infty^n \longrightarrow L_2^n)$ since

$$n = tr(P_n) \leq \|P_n : L_2^n \longrightarrow L_\infty^n\| \; \nu(P_n : L_\infty^n \longrightarrow L_2^n)$$

and thus, using $(1-\rho) P_n = P_n T_n$,

$$\begin{aligned}(1-\rho)\sqrt{n} &\leq \nu((1-\rho) P_n : L_\infty^n \longrightarrow L_2^n) = \pi_1((1-\rho) P_n : L_\infty^n \longrightarrow L_2^n) \\ &\leq \| T_n : L_\infty^n \longrightarrow L_1^n \| \; \pi_1(P_n : L_1^n \longrightarrow L_2^n) \\ &\leq K_G(n) \sqrt{n} \; \| T_n : L_\infty^n \longrightarrow L_1^n \|\end{aligned}$$

Thus with $M_\rho^n := \|T_n(\rho) : L_\infty^n \longrightarrow L_1^n\|$, $K_G(n) \geq \sup_{0 \leq \rho \leq 1} \frac{1-\rho}{M_\rho^n}$.

Taking $\rho = 0$, one gets for $n \longrightarrow \infty$ the classical lower bounds $K_G^{\mathbb{R}} \geq \pi/2$, $K_G^{\mathbb{C}} \geq 4/\pi$. Explicitely

$$M_\rho^n = \sup_{|f|,|g| \leq 1} \left| \int_{S^{n-1}} \int_{S^{n-1}} n <x,y> d\omega(x)\, d\omega(y) - \rho \int_{S^{n-1}} f(x)\, g(x)\, d\omega(x) \right|.$$

Davie ingeneously estimates this: the projection P_n leads to a linear function $h_e(x) = \sqrt{n} <x,e>$ with $e \in S^{n-1}(\mathbb{C})$ such that in the <u>complex</u> case

$$M_\rho^n \leq \sup_{0<\mu<\sqrt{n}} \left\{ \left(\int_{S^{n-1}(\mathbb{C})} \psi_\mu(x)\, |h_e(x)|\, dx \right)^2 + \rho - 2\rho \int_{S^{n-1}(\mathbb{C})} |\psi_\mu(x)|^2\, dx \right\}$$

where $\psi_\mu(x) = \begin{cases} |h_e(x)|/\mu & |h_e(x)| < \mu \\ 1 & |h_e(x)| \geq \mu \end{cases}$. Actually, equality holds.

This is clearly independent of $e \in S^{n-1}(\mathbb{C})$. The integrals can be calculated by using the formula in (v) of section 2. Rescaling $\lambda = \mu/\sqrt{n}$ and differentiating with respect to λ shows that the maximum is attained for $0 < \lambda < 1$ determined by

$$\rho = \rho(\lambda) = \frac{1-(1-\lambda^2)^{n-1}}{2} - \frac{n-1}{2}\lambda^2(1-\lambda^2)^{n-1} + n(n-1)\lambda\int_\lambda^1 t^2(1-t^2)^{n-2}dt. \quad (15)$$

(Davie takes $n \longrightarrow \infty$, working with error integrals.) The last equation is uniquely solvable for $0 < \lambda < 1$ provided that $0 < \rho < 1/2$; only this range of ρ's turns out to be of interest. Inserting this value of $\rho = \rho(\lambda)$, one finds

$$M^n_{\rho(\lambda)} = \rho(\lambda)\left(1-2(1-\lambda^2)^{n-1} + \frac{4(n-1)}{\lambda}\int_\lambda^1 t^2(1-t^2)^{n-2}dt\right). \quad (16)$$

Now $K_G^{\mathbb{C}}(n) \geq \sup_{0<\rho<1/2} \frac{1-\rho}{M^n_\rho} = \sup_{0<\lambda<1} \frac{1-\rho(\lambda)}{M^n_{\rho(\lambda)}}$. Using formulas (15), (16), the expression $\frac{1-\rho(\lambda)}{M^n_{\rho(\lambda)}}$ can be calculated for any $0 < \lambda < 1$. Numerical evaluation yields the bounds given in the table in section 1. The optimal value of $\rho = \rho(\lambda)$ does not depend very strongly on $n \in \mathbb{N}$ and is close to $\rho \approx .23$.

References

[Da] A. M. Davie; Unpublished notes.

[DGS] P. Delsarte, J. M. Goethals, J. J. Seidel; Bounds for systems of lines and Jacobi polynomials, Philips Research Reports 30(1975), 91-105.

[Ha] U. Haagerup; A new upper bound for the complex Grothendieck constant, Israel J. Math. 60 (1987), 1999-224.

[Ko] T. H. Koornwinder; The addition formula for Jacobi polynomials and spherical harmonics, SIAM J. Appl. Math. 2 (1973), 236-246.

[Kr] J. L. Krivine; Constantes de Grothendieck et fonctions de type positif sur les sphères, Advances in Math. 31(1979), 16-30.

[LP] J. Lindenstrauss, A. Pelczyński; Absolutely summing operators in $\mathcal{L}_p$-spaces and their applications, Stud. Math. 29(1968), 275-326.

[LT] J. Lindenstrauss, L. Tzafriri; Classical Banach Spaces II, Springer, 1979.

[Mü] C. Müller, Spherical Harmonics, Lect. Notes in Math. 17, Springer, 1966.

[Pi] G. Pisier; Grothendieck's theorem for non-commutative C^*-algebras with an appendix on Grothendieck's constant, J. Funct. Anal. 29 (1978), 379-415.

[Sz] G. Szegö; Orthogonal Polynomials, AMS 1959.

[TJ] N. Tomczak-Jaegermann; Banach-Mazur distances and finite-dimensional operator ideals, Pitman, 1989.

Mathematisches Seminar
Universität Kiel
Olshausenstr. 40
2300 KIEL
W-Germany

Pathological properties and dichotomies for random quotients of finite-dimensional Banach spaces

PIOTR MANKIEWICZ
Institute of Mathematics
Polish Academy of Sciences
Warsaw, POLAND

NICOLE TOMCZAK-JAEGERMANN [1]
Department of Mathematics
University of Alberta
Edmonton, Alberta, CANADA

1 Introduction and preliminaries

In this note we study Banach–Mazur distances and basic constants of proportional dimensional quotients of an arbitrary n-dimensional Banach space X. The main result establishes a dichotomous behaviour of these invariants for families of random quotients. It is shown that every space X has a quotient X_1 such that either X_1 is Euclidean or otherwise basis constants of random quotients of X_1 are large and, for a random pair of quotients, the Banach–Mazur distances are also large. This provides a strengthening of some results from [M-T].

Our notation follows [M-T]. In particular let $(X, \|\cdot\|_X)$ be an n-dimensional Banach space. The unit ball of X is denoted by B_X. Let $|\cdot|_2$ be a Euclidean norm on X and let $(\cdot,\cdot)$ be the associated inner product. Identify $(X, |\cdot|_2)$ with $\mathbb{R}^n$ under the natural Euclidean norm.

Let $1 \leq l \leq n$. Denote by $G_{n,l}$ the Grassmann manifold of all l-dimensional subspaces of $\mathbb{R}^n$ with the Haar measure $h_{n,l}$. Fix $0 < \gamma < 1$. For a family of properties $\mathcal{M} = \{\mathcal{M}_n\}$ of γn-dimensional spaces we say that $\mathcal{M}$ is satisfied for a random subspace, if there exists $0 < \delta < 1$ such that

$$h_{n,\gamma n}\{E \in G_{n,\gamma n} \mid E \text{ has } \mathcal{M}_n\} \geq 1 - \delta^n,$$

[1] Supported in part by National Sciences and Engineering Research Council Grant A8854.

for every $n = 1, 2, \ldots$. We say that $\mathcal{M}$ is satisfied for a random quotient if the set of $E \in G_{n,(1-\gamma)n}$ such that X/E satisfies $\mathcal{M}$ has measure larger than or equal to $1 - \delta^n$. Similarly we define a random pair of subspaces and a random pair of quotients.

Let P be an orthogonal projection on $(X, |\cdot|_2)$. The quotient space $Y = X/\ker P$ can be identified with the space $P(X)$ under the norm

$$\|y\|_{P(X)} = \inf\{\|y+z\|_X \mid z \in \ker P\} \quad \text{for} \quad y \in P(X).$$

Notice that with this identification Y has a Euclidean norm inherited from X, namely $|\cdot|_2$. Another representation of quotient spaces will be described in Section 2.

Define the norm $\|\cdot\|_*$ on X by

$$\|z\|_* = \sup\{|(x,z)| \mid x \in B_X\} \quad \text{for } z \in X. \tag{1.1}$$

Then the space $(X, \|\cdot\|_*)$ is clearly isometric to the dual space X^*.

Finally, for detailed information on random subspaces and measure concentration phenomena we refer the reader to [M-S]; on various ideal norms and Banach–Mazur distances—to [T]; and on recent geometric and related random methods in the local theory of Banach spaces—to [P].

The second named author would like to thank Bill Johnson for valuable conversations by E-mail concerning Lemma 3.4.

2 Estimates for spaces in special positions

In this section we discuss the space $\mathbb{R}^n$ with an arbitrary norm $\|\cdot\|$ and we denote this space by X. The unit ball of X is denoted by B_X or by B_n, to emphasize the dimension. Recall that the norm from l_n^2 is denoted by $\|\cdot\|_2$, the Euclidean unit ball is denoted by B_n^2 and $\{e_1, \ldots, e_n\}$ denotes the standard unit vector basis in $\mathbb{R}^n$.

Let $1 \le k \le n$. By $\mathbb{R}^k \subset \mathbb{R}^n$ we denote span$\{e_1, \ldots, e_k\}$. By $L(\mathbb{R}^k)$ we denote the space of all linear operators on $\mathbb{R}^k$. For every $(n-k)$-dimensional subspace E of $\mathbb{R}^n$ by P_E we denote the orthogonal projection on $E^\perp$. For every $(n-k)$-dimensional subspace E of $\mathbb{R}^n$, which does not intersect $\mathbb{R}^k$, by $Q_E : \mathbb{R}^n \to \mathbb{R}^k$ we denote the projection onto $\mathbb{R}^k$ with the kernel E.

By $Q_E(X)$ we denote $\mathbb{R}^k$ with the norm for which the unit ball is $Q_E(B_n)$. That is, $\|y\|_{Q_E(X)} = \inf\{\|y+z\| \mid z \in E\}$, for $y \in \mathbb{R}^k$. In particular, in this section we will identify the quotient space X/E (where $E \cap \mathbb{R}^k = \{0\}$) with $Q_E(X)$.

For $1 \leq k \leq m < n$ set

$$\begin{aligned} V_{m,k} &= V_{m,k}(X) \qquad (2.1) \\ &= \sup_{E \in G_{n,n-m}} \inf_{F \subset E^\perp, \dim F = k} \left(\mathrm{vol}_k(F \cap P_E(B_n))/\mathrm{vol}_k(F \cap P_E(B_n^2))\right)^{1/k}. \end{aligned}$$

We write V_k for $V_{k,k}$.

The quantity $V_{m,k}$ is closely related to the notion of the volume ratio numbers of operators, which will be of importance later on. Let us recall ([M-P], [P-T], [P]) that for an operator $u : X \to l_n^2$ and $1 \leq j \leq n$, the j-th volume ratio number is defined by

$$\mathrm{vr_j}(u) = \sup\{(\mathrm{vol}_k(P_E(B_X))/\mathrm{vol}_k P_E(B_n^2))^{1/k} \mid E \subset \mathbb{R}^n, \mathrm{codim} E < j\}. \qquad (2.2)$$

In particular, $V_k(X) = \mathrm{vr}_{n-k}(i)$, where $i : X \to l_n^2$ is the identity operator.

Random subspaces of $\mathbb{R}^n$ enjoy a number of good geometric properties which are fundamental in this investigation. These properties were discussed in details in [M-T]; here we only list the ones which are directly used in our arguments.

Fact 2.1 *Let $0 < \gamma < \alpha < 1$. There exist $c_1 = c_1(\alpha, \gamma)$, $c_2 = c_2(\alpha)$ and $\varepsilon = \varepsilon(\gamma) > 0$ such that a random subspace $E \in G_{n,(1-\alpha)n}$ has the property:*

(A) *The operator $Q_E \mid E^\perp : E^\perp \to \mathbb{R}^{\alpha n}$ has at most γn s-numbers larger than c_1.*

(B) $\|Q_E e_i\|_2 \leq c_2$, *for $i = \alpha n + 1, \alpha n + 2, \ldots, n$.*

(C) dist $(Q_E e_i, \mathrm{span}\,\{Q_E e_k \mid (1-\gamma)n \leq k < i\}) > \varepsilon$, *for $(1-\gamma)n < i \leq n$.*

Let $(\Omega, \mathbf{P})$ be a probability space. We shall say that a Gaussian random variable in a k-dimensional subspace E of $\mathbb{R}^n$ has distribution $\mathrm{N}(0, 1, E)$ if its density (with respect to the k-dimensional Lebesgue measure on E) is equal to

$$f(x) = (\frac{k}{2\pi})^{k/2} e^{-k\|x\|_2^2/2} \quad \text{for} \quad x \in E.$$

In the sequel we shall need the following well known basic properties of such random variables (*cf. e.g.* [Sz]).

Fact 2.2 *Let E be a k-dimensinal subspace of $\mathbb{R}^n$ and let $g(\omega)$ be a Gaussian random variable with distribution $N(0,1,E)$. Then*

(D) *For every orthogonal projection P in E onto an m-dimensinal subspace F, $\sqrt{k/n}P(g(\omega))$ is a Gaussian random variable with distribution $N(0,1,F)$.*

(E) *For every pair F_1, F_2 of orthogonal subspaces of E, the random variables $P_{F_1^\perp}(g(\omega))$ and $P_{F_2^\perp}(g(\omega))$ are independent.*

(F) *There is a universal constant a such that*

$$\mathbf{P}\left(\{\omega \in \Omega \,|\, 1/2 \leq \|g(\omega)\|_2 \leq 2\}\right) > 1 - e^{-an},$$

for every Gaussian random variable g with distribution $N(0,1,\mathbb{R}^n)$.

The following two lemmas are well known to specialists.

Lemma 2.3 *Let B be a centrally symmetric convex body in $\mathbb{R}^n$ and let g be a Gaussian random variable with distribution $N(0,1,\mathbb{R}^n)$. For every operator $T \in L(\mathbb{R}^n)$ we have*

$$\mu_{S^{n-1}}\Big(\{x \in S^{n-1} \,|\, Tx \in B\}\Big) \leq 2\mathbf{P}\left(\{\omega \in \Omega \,|\, Tg(\omega) \in 2B\}\right).$$

Here $\mu_{S^{n-1}}$ denotes the normalized Haar measure on the unit sphere S^{n-1} in $\mathbb{R}^n$.

Proof Observe that the measure ν defined for a Borel subset $C \subset S^{n-1}$ by

$$\nu(C) = \frac{1}{A}\mathbf{P}\left(\{\omega \in \Omega \,|\, \|g(\omega)\|_2 \leq 2 \quad \text{and} \quad g(\omega)/\|g(\omega)\|_2 \in C\}\right),$$

where $A = \mathbf{P}(\{\omega \in \Omega \,|\, \|g(\omega)\|_2 \leq 2\})$, is normalized and rotation invariant. Therefore, by the uniquness of Haar measure on S^{m-1} , Fact 2.2 (F) and the convexity of B, we have

$$\begin{aligned}
&\mu_{S^{m-1}}\Big(\{x \in S^{m-1} \,|\, Tx \in B\}\Big) \\
&= \nu\Big(\{x \in S^{m-1} \,|\, Tx \in B\}\Big) \\
&\leq 2\mathbf{P}\left(\{\omega \in \Omega \,|\, \|g(\omega)\|_2 \leq 2 \quad \text{and} \quad Tg(\omega) \in \|g(\omega)\|_2 B\}\right) \\
&\leq 2\mathbf{P}\left(\{\omega \in \Omega \,|\, Tg(\omega) \in 2B\}\right).
\end{aligned}$$

This concludes the proof. □

Lemma 2.4 *Let E be a k-dimensional subspace of $\mathbb{R}^n$ and g be a Gaussian random variable with distribution $N(0,1,E)$. Then for every random variable $y = y(\omega)$ independent of g and taking values in $\mathbb{R}^n$, and for every centrally symmetric convex body $B \subset \mathbb{R}^n$ we have*

$$\mathbf{P}\left(\{\omega \in \Omega \,|\, g(\omega) \in B + y(\omega)\}\right) \le a_1^k \mathrm{vol}_k(E \cap B)/\mathrm{vol}_k B_k^2.$$

where $a_1 > 1$ is a universal constant.

Proof It is enough to prove the lemma with y being a constant variable; the general case is then obtained by an appropriete integration. Clearly,

$$\mathbf{P}\left(\{\omega \in \Omega \,|\, g(\omega) \in B + y\}\right) = \mathbf{P}\left(\{\omega \in \Omega \,|\, g(\omega) \in E \cap (B + y)\}\right). \quad (2.3)$$

By Lemma 6.6 in [Sz], the latter probability is smaller than or equal to

$$a_1^k \mathrm{vol}_k(E \cap (B + y))/\mathrm{vol}_k B_k^2,$$

where $a_1 > 1$ is a universal constant. Obviously, $\mathrm{vol}_k(E \cap (B + y)) = \mathrm{vol}_k((E - y) \cap B)$. It is a well-known consequence of the classical Brunn inequality (*cf. e.g.* [H]) that the latter volume is the largest if $y = 0$. That is, the latter quantity in (2.3) is smaller than or equal to $\mathrm{vol}_k(E \cap B)/\mathrm{vol}_k B_k^2$, completing the proof. □

The next proposition gives an upper estimate for probability of certain sets in terms of the invariant $V_{m,k}(X)$.

Proposition 2.5 *Let E be a βn-dimensional subspace of $\mathbb{R}^n$ with $0 < \beta \le 1$ and let g be a Gaussian random variable with distribution $N(0,1,E)$. Let B be a centrally symmetric convex body in $\mathbb{R}^n$. Let $S : E \to S(E) \subset \mathbb{R}^n$ be an operator with all s-numbers greater than or equal to $a > 0$. Let $0 < \delta < 1$. For every k with $\delta\beta n \le k \le \beta n$ we have*

$$\mathbf{P}\left(\{\omega \in \Omega \,|\, Sg(\omega) \in P_{S(E)^\perp}(B)\}\right) \le \left(ca^{-1} V_{\beta n,k}(B)\right)^k,$$

where $c = c(\delta)$.

Proof Fix k satisfying $\delta\beta n \le k \le \beta n$. By the definition of $V_{\beta n,k}$ there exist a subspace $F \subset S(E)$ with $\dim F = k$ such that

$$\left(\mathrm{vol}_k(F \cap P_{S(E)^\perp}(B))/\mathrm{vol}_k B_k^2\right)^{1/k} \le V_{\beta n,k}(B). \quad (2.4)$$

Set $F_1 = S^{-1}(F)$. Let F_2 be the orthogonal complement of F_1 in E and let P_1 and P_2 be the orthogonal projections in E onto F_1 and F_2 respectively. Write $g = P_1 g + P_2 g$. Note that, by Fact 2.2 (E), the random variables $P_1 g$ and P_2 are independent. Moreover, by Fact 2.2 (D), $\sqrt{\beta n/k}\, P_1 g$ is a Gaussian random variable with distribution $\mathrm{N}(0,1,F_1)$. Therefore, by Lemma 2.4 we have

$$\begin{aligned}
&\mathbf{P}\left(\{\omega \in \Omega \mid Sg \in P_{S(E)^\perp}(B)\}\right)\\
&= \mathbf{P}\left(\{\omega \in \Omega \mid g \in S^{-1}P_{S(E)^\perp}(B)\}\right)\\
&= \mathbf{P}\left(\{\omega \in \Omega \mid P_1 g \in S^{-1}(P_{S(E)^\perp}(B)) - P_2 g\}\right) \qquad (2.5)\\
&= \mathbf{P}\left(\{\omega \in \Omega \mid \sqrt{\frac{\beta n}{k}} P_1 g \in \sqrt{\frac{\beta n}{k}}\left(F_1 \cap [S^{-1}(P_{S(E)^\perp}(B)) - P_2 g]\right)\}\right)\\
&\leq a_1^k \mathrm{vol}_k\left(\sqrt{\frac{\beta n}{k}} S^{-1}(F \cap P_{S(E)^\perp}(B))\right) / \mathrm{vol}_k B_k^2,
\end{aligned}$$

where a_1 is a universal constant. On the other hand, by (2.4), easy computation yields

$$\begin{aligned}
&\mathrm{vol}_k\left(\sqrt{\frac{\beta n}{k}} S^{-1}(F \cap P_{S(E)^\perp}(B))\right) / \mathrm{vol}_k B_k^2\\
&\leq (a^{-1}\sqrt{1/\delta}\,)^k \mathrm{vol}_k(F \cap P_{S(E)^\perp}(B)) / \mathrm{vol}_k B_k^2\\
&\leq (a^{-1}\sqrt{1/\delta}\, V_{\beta n,k})^k, \qquad (2.6)
\end{aligned}$$

which combined with (2.5) concludes the proof. □

We come now to a crucial point of this section which is a refinement of Lemma 3.3 in [M-T]. To state this very technical estimate some more notation from [M-T] is required.

Fix $0 < \alpha < 1$. Set

$$\mathcal{F} = \{E \in G_{n,(1-\alpha)n} \mid E \text{ satisfies (A), (B), (C)}\}.$$

Let $0 < \beta < \alpha$, and $\gamma \leq \min(\alpha, 1-\alpha)$, and $\eta > 0$ satisfy $\alpha - (\eta + 2\gamma) \geq \beta$. Denote $\mathrm{span}\{e_1, \ldots, e_{\gamma n - 1}\}$ by $\mathbb{R}^{\gamma n - 1}$ and let $\mathcal{G}$ denote the group of all

$W \in \mathcal{O}(n)$ such that

$$W \,|\, \mathbb{R}^{\gamma n-1} = Id \quad \text{and} \quad W \,|\, (\mathbb{R}^{\alpha n})^{\perp} = Id. \tag{2.7}$$

By $h_{\mathcal{G}}$ denote the normalized Haar measure on $\mathcal{G}$.

Lemma 2.6 *Let $0 < \delta < 1$ and let $A > 0$. Let X be an n-dimensional Banach space. Let $E_0 \in G_{n,(1-\alpha)n}$ satisfy* (A) *and let $B_{\alpha n} = Q_{E_0} B_n$. Fix an arbitrary $F_0 \in \mathcal{F}$ and arbitrary $U \in \mathcal{O}(n)$ such that $U(F_0) \in \mathcal{F}$. Then we have*

$$h_{\mathcal{G}}\{W \in \mathcal{G} \,|\, TWQ_{U(F_0)}e_n \in AB_{\alpha n})\} \leq (c\varepsilon^{-1}AV_{\beta n,k})^k,$$

for every k with $\delta\beta n \leq k \leq \beta n$ and for every $T \in L(\mathbb{R}^{\alpha n})$ which has exactly ηn s-numbers smaller than 1; where ε satisfies condition (C) *and $c = c(\alpha, \beta, \delta)$.*

Proof By condition (A), $E_0^{\perp} = E_1 \oplus E'_1$ with $E_1 \perp E'_1$, $Q_{E_0}(E_1) \perp Q_{E_0}(E'_1)$ and $\dim E_1 \leq \gamma n$, and such that $Q_{E_0} \,|\, E_1$ (resp. $Q_{E_0} \,|\, E'_1$) has all s-numbers larger than c_1 (resp. smaller than c_1). Set $F_1 = Q_{E_0}(E_1) \subset \mathbb{R}^{\alpha n}$.

Fix an operator $T \in L(\mathbb{R}^{\alpha n})$ which has exactly ηn s-numbers smaller than 1. Write $\mathbb{R}^{\alpha n} = E_2 \oplus E'_2$ with $E_2 \perp E'_2$, $T(E_2) \perp T(E'_2)$ and $\dim E_2 = \eta n$, and such that $T \,|\, E_2$ (resp. $T \,|\, E'_2$) has all s-numbers smaller than 1 (resp. larger than or equal to 1). Set $F_2 = T(E_2) \subset \mathbb{R}^{\alpha n}$.

Set $F_3 = T(\mathbb{R}^{\gamma n-1})$, so that $\dim F_3 < \gamma n$. Let $P : \mathbb{R}^{\alpha n} \to \mathbb{R}^{\alpha n}$ be any orthogonal projection with $\operatorname{rank} P = \beta n$ such that

$$F_j \subset \ker P \quad \text{for} \quad j = 1, 2, 3.$$

The existence of P like this follows from the condition $\alpha - (\eta + 2\gamma) \geq \beta$.

Set

$$\mathcal{A}' = \{W \in \mathcal{G} \mid PTWQ_{U(F_0)}e_n \in APB_{\alpha n}\}.$$

Clearly, if $\mathcal{A}$ denotes the set in the conclusion of the lemma, then

$$\mathcal{A} \subset \mathcal{A}'. \tag{2.8}$$

Denote by H the orthogonal complement of $\mathbb{R}^{\gamma n-1}$ in $\mathbb{R}^{\alpha n}$ and by P_1— the orthogonal projection on H. Since $\ker P_1 \subset \ker PT$ then $PT = PTP_1$. Also, $WP_1 = P_1 W$ for every $W \in \mathcal{G}$. Set

$$z = Q_{U(F_0)}e_n$$

and

$$z_0 = P_1(z)/\|P_1(z)\|_2.$$

We have $PTWQ_{U(F_0)}e_n = PTW(z_0)\|P_1(z)\|_2$. Recall that $U(F_0) \in \mathcal{F}$ and in particular it satisfies condition (C). Let $\varepsilon > 0$ be a constant as in (C). Then $\|P_1(z)\|_2 \geq \varepsilon$. Set

$$\mathcal{A}'' = \{W \in \mathcal{G} \mid PTW(z_0) \in \varepsilon^{-1}APB_{\alpha n}\}. \tag{2.9}$$

Thus

$$\mathcal{A}' \subset \mathcal{A}''. \tag{2.10}$$

Since $\mathcal{G}$ can be identified with the orthogonal group acting on H, the measure of $\mathcal{A}''$ is equal to the measure of a suitable subset of the sphere S_H in H. We have

$$h_{\mathcal{G}}(\mathcal{A}'') = \mu_{S_H}\left(\{x \in S_H \mid PTx \in \varepsilon^{-1}APB_{\alpha n}\}\right). \tag{2.11}$$

Denote by $\tilde{\mathcal{A}}$ the subset of S_H which appears on the right hand side of (2.11). Let P_2 be the orthogonel projection in H with $\ker P_2 = \ker(PT \mid H)$. By Lemma 2.3 we have

$$\begin{aligned}\mu_{S_H}(\tilde{A}) &\leq 2\mathbf{P}\left(\{\omega \in \Omega \mid PTg' \in 2\varepsilon^{-1}APB_{\alpha n}\}\right)\\ &= 2\mathbf{P}\left(\{\omega \in \Omega \mid PTP_2 g' \in 2\varepsilon^{-1}APB_{\alpha n}\}\right),\end{aligned} \tag{2.12}$$

where g' is a Gaussian random variable with distribution $\mathrm{N}(0,1,H)$. Set $E = P_2(H)$ and $g = \sqrt{(\alpha-\gamma+1)/\beta}P_2 g'$. Clearly, $\dim E = \beta n$ and, by Fact 2.2 (H), g is a Gaussian random variable with distribution $\mathrm{N}(0,1,E)$. By (2.12), we infer that

$$\mu_{S_H}(\tilde{A}) \leq 2\mathbf{P}\left(\{\omega \in \Omega \mid PTg \in c'\varepsilon^{-1}APB_{\alpha n}\}\right), \tag{2.13}$$

where $c' = 2\sqrt{\alpha-\gamma+1}$.

Consider now the operator $Q = Q_{E_0} | E_0^{\perp} : E_0^{\perp} \to \mathbb{R}^{\alpha n}$. Recall that P_{E_0} denotes the orthogonal projection onto $E_0^{\perp}$. Therefore $Q_{E_0} = QP_{E_0}$ and $PB_{\alpha n} = PQ_{E_0}B_n = PQP_{E_0}B_n$. Clearly, Q is an isomorphism. There is a projection P_3 in $\mathbb{R}^{\alpha n}$ such that $\ker P_3 = \ker P$ and that $P' = Q^{-1}P_3Q$ is an orthogonal projection in $E_0^{\perp}$ of rank βn. In fact, P_3 is the projection onto

$Q(Q^{-1}(\ker P))^{\perp}$. In particular, $P_3(\mathbb{R}^{\alpha n}) \subset Q(E'_1)$. Clearly, $P_3 = P_3 P$, so that

$$\tilde{P} = Q^{-1} P_3 P Q P_{E_0} \tag{2.14}$$

is an orthogonal projection of rank βn with $\ker \tilde{P} \supset E_0$.

Set $S = Q^{-1} P_3 T \,|\, E : E \to S(E) \subset E_0^{\perp}$. Then

$$\begin{aligned} &\{\omega \in \Omega \,|\, PTg \in c'\varepsilon^{-1} A P B_{\alpha n}\} \\ &\subset \{\omega \in \Omega \,|\, Sg \in c'\varepsilon^{-1} A \tilde{P} B_n\}. \end{aligned} \tag{2.15}$$

Observe that $\tilde{P}S = S$, so since $\operatorname{rank} S = \beta n$, $\tilde{P}$ is the orthogonal projection in $\mathbb{R}^n$ onto $S(E)$. We will show that all s-numbers of S are larger than or equal to c_1^{-1}, where c_1 is the constant from condition (A). By Proposition 2.5 this will imply that the measure of the latter set in (2.15) is smaller than or equal to $(ca^{-1} V_{\beta n,k}(B))^k$. We will then conclude the proof combining this estimate with (2.13) and (2.15).

To estimate s-numbers of S, let $R = T \,|\, E$. Since $E \subset E_1'$, all s-numbers of R are larger than or equal to 1. We have $PT \,|\, E = R(R^{-1}PR) = RP''$, where $P'' = R^{-1}PR$ is a projection. Clearly, $\ker P'' = \ker P_2 \perp E$, hence $P'' \,|\, E$ has all s-numbers larger than or equal to 1. Similarly, since $P(\mathbb{R}^{\alpha n}) \perp \ker P_3$, then all s-numbers of $P_3 \,|\, P(\mathbb{R}^{\alpha n})$ also are larger than or equal to 1. Finally, since $P(\mathbb{R}^{\alpha n}) \subset Q_{E_0}(E_1')$ then all s-numbers of $Q^{-1}|P(\mathbb{R}^{\alpha n})$ are larger than or equal to c_1.

Now recall that for any operator $U : \mathbb{R}^m \to \mathbb{R}^m$, all s-numbers of U are larger than or equal to $c > 0$ if and only if the inequality $|Ux| \geq c|x|$ is satisfied for every $x \in \mathbb{R}^m$. (Here $|\cdot|$ denotes the Euclidean norm on $\mathbb{R}^m$.) This implies that all s-numbers of the operator $S = Q^{-1} P_3 R P''$ are larger than or equal to c_1^{-1}, completing the proof. □

Lemma 2.6 is the only modification required to get strengthenings of the main technical results of [M-T], Sections 3 and 6.

Similarly as in [M-T], we are able to obtain estimates for distances and basis constant for random quotients of n-dimensional spaces in a special position. In fact we shall state the result for spaces which additionally admit a good control of volume ratio, leaving a general case to an interested reader.

Let us recall that a space $X = (\mathbb{R}^n, \|\cdot\|)$ is said to be in a special position if

$$a\|x\|_2 \leq \|x\| \leq \|x\|_1 \quad \text{for} \quad x \in X, \tag{2.16}$$

for some constant $a > 0$. (Here $\|\cdot\|_1$ and $\|\cdot\|_2$ denote the norms in l_n^1 and l_n^2 respectively.)

Recall that the volume ratio of X, $\mathrm{vr}(X)$, is defined by

$$\mathrm{vr}(X) = (\mathrm{vol}\, B_X / \mathrm{vol}\, \mathcal{E})^{1/n}, \tag{2.17}$$

where $\mathcal{E} \subset B_X$ is the ellipsoid of maximal volume contained in the unit ball B_X of X.

The main result of this section is a refinement of Theorems 3.5 and 6.5 from [M-T]. Its proof repeates the arguments from Proposition 3.2, Theorems 3.5, 3.7, Proposition 6.1, Theorem 6.5 and Proposition 4.2 in [M-T], and combines them with Lemma 2.6 above.

Theorem 2.7 *Let X be an n-dimensional Banach space satisfying (2.16) with the volume ratio $\mathrm{vr}(X) \leq A$, for some constant $A \geq 1$. Let $0 < \alpha < 1$, let $0 < \beta < \alpha/2$ and let $k = \delta\beta n$, for some $0 < \delta < 1$. For a random pair (E_1, E_2) of $(1-\alpha)n$-dimensional subspaces of $\mathbb{R}^n$ one has*

$$\mathrm{d}(Q_{E_1}(X), Q_{E_2}(X)) \geq cA^{-c'} V_{\beta n,k}^{-2}. \tag{2.18}$$

Furthermore, set $m = \min(3\alpha/2 - \beta, 1)n$ and let $X_m = P_m(X)$, where $P_m : \mathbb{R}^n \to \mathbb{R}^m$ is the orthogonal projection onto $\mathbb{R}^m$. Then for a random $(m - \alpha n)$-dimensional subspace E of $\mathbb{R}^m$ the following basis constant estimate holds

$$\mathrm{bc}(Q_E(X_m)) \geq cA^{-c'} V_{\beta n,k}^{-1}. \tag{2.19}$$

In general one has $c = c(\alpha, \beta, \delta) > 0$ and $c' = c'(\alpha, \beta, \delta) > 1$. If for some $0 < \kappa < 1$ one has $\alpha < \kappa$, $\beta < \kappa\alpha/2$ and $\delta < \kappa$, then $c = c(\kappa)$ and $c' = c'(\kappa)$.

3 Dichotomies for random quotients

Now we are ready to study proportional dimensional quotients of an arbitrary n-dimensional Banach space X. Actually, we will be mainly interested in properties shared by random quotients with respect to some Euclidean norm on X. The main result of this note says that every space X has a quotient X_1 which satisfies the dichotomy: either X_1 is Euclidean or ¿otherwise basis constants of random quotients of X_1 are large and, for a random pair of quotients, the distances are also large.

The notion of random quotient obviously depends on the Euclidean norm on X, and the result will be established for a large class of such norms. Denote a Euclidean norm on X by $|\cdot|_2$ and let $[\cdot,\cdot]$ be the associated inner product. Assume that

$$|x|_2 \leq \|x\| \quad \text{for } x \in X. \tag{3.1}$$

Moreover, in the present context it is natural to assume that the Euclidean unit sphere and the unit ball of X are close to each other in some directions, and that these directions are equally distributed in the space. Perhaps the weakest assumption of this type is the following. Fix absolute constants $0 < \delta, \delta_1 < 1$. Assume that there exist vectors $y_1, \ldots, y_l$, with $l = \delta n$, which are orthonormal in $|\cdot|_2$ and such that

$$\|y_j\|_* = \sup\{|[x, y_j]| \mid \|x\| \leq 1\} \geq \delta_1 \quad \text{for } j = 1, \ldots, l. \tag{3.2}$$

Note that if $|\cdot|_2$ is the norm determined by the ellipsoid of minimal volume containing the unit ball of X, then for every κn-dimensional subspace E of X there is $y \in E$ with $|y|_2 = 1$ satisfying (3.2) with $a_1 = \sqrt{\kappa}$.

Now our result states.

Theorem 3.1 *There is a constant $0 < \alpha_0 < 1/2$ such that the following holds, for every $0 < \alpha < \alpha_0$. Let X be an n-dimensional Banach space, and let $|\cdot|_2$ be a Euclidean norm on X satisfying (3.1) and (3.2). There is an integer $2\alpha_0 n \leq m < n$ and an m-dimensional quotient X_1 of X such that for every $K > 10$ exactly one of the following two conditions holds.*

(i) *A random pair (F_1, F_2) of αn-dimensional quotients of X_1 satisfies*

$$\mathrm{d}(F_1, F_2) \geq K^2; \tag{3.3}$$

and a random αn-dimensional quotient F of X_1 satisfies

$$\mathrm{bc}(F) \geq K. \tag{3.4}$$

(ii) *There exist constants $c_1, c_2 > 0$ such that $c_1 c_2 < K$ and that*

$$(1/c_1)|x|_2 \leq \|x\|_{X_1} < c_2|x|_2 \quad \text{for} \quad x \in X_1. \tag{3.5}$$

The proof of the theorem is based on Theorem 2.7 and powerful methods from the local theory of Banach spaces. It requires some additional results. The first one is analogous to [M-T] Proposition 4.3. It ensures that the assumptions of Theorem 2.7 are satisfied for some quotient of X.

Proposition 3.2 *There exist constants $0 < a_0 < 1$ and $1 < \tilde{a}, a', a''$ with the following property. Let X be an n-dimensional Banach space, and let $|\cdot|_2$ be a Euclidean norm on X satisfying (3.1) and (3.2). There exists a quotient of X, say $Y = P(X)$, for some orthogonal projection P, with $\dim Y = k \geq a_0 n$, such that*

$$\mathrm{vr}\,(Y) \leq \tilde{a}. \tag{3.6}$$

Furthermore, there exist a Euclidean norm $\|\cdot\|_2$ on Y, and an orthonormal basis $\{e_1, \ldots, e_k\}$ in $(Y, \|\cdot\|_2)$, such that

$$(1/a')\|y\|_2 \leq \|y\| \leq \sum_i |t_i| \quad \text{for} \quad y = \sum_i t_i e_i \in Y, \tag{3.7}$$

and

$$(1/a'')|y|_2 \leq \|y\|_2 \leq a''|y|_2 \quad \text{for} \quad y \in Y. \tag{3.8}$$

Remark If $|\cdot|_2$ is the norm determined by the ellipsoid of minimal volume containing the unit ball of X, then the above proposition holds for every $1 \leq k < n$. Constants $\tilde{a}$, a' and a'' depend then on $\delta = k/n$.

Proof For every $j = 1, \ldots, l$, pick $x_j \in X$ such that $\|x_j\| = 1$ and that $\|[x_j, y_j]\| = \sup_{\|x\| \leq 1} \|[x, y_j]\|$. Using Bourgain–Tzafriri's method ([B-T] Theorem 7.2, *cf.* *also* [B-S]) one can choose a subset $\sigma \subset \{1, \ldots, l\}$, with $|\sigma| \geq 2^{-10}\delta_1^2 l$ such that

$$\sum_{i\in\sigma} |t_i| \geq \|\sum_{i\in\sigma} t_i x_i\| \geq |\sum_{i\in\sigma} t_i x_i|_2 \geq (\delta_1/4)\left(\sum_{i\in\sigma} |t_i|^2\right)^{1/2}, \tag{3.9}$$

for any sequence (t_i) of scalars.

Denote by Y_1 the space $\mathrm{span}\,\{x_i\}_{i\in\sigma}$ with the quotient norm given by the orthogonal projection from X onto Y_1. Clearly, the norm $\|\cdot\|_{Y_1}$ also satisfies (3.9). It was proved by V. Milman, as a consequence of his inverse Brunn–Minkowski inequality [M.1], that Y_1 has a quotient Y_2 of dimension $\dim Y_2 = \dim Y_1/2$, say, such that $\mathrm{vr}\,(Y_2) \leq \tilde{a}$.

Represent Y_2 as an image of Y_1 (and also of X), by some orthogonal projection, say R. Now proceed as in Proposition 4.3 of [M-T]. Let S be an orthogonal projection in Y_2 with rank $S \geq a_2 \dim Y_2$ such that

$$|\sum_{i\in\sigma} t_i SRx_i|_2 \leq a(\sum_{i\in\sigma} |t_i|^2)^{1/2},$$

where $0 < a_2 < 1 < a$ are absolute constants. Required vectors e_i's can be chosen among the SRx_i's. The space Y is the span of the e_i's under the quotient norm, and the new Euclidean norm $\|\cdot\|_2$ on Y is set to be $\|\sum_i t_i e_i\|_2 = (\sum_i |t_i|^2)^{1/2}$. Then (3.7) and (3.8) follow from Lemmas D and C in [B-S] by the same argument as in [M-T]. □

We also need some estimates for Gelfand numbers of operators from a Banach space X to a Hilbert space, in terms of volume ratio numbers of u. Let us recall that for $u : X \to Y$ the m-th Gelfand number is defined by

$$c_m(u) = \inf\{\|u \,|\, E\| \mid E \subset X, \ \operatorname{codim} E < m\}.$$

Recall that the volume ratio numbers were defined in (2.2). The following result was observed by Pajor and Tomczak in [P-T]. It is close in spirit to the volume ratio result of Szarek.

Proposition 3.3 *Let X be an n-dimensional Banach space and let u : $X \to l_2^n$. For every $m = 1, \ldots, n$ one has*

$$c_{2m}(u) \leq a_2(n/m)\log(1+n/m)\mathrm{vr_m}(u), \tag{3.10}$$

where $a_2 \geq 1$ is a universal constant.

Let us provide an outline of the proof of (3.10).

Proof Define the Euclidean norm on X by $|x|_2 = \|ux\|$ for $x \in X$. It is clearly enough to prove (3.10) for the formal identity operator $i : X \to (X, |\cdot|_2)$. Fix m. By Milman's theorem on quotients of subspaces [M.2] (*cf. also* [M.1], [P]), there is an $m/2$-codimensional subspace $E \subset X$ and an $(n-m)$-dimensional quotient F of E such that $\mathrm{d}(F, l_{n-m}^2) \leq D$, where $D = \tilde{a}(n/m)\log(1+n/m)$, for some universal constant $\tilde{a}$.

Represent F as $P(E)$, for some orthogonal projection P, and denote its quotient norm by $\|\cdot\|_{P(E)}$. The required inequality is easy to establish for operators acting between Hilbert spaces. In particular (*cf. e.g.* [P-T]), there is a subspace $H \subset P(E)$ with $\dim H = n - 2m$ such that

$$\|x\|_2 \leq D \sup \frac{\operatorname{vol} QP(B_E)}{\operatorname{vol} QP(B^2_{n-m/2})} \|x\|_{P(E)} \quad \text{for } x \in H, \tag{3.11}$$

with the supremum taken over all orthogonal projections $Q : P(E) \to P(E)$ of rank $n - 2m$. Consider now $F = P(E)$ as a subspace of E, that is, F with the norm $\|\cdot\|$, inherited from E. Clearly, $\|x\|_{P(E)} \leq \|x\|$, for $x \in F$. Furtheremore, if R is the orthogonal projection on E, then

$$\operatorname{vol} QP(B_E) \leq \operatorname{vol} QPR(B_X).$$

Substituting these estimates into (3.11) and observing that $(H, \|\cdot\|)$ is a subspace of X and QPR is an orthogonal projection in X, we conclude the proof of (3.10) by the definition of volume ratio numbers. □

Finally, we also require the following simple lemma, known to specialists.

Lemma 3.4 *Let X be a k-dimensional Banach space such that $\mathrm{d}(X, l_2^k) \leq D$, for some $D \geq 10$. There exists a $(0.01k)$-dimensional subspace $E \subset X$ such that $\mathrm{d}(E, l^2_{0.01k}) \leq (1/2)D$.*

Proof Let $\|\cdot\|_2$ be a Euclidean norm on X satisfying

$$\|x\|_2 \leq \|x\| \leq D\|x\|_2 \quad \text{for } x \in X. \tag{3.12}$$

Let M_r denotes the median of $\|\cdot\|$ on the unit shere $\{x \in X \mid \|x\|_2 = 1$. The classical proof of Dvoretzky's theorem (*cf. e.g* [M.3], [F-L-M] or [M-S] Theorem 4.2) shows that for a fixed $\varepsilon > 0$, whenever

$$m < (\ln 9)^{-1}[(1/2)\ln(\pi/2) + \varepsilon^2(k-3)/2], \tag{3.13}$$

there exists an m-dimensional subspace $E \subset X$ such that for $x \in E$ one has

$$[(2/3)M_r - (4/3)D\varepsilon]\|x\|_2 \leq \|x\| \leq (4/3)(M_r + D\varepsilon)\|x\|_2. \tag{3.14}$$

Now, set *e.g.* $\varepsilon = 3/40$, so that $m = 0.01k$ satisfies (3.13). If $M_r \geq 0.3D$ then (3.14) implies that for some m-dimensional subspace $E \subset X$ one has

$$\mathrm{d}(E, l^2_m) \leq 2(1 + \kappa\varepsilon)/(1 - 2\kappa\varepsilon) \leq 5,$$

where $\kappa = D/M_r$. Otherwise, combining the upper estimate in (3.14) with the lower estimate in (3.12) we get that some m-dimensional subspace $E \subset X$ satisfies

$$d(E, l_m^2) \leq (4/3)(M_r + D\varepsilon) \leq 0.5D,$$

completing the proof. □

Proof of Theorem 3.1 Let a_0 be the constant from Proposition 3.2. Fix $\alpha_0 < a_0/2$ to be defined later. Let Y be a quotient space of X, and let $\|\cdot\|_2$ be a Euclidean norm on Y such that $\dim Y = a_0 n$ and (3.6), (3.7) and (3.8) are satisfied.

Fix an integer m with $2\alpha_0 n \leq m < a_0 n$ to be defined later. Let $0 < \alpha < \alpha_0$. Assume that for every m-dimensional quotient Y_1 of Y, condition (i) is violated. For a given Y_1 and a fixed $\varepsilon > 0$ small enough this means the following. In case of the distance estimate, the measure of the set of pairs (F_1, F_2) which satisfy (3.3), is smaller than $1 - \varepsilon^n$; in case of the basis constant estimate, for $m' = 5\alpha n/4$ and $P_{m'}$ being the orthogonal projection on the first m' coordinates, the measure of the set of quotients F's of $P_{m'}(Y_1)$ which satisfy (3.4), is smaller than $1 - \varepsilon^n$. By the concentration of measure phenomenon, for any Y_1 we have: either a random pair (F_1, F_2) of αn-dimensional quotients of Y_1 actually satisfy

$$d(F_1, F_2) \leq aK^2, \tag{3.15}$$

or a random αn-dimensional quotient F of $P_{m'}(Y_1)$ satisfies

$$\mathrm{bc}(F) \leq aK. \tag{3.16}$$

We will show that there exist an m-dimensional quotient X_1 of Y for which (ii) holds.

Define α' by $\alpha' m = \alpha n$. Set $\beta' = \alpha'/4$ and $k = 3\beta' m/4$.

Now the proof is similar as in [M-T] Theorem 5.3, in which the use of Szarek's volume ratio argument is replaced by Proposition 3.3. It shows that if any of (3.15) or (3.16) holds then every m-dimensional subspace $Z = (Y_1)^*$ of Y^* contains a subspace H with $\dim H = \beta' m/2$ such that

$$d(H, l_{\dim H}^2) \leq a_1 K.$$

In general, a_1 depends on α', β' and $k/\beta' m$. However, as we have $\alpha' \le 1/2$, $\beta' = (1/2)(\alpha'/2)$ and $k \le (3/4)\beta' m$, then a_1 may be taken as an absolute constant.

By Proposition 5.5 from [M-T], this implies the existence of a subspace S of Y^* such that

$$d(S, l_k^2) \le a_1 K \tag{3.17}$$

and $k = \dim S \ge \dim Y^* - m = a_0 n - m$.

We want to show that on a certain m-dimensional subspace of S, the original norm and $|\cdot|_2$ are equivalent, with proper constants. If $a_1 > 1$, let s be the smallest integer larger than $\log_2 a_1$. Applying Lemma 3.4 s times we get a subspace Z of S, with $\dim Z \ge 0.01^s(a_0 n - m)$ such that $d(Z, l^2_{\dim Z}) < K$. It is well-known that given two ellipsoids on an l-dimensional space, there exists a subspace of dimension $l/2$ on which the ellipsoids are proportional. Comparing a distance ellipsoid on Z with the ellipsoid defined by $|\cdot|_2$, and passing to a $(\dim Z/2)$-dimensional subspace of Z if neccessary, we get a subspace $Z_1 \subset Z$ on which (3.5) holds. Let m be the largest integer such that $m \le 0.01^s(a_0 n - m)/2$ and set $\alpha_0 = 0.01^s a_0/2(1 + 0.01^s)$. Since $\dim Z_1 \ge m$, $X_1 = Z_1^*$ is the required quotient of X. If $a_1 < 1$, the proof can be completed by applying the latter step alone. □

The following corollary is very close to Theorem 3.1, and it is its immediate consequence.

Corollary 3.5 *There is a constant $0 < \alpha_0 < 1/2$ such that the following holds, for every $0 < \alpha < \alpha_0$. Let X be an n-dimensional Banach space, and let $|\cdot|_2$ be a Euclidean norm on X satisfying (3.1) and (3.2). There is an integer m with $2\alpha_0 n \le m < n$, and an m-dimensional quotient X_1 of X such that whenever $d(X_1, l_m^2) > 10$ then a random pair (F_1, F_2) of αn-dimensional quotients of X_1 satisfies*

$$d(F_1, F_2) > d(X_1, l_m^2)^2 \tag{3.18}$$

and a random αn-dimensional quotient F of X_1 satisfies

$$\mathrm{bc}(F) > d(X_1, l_m^2). \tag{3.19}$$

If $|\cdot|_2$ is the norm determined by the ellipsoid of minimal volume containing the unit ball of X, then a version of the corollary, with additional universal constants smaller than 1 on the right hand sides of (3.18) and (3.19), is satisfied for an arbitrary $\alpha < 1/2$ and any m with $\alpha < m < 1/2$ (*cf.* Theorem 5.3 in [M-T]).

Another consequence of Theorem 3.1 is the following infinite-dimensional result, rather weaker than the theorem itself. It is more natural to state it for subspaces of a given space rather than for quotients.

Corollary 3.6 *There is a constant* $0 < \alpha_0 < 1$ *such that the following holds, for every* $0 < \alpha < \alpha_0$. *Let* Z *be a Banach space. Let* $f : N \to \mathbb{R}_+$ *be a non-decreasing function,* $\lim f(n) = \infty$. *Exactly one of the following two conditions holds.*

(i) *There exist a sequence* $\{Z_n\}$ *of subspaces of* Z, *with* $k_n = \dim Z_n \to \infty$, *as* $n \to \infty$, *and Euclidean norms* $|\cdot|_n$ *on the* Z_n*'s so that the following holds. There is a family* $\mathcal{F}'$ *of random pairs* (E_1, E_2) *of* αk_n*-dimensional subspaces of* Z_n, *such that*

$$\inf_{(E_1,E_2)\in\mathcal{F}'} \mathrm{d}(E_1, E_2)/f(k_n) \to \infty \tag{3.20}$$

and there is a family $\mathcal{F}$ *of random* αk_n*-dimensional subspaces* E *of* Z_n *such that*

$$\inf_{E\in\mathcal{F}} \mathrm{bc}(E)/f(k_n) \to \infty. \tag{3.21}$$

(ii) *For every* k*-dimensional subspace* $\{Z'\}$ *of* Z, *and every Euclidean norm* $|\cdot|$ *on* Z', *there is a family* $\mathcal{F}$ *of random* αk*-dimensional subspaces* E *of* Z' *such that*

$$\mathrm{d}(E, l^2_{\alpha k}) \le f(k) \quad \text{for } E \in \mathcal{F}. \tag{3.22}$$

To conclude, let us mention that some other results of [M-T] also admit suitable strengthenings, in which an $\varepsilon > 0$, which appeared originally, can be taken to be 0. For example, Theorem 5.6 in [M-T] can be modified as follows.

Theorem 3.7 *There is a constant* $0 < \alpha_0 < 1$ *such that the following holds, for every* $0 < \alpha < \alpha_0$. *Let* X *be an* n*-dimensional Banach space such that arbitrary quotients of* X, *say* F_1 *and* F_2, *with* $\dim F_1 = \dim F_2 = \alpha n$ *satisfy* $\mathrm{d}(F_1, F_2) \le K$, *for some* K. *Then*

$$\mathrm{d}(X, l^2_n) \le aK^{3/2},$$

where $a \geq 1$ is a universal constant.

References

[B-S] BOURGAIN, J. & SZAREK, S. J., The Banach–Mazur distance to the cube and the Dvoretzky–Rogers factorization. *Israel J. of Math.* 62 (1988), 169–180.

[B-T] BOURGAIN, J. & TZAFRIRI, L., Invertibility of "large" submatrices with applications to the geometry of Banach spaces and harmonic analysis. *Israel J. of Math.* 57 (1987), 137–224.

[H] HADWIGER, H., *Vorlesungen Über Inhalt: Oberfläche und Isoperimetrie.* Springer Verlag, 1957.

[F-L-M] FIGIEL, T., INDENSTRAUSS, J., & MILMAN, V. D., The dimension of almost spherical sections of convex bodies. *Acta Math.* 139 (1977), 56–94.

[M-T] MANKIEWICZ, P. & TOMCZAK-JAEGERMANN, N., Random subspaces and quotients of finite-dimensional Banach spaces. *Preprint*, Odense University, 1989.

[M.1] MILMAN, V. D., Inégalité de Brunn–Minkowski inverse et applications à le théorie local des espaces normés. *C. R. Acad. Sci. Paris.*, 302 Sér 1.(1986), 25–28.

[M.2] MILMAN, V. D., Almost Euclidean quotient spaces of subspaces of finite dimensional normed spaces. *Proc. Amer. Math. Soc.*, 94 (1985), 445–449.

[M.3] MILMAN, V. D., A new proof of the theorem of A. Dvoretzky on sections of convex bodies. *Func. Anal. Appl.*, 5 (1971), 28–37. (translated from Russian).

[M-P] MILMAN, V. D. & PISIER, G., Gaussian processes and mixed volumes. *Ann. of Prob.*, 15 (1987), 292–304.

[M-S] MILMAN, V. D. & SCHECHTMAN, G., *Asymptotic theory of finite dimensional normed spaces.* Springer Lecture Notes No. 1200 (1986).

[P-T] PAJOR, A. & TOMCZAK-JAEGERMANN, N., Volume ratio and other s-numbers of operators related to local properties of Banach spaces. *J. of Func. Anal.*, 87 (1989), 273–293.

[P] PISIER, G., *Volumes of Convex Bodies and Banach Spaces Geometry.* Cambridge Univ. Press, 1989

[Sz] SZAREK, S. J., The finite dimensional basis problem with an appendix on nets of Grassmann manifolds. *Acta Math.*, 151 (1983), 153–179.

[T] TOMCZAK-JAEGERMANN, N., *Banach–Mazur Distances and Finite Dimensional Operator Ideals.* Pitman Monographs and Surveys in Pure and Applied Mathematics, Longman Scientific & Technical, Harlow and John Wiley, New York, 1989.

A NOTE ON A LOW M^*-ESTIMATE

V. Milman*

Raymond and Beverly Sackler
Faculty of Exact Sciences
Tel Aviv University
Tel Aviv, Israel

1. Introduction and Notations

Consider a finite dimensional normed space $X = (\mathbf{R}^n, \|\cdot|, |\cdot|)$ equipped with a norm $\|\cdot\|$ and a euclidean norm $|\cdot|$. Then the dual space $X^* = (\mathbf{R}^n, \|\cdot\|^*)$ equipped with the dual norm $\|x\|^* = \sup\left\{|(x,y)| \mid \|y\| \le 1\right\}$ arises naturally (here (x,y) is the inner product which produces the euclidean norm $|x|^2 = (x,x)$). Let $S = \{x \in \mathbf{R}^n \ , \ |x| = 1\}$ be the euclidean sphere equipped with the rotation invariant probability measure μ. The following two parameters play an important role in the study of the linear structure of the space X:

$$M(=M_X) = \int_{x\in S} \|x\| d\mu(x) \quad \text{and} \quad M^*(=M_{X^*}) = \int_{x\in S} \|x\|^* d\mu(x) \ .$$

The discussion in this note will center around the following (known) proposition.

Main Proposition 1. *There is a function $f(\lambda)$, $0 < \lambda < 1$, such that for all integers n and $k < n$ and any $X = (\mathbf{R}^n, \|\cdot\|, |\cdot|)$ there is a subspace E^k of* codim $E^k = k$ *(i.e., $E^k \in G_{n,n-k}$ - the Grassmann manifold of $(n-k)$-dimensional subspaces of $\mathbf{R}^n$) and for every $x \in E^k$*

$$\frac{f\left(\frac{k}{n}\right)}{M^*}|x| \le \|x\| \ . \tag{1}$$

This proposition was presented for the first time at the Laurent Schwartz Colloquium (Summer 1983) where (see [M1]) it was shown that it plays a central role in the developing of a "proportional" theory (as it was called there). Since that time it has been used in different forms in most publications on Local Theory (in the study of finite dimensional spaces or operators). In the above form (1), different estimates on a function $f(\lambda)$ may be important and we will discuss this below. However, even the roughest and easiest estimate with an exponential dependence on λ (for $\lambda \to 0$)

* Supported in part by an G.I.F. Research Grant

already had unexpected and, probably, the most interesting applications, as can be seen in [M2]. In addition, a completely different form of inequality (1) with the spherical (Gaussian type) average M^* substituted by Rademacher type average (i.e. average over signs), was suggested by the author (unpublished) and used very successfully in [B.Tz].

Today, a few different approaches to the Proposition are known, leading to different functions $f(\lambda)$:

(i) the original approach (1983) used Urysohn's inequality (see [M1]) and gave $f(\lambda) = c^{1/\lambda}$ for a number $0 < c < 1$;

(ii) in the next step (1984), we returned to the use of isoperimetric inequality on the euclidean sphere S^n to show that (see [M3])

$$f(\lambda) \geq c\lambda \ .$$

(It was later shown ([MP], Appendix) that this is an easy consequence of pure entropy considerations.) Moreover, approach [M3] gives

$$f(\lambda) \geq \lambda + o(1-\lambda) \quad \text{when} \quad \lambda \to 1 \ .$$

(We note this because inequality $\lim_{\lambda\to 1} f(\lambda)/\lambda \geq 1$ was recently used.)

(iii) Asymptotically ($\lambda \to 0$) the best dependence

$$f(\lambda) \geq c\sqrt{\lambda}$$

was received by Pajor-Tomczak [PT] in 85-86 by adapting Gluskin's method for ℓ_1^n to a general normed space and joining it with Sudakov minoration estimate on covering numbers.

(iv) Y. Gordon [G] showed, using his minimax comparison principle, that, in fact, for $0 < \lambda < 1$,

$$f(\lambda) \geq \sqrt{\lambda}\left(1 + O\left(\frac{1}{\lambda n}\right)\right)$$

(i.e., a constant c above may be taken $\simeq 1$).

In this note, we give another proof of the Proposition with an estimate (iii) as in Pajor-Tomczak. For this purpose, we adapt to a general normed space a new approach of Makovoz [Ma] to prove Garnaev-Gluskin's estimates on the relative position of the Euclidean ball of ℓ_2^n in the unit ball of ℓ_1^n.

I would like to emphasize that I found the Makovoz arguments very interesting (they are, in fact, a "right" organization of Garnaev-Gluskin's arguments) and a way of adapting them to a general normed space could be of independent interest.

Some new results in this direction are presented in section 4.

General Remark. All proofs of Proposition 1 have a probabilistic nature. Therefore, in fact, we obtain a set $A \subset G_{n,n-k}$ of subspaces which satisfy (1) of "almost" full measure. We mean by this that $\mu(A) \geq 1 - \alpha(n, \lambda)$, where $\alpha(n, \lambda) \to 0$ $(n \to \infty)$ and $\lambda \geq$ fixed $\varepsilon > 0$; here μ is the Haar probability measure on the Grassmann manifold $G_{n,n-k}$. It is exceptable jargon to call an individual element of A a "random" subspace and to state Proposition 1 for a "random" k-codimensional subspace.

Usually, we automatically derive an estimate on $\alpha(n, \lambda)$ which, in all cases, has a form

$$\alpha(n, \lambda) \leq c(\lambda)^k = c(\lambda)^{\lambda \cdot n}$$

where $c(\lambda) < 1$ is a constant depending on $\lambda > 0$ only. However, in case i) (the worst estimate), we compensate for a bad estimate of $f(\lambda)$ by a very strong improvement in our knowledge of the behavior of the tail, i.e. about decay of $\alpha(n, \lambda)$; in this case, for *any* $0 < c < 1$, there is a $c_1 = c_1(c)$ such that a set of subspaces, which does not satisfy (1) with $f(\lambda) = c_1^{1/\lambda}$, has a measure

$$\alpha(n, \lambda) \leq c^{\lambda \cdot n} .$$

(In some applications we have to use a super-large set of subspaces satisfying some "good" properties, then such a strong estimate on $\alpha(n, \lambda)$ may be important; also most applications of the inequality (1) do not use $\lambda \to 0$ but, say, $\lambda = \frac{1}{4}$, then an exponential form of function $f(\lambda)$ becomes irrelevant.)

We denote by $K(= K_X) = \{x \in \mathbf{R}^n \mid \|x\| \leq 1\}$ - the unit ball of space X, and $D = \{x \in \mathbf{R}^n \mid |x| \leq 1\}$ - the euclidean unit ball. Also $D(\xi)$ is the unit euclidean ball of a subspace ξ. Starting from a centrally symmetric convex compact body $K \subset \mathbf{R}^n$, we consider a normed space X $(= X_K)$ such that $K_X = K$. Then the polar body $K^\circ = \{x \in \mathbf{R}^n \mid (x, y) \leq 1 \text{ for any } y \in K\}$ is the unit ball of the dual space X^*. We write $M^*(K)$ instead of M^* of the space X_K when different norms can be considered in $\mathbf{R}^n$. We will also write $a \lesssim b$ instead of stating that for some universal constant c we have $a \leq c \cdot b$.

Remark.

Most proofs of Proposition 1 do not use the central symmetry of norm. This means that a convex set $K = \{x \mid \|x\| \leq 1\}$ may not be centrally symmetric; however, to define K° properly, we assume that $0 \in \overset{\circ}{K}$.

Acknowledgement. I am grateful to Professor G.G. Lorentz who brought my attention to Makovoz's paper.
I would like to thank IHES where this paper was written.

2. Preliminary Lemmas

The following two lemmas are close enough to well-known results. So we will not develop detailed proofs. We denote below by P_ξ the orthogonal projection onto subspace ξ.

Lemma 2.1. *Let $|x| \le \|x\|$ (i.e., $K \subset D$). Then for every integer $1 \le k \le n$, for a "random" subspace $\xi \in G_{n,k}$ (of dimension k)*

$$P_\xi K \subseteq \begin{cases} C\sqrt{k/n}D(\xi) & \text{if } \sqrt{k/n} \ge M^* \\ CM^*D(\xi) & \text{if } \sqrt{k/n} < M^* \end{cases}$$

(here, as usual, C is a universal constant).

Proof: First consider a case $\sqrt{k/n} \simeq M^*$, i.e., $k \simeq (M^*)^2 n$. The standard estimate of the dimension of euclidean sections in Dvoretzky Theorem (see [M.Sch], 4.2) for the dual space X^* shows that, for a "random" subspace ξ, $P_\xi K \simeq M^* D(\xi)$. Then of course, the same is true for $k < n(M^*)^2$ because $P_\xi D(\eta) = D(\xi)$ for $\xi \subset \eta$ (ξ and η being subspaces of $\mathbf{R}^n$). In the case of $k > n(M^*)^2$, note that we want to prove now only embedding $P_\xi K \subset C\sqrt{k/n}D(\xi)$ (in the previous case we had, in fact, equivalence). For the dual sets, this means the following inequality for norms

$$\|x\|^* \le C\sqrt{\frac{k}{n}}|x| , \qquad \forall\, x \in \xi .$$

The standard "concentration phenomenon" approach to Dvoretzky's Theorem (as described in [M.Sch], Chapters 2-4) gives this upper bound for a random k-dimensional subspace of X^* (under condition $\|x\|^* \le |x|$).

Remarks.

1. I realize that the above sketch of the proof may suffice only for readers well versed with the methods of the first part of the book [M.Sch].
2. The interesting point of Lemma 2.1 is that it combined the modern form of Dvoretsky's Theorem (as was put forward in [M4] and later developed in [FLM]) and Johnson-Lindenstrauss' Lemma [J-L] which states:

Let $A \subset S^n$ be a set of at most 2^k points (i.e., $\#A \leq 2^k$) on the euclidean sphere S^n; then for a random orthogonal projection P_ξ onto a k-dimensional subspace ξ for any $x \in A$

$$|P_\xi x| \simeq C\sqrt{k/n} \ .$$

We derive from Lemma 2.1 only an upper bound: $|P_\xi x| \leq C\sqrt{k/n}$. Let $K = \operatorname{Conv} A \cup (-A)$ and $X = X_K$. Then M^* of X is at most $C\sqrt{k/n}$ (for a universal constant C) which implies the upper bound $P_\xi K \subset C\sqrt{k/n}D(\xi)$.

Lemma 2.2. *There are constants C and $c > 0$ such that for every integer k, $\log n < k \leq n$, there is a set $A_k \subset S^{n-1}$, $\#A_k = C^k$, producing the body $T_k = \operatorname{Conv} A_k \cup (-A_k)$ with the following property:*

$$P_\xi T_k \supset c\sqrt{k/n}D(\xi)$$

for the orthogonal projection P_ξ onto a random k-dimensional subspace $\xi \subset \mathbf{R}^n$.

Sketch. The Lemma may be verified by a probabilistic argument (althoug I would prefer to see here the construction of a set A_k). For that, note that for a set $T = \{x_i\} \subset S^{k-1}$, $\#T = 10^k$, of randomly and indipendently chosen points $\{x_i\}$ on the sphere and any numbers a_i, $\frac{1}{2} < a_i < 2$, with high probability

$$\operatorname{Conv}\{\pm a_i x_i\} \supset cD$$

for some universal constant $c > 0$. Combine this observation with Johnson-Lindenstrauss' lemma from Remark 2 above.

Remark 3. Normed spaces $X_k = X_{T_k}$ give a certain scaling which controls, as we will see, the behavior of a general normed space. Note that spaces ℓ_1^n and ℓ_∞^n belong to this scale: $\ell_1^n = X_{\log n}$ and $\ell_\infty^n = X_{cn}$ (the last fact follows from Kašin's well-known result [K] stating that a random orthogonal projection of a n-dimensional cube onto a $\frac{n}{2}$-dimensional subspace is uniformly isomorphic to a euclidean ball).

Problem. A general problem which arises at this point is to construct, for every $n \geq k > \log n$, a symmetric set A_k, $\#A_k = 2^k$, such that for a set of k-dimensional subspaces $\mathfrak{A} \subset G_{n,k}$ of measure $\mu(\mathfrak{A}) \geq \frac{1}{10^k}$ (so, *very small* but anyway not too small)

$$P_\xi(\operatorname{Conv} A_k) \supset \delta D(\xi) \qquad (\forall\, \xi \in \mathfrak{A})$$

for a *maximal possible* $\delta > 0$.

It is, of course, enough for applications to prove an existence of a set A_k which gives a large enough δ. But how large can this δ be? As stated in Lemma 2.2, a set $\mathfrak{A}$ of subspaces is of almost full measure if $\delta \sim \sqrt{k/n}$. Could this be improved for a set of small measure?

Remark 4. For a small $\delta \sim \sqrt{\frac{k}{n}}/\sqrt{\log \frac{2en}{k}}$, Lemma 2.2 can be improved:

Take integer ℓ such that $\left[\ell \log \frac{2en}{\ell}\right] = k > \log n$ and consider the set $\mathfrak{N} \subset S^{n-1}$ of all n-dimensional vectors with 0 and $\pm\frac{1}{\sqrt{\ell}}$ enters (so, 0's appear exactly $n - \ell$ times). Then $\#\mathfrak{N} = 2^\ell \binom{n}{\ell} \simeq 2^{\ell \log \frac{2en}{\ell}} \simeq 2^k$. Consider $T = \operatorname{Conv} \mathfrak{N}$. Then $\sqrt{\frac{\ell}{n}} D \subset T$ which means that

$$\frac{\sqrt{k/n}}{\sqrt{\log \frac{cn}{k}}} D \subset T .$$

So, we don't need to take any projections on subspaces at all.

3. Proof of Proposition 1

We will now prove Proposition 1 with estimate iii) using some ideas from Makovoz paper [Ma]. Let $\overline{y} = (y_1, \dots, y_k) \subset S^{n-1}$. Consider

$$F(x, \overline{y}) = \frac{1}{k} \sum_{i=1}^{k} |(x, y_i)| .$$

Lemma 3.1 [Ma]. *Denote by P the probability porduct-measure on the product of k spheres $\prod_1^k S^{n-1}$ (and every sphere equipped with the rotation invariant probability measure). For every $x \in S^{n-1}$*

$$P\left(\overline{y} \in \prod_1^k S^{n-1} \,\middle|\, \frac{1}{100\sqrt{n}} \le F(x, \overline{y}) \le \frac{3}{\sqrt{n}}\right) > 1 - e^{-k/2} .$$

(Therefore, for any set $\mathfrak{N} \subset \mathbf{R}^n$, $\#\mathfrak{N} < e^{k/2}$, there is $\overline{y} = (y_1, \dots, y_k)$ such that $\frac{|x|}{100\sqrt{n}} \le F(x, \overline{y}) \le \frac{3|x|}{\sqrt{n}}$ for any $x \in \mathfrak{N}$.)

Lemma 3 may be proved by direct computation (see [Ma]). In the framework of Local Theory, it is an immediate consequence of Bernstein's inequality (see, e.g. [H] or [BLM] for exactly such a use): consider a function $\varphi(y) = |(x, y)|$. Then the ψ_2-Orlicz norm $\|\varphi(y)\|_{L_{\psi_2}} \le c_1 \|\varphi(y)\|_{L_1} = \frac{c_2}{\sqrt{n}}$ (here $\psi_2(t) = e^{t^2} - 1$). Therefore, by the standard sampling method for ψ_2-distribution (see [BLM], Prop. 1, (iii))

$$\operatorname{Prob}\left\{\overline{y} \in \prod_1^k S^{n-1} \,\middle|\, \left|\frac{1}{k}\sum_1^k |\varphi(y_i)| - \frac{c}{\sqrt{n}}\right| < \varepsilon\right\} \ge 1 - e^{-\varepsilon^2 k / c\|\varphi\|^2_{L_{\psi_2}}} \ge$$

$$\ge 1 - e^{-c\varepsilon^2 kn} .$$

Take $\varepsilon \sim \varepsilon_0/\sqrt{n}$ and we derive Lemma 3.1 with some constants C_1 and C_2 instead of 100 and 3 above.

Remarks.

1. If we are interested only in an upper estimate like $F(x,\overline{y}) \leq \frac{T}{\sqrt{n}}|x|$ (for a large T), we can take $\varepsilon = \frac{T}{\sqrt{n}}$ to be large and improve the estimate of the probability of $\overline{y}$'s.
2. Taking $T = c\sqrt{n/k}$, we have

$$\text{Prob}\left\{\overline{y} \in \prod_1^k S^{n-1} \;\middle|\; F(x,\overline{y}) \lesssim \frac{|x|}{\sqrt{k}}\right\} \geq 1 - e^{-cn}$$

and, therefore, we may satisfy the inequality

$$F(x,\overline{y}) \leq c\frac{|x|}{\sqrt{k}}$$

for any x from some ε-net on S^{n-1} (for fixed $\varepsilon > 0$) and, consequently, for any $x \in S^{n-1}$.

We are now going to prove (1) with $f(\lambda)$ given by iii). The inequality is homothetically invariant and so we may choose homothetical normalization of the euclidean norm in such a way that $M^* = \sqrt{k/n}$.

We want to prove then that for some universal constant C and a random k-codimensional subspace ξ

$$|x| \leq C\|x\|\,, \qquad \forall\, x \in \xi\,.$$

We use a notion of the covering number $N(K,D) = \min\left\{N \mid \exists\{x_i\}_1^N \text{ such that } K \subset \bigcup_1^N (x_i + D)\right\}$. It is well-known (Sudakov's inequality - see, e.g. [LT]) that (K below is the unit ball K_X of X)

$$N(K,D) \leq e^{n(M^*)^2} = e^k$$

for our normalization. Let $\mathfrak{N}_0 = \{x_i\}$ be a D-net of K such that $\#\mathfrak{N} \leq e^k$ and let

$$\mathfrak{N} = \mathfrak{N}_0 \cup A_{k_1}$$

where A_{k_1} is a set on S^{n-1} from Lemma 2.2 (and $k_1 = c_1 k$ will be chosen next). Therefore, $\#\mathfrak{N} \leq C^k$ for a universal constant C.

Use Lemma 3.1 for $k_1 = c_1 k$ ($c_1 = 2\log C$ - universal constant). Then there is $\overline{y} = (y_i \in S^{n-1})_{i=1}^{k_1}$ such that for every $x_j \in \mathfrak{N}$

$$F(x_j;\overline{y}) \simeq \frac{|x_j|}{\sqrt{n}}\,. \tag{3.1}$$

Moreover, we have a large measure of such $\{y_i\} \in \prod S^{n-1}$ which means a large measure of subspaces $\xi = \operatorname{span}\{y_i\}_1^{k_1} \in G_{n,k_1}$.

By Lemma 2.1 (which we apply to a body $K \cap D$ noting that $M^*(K \cap D) \leq M^*(K) = \sqrt{k/n}$), there is a set of large measure of k_1-dimensional subspaces such that the orthogonal projection on any ξ from this set $P_\xi(K \cap D) \subseteq c\sqrt{\frac{k}{n}} \cdot D(\xi)$. We now combine all 3 conditions (from Lemmas 2.1, 2.2 and 3.1) and find a subspace $\xi \in G_{n,k_1}$ from Lemma 2.1 for the set $K \cap D$, and, at the same time, from Lemma 2.2 for k_1, and $\xi = \operatorname{span}(y_i)_{i=1}^{k_1}$ satisfying (3.1). We put (briefly) $F(x) \equiv F(x, \bar{y})$.

Consider $E = \xi^\perp (= \{y_i\}^\perp)$. We will show that E is a subspace which we are looking for. Clearly $\operatorname{codim} E = k_1 \sim k$ and $F(x) = 0$ for any $x \in E$. We have to show that for every $x \in E$, $\|x\| \leq 1$,

$$|x| \leq \text{Const}.(?) .$$

Write $x = x'' + x'$ where $x'' \in \mathfrak{N}_0$ and $|x'| \leq 1$. We will estimate $|x''|$ from above by estimating $F(x'') \simeq \frac{|x''|}{\sqrt{n}}$, Because our $x \in K \cap (x'' + D)$ we have $x' = x - x'' \in 2(K \cap D)$ (note that $K \cap (x'' + D) \subset x'' + (2K) \cap D \subset x'' + 2(K \cap D)$). By Lemma 2.1, for some constant c_2,

$$P_\xi\big(2(K \cap D)\big) \subset c_2\sqrt{\frac{k_1}{n}} D(\xi) .$$

Then there exists $x_1' \in 2[K \cap D]$, $P_\xi x' = P_\xi x_0'$, and

$$c_3 x_0' \in \operatorname{Conv}\{z_i\}_{z_i \in A_{k_1}} = T_{k_1}$$

(again, for some universal constant $c_3 > 0$). It shows that

$$F(x_0') \leq \frac{1}{c_3} \max F(z_j) \simeq \frac{1}{\sqrt{n}}$$

(recall that $z_j \in \mathfrak{N}$ and therefore $F(z_j) \simeq \frac{|z_j|}{\sqrt{n}}$). It remains to note that $F(x'') = F(x_0') \leq c/\sqrt{n}$ which implies $|x''| \leq$ Const. and $|x| \leq |x''| + |x'|$. □

4. Some New Connected Results and Problems

In fact, a stronger statement was proved above. First, define entropy numbers (the inverse function to covering numbers): for every integer $k \geq 1$,

$$e_k \equiv e_k(K, D) \equiv \inf\left\{\varepsilon \mid N(K, \varepsilon D) \leq 2^{k-1}\right\} .$$

Assume that $e_k = 1$. Then it was proved that, for a random subspace E^k of $\operatorname{codim} E^k \simeq k$ and any $x \in E^k$,

$$\frac{\sqrt{k/n}}{M^*(K \cap D)}|x| \lesssim \|x\| \ .$$

Note, that the expression $M^*(K \cap e_k D)$ is independent of homothetic change of the euclidean norm. Therefore we obtain

Proposition 2. *For any $X = (\mathbf{R}^n, \|\cdot\|, |\cdot|)$ and any integer $k < n$ for a random subspace E^k of* $\operatorname{codim} E^k = k$

$$c\frac{\sqrt{k/n}}{M^*(K \cap e_k D)}|x| \leq \|x\| \qquad (\forall\, x \in E^k)$$

where c is a universal constant.

The above inequality is stronger than Proposition 1 with $f(\lambda)$ given by iii), because $M^*(K \cap e_k D) \leq M^*(K)$.

For the following result we introduce the so called "Sudakov" numbers.

Let K be, as above, the unit ball of a space $X = (\mathbf{R}^n, \|\cdot\|)$ and let e_i be the entropy numbers of K. Consider

$$S(K) = \sup_j \sqrt{j} e_j \ .$$

By Sudakov's minoration theorem $S(K) \lesssim \sqrt{n} M^*$. Therefore, to substitute inequality (1) in the framework of Proposition 1 by the following one

$$c\frac{\sqrt{k}}{S(K)}|x| \leq \|x\|$$

would be an improvement over all known estimates. Unfortunately, we cannot prove this fact but will prove one close to it.

Theorem 3. *For any $X = (\mathbf{R}^n, \|\cdot\|, |\cdot|)$ and any integer $k < n$ for a random subspace E^k of* $\operatorname{codim} E^k = k$

$$c\frac{\sqrt{k}}{\log\left(\frac{n}{k}\right) S(K)}|x| \leq \|x\| \qquad (\forall\, x \in E^k)$$

(c is a universal constant).

Proof: Let $\mathfrak{N}_j$ be the smallest e_{j+1}-net of K. Then $\#\mathfrak{N}_j = 2^j$. Let k_i be such that $e_{k_i} = \frac{1}{2^i}$. Then, by Remark 1 to Lemma 3.1, we have a large measure of $\overline{y}$'s (i.e., of $\operatorname{span} \overline{y} = \xi \in G_{n,k}$) such that

$$\sqrt{n} F(x, \overline{y}) \lesssim \sqrt{\frac{k_i}{k}}|x| \tag{4.1}$$

for any $x \in \mathfrak{N}_{k_i} - \mathfrak{N}_{k_{i-1}}$. We achieve this first for a fixed i but choose next the same $\overline{y}_0$ for all i, $0 \le i \le t$. (We choose later $t \sim \log n/k$.) As in the proof of Proposition 1 a subspace which we are looking for is $E^k = [\overline{y}_0]^\perp$. To prove this, take any $x \in E^k$, $\|x\| \le 1$, and decompose it (for every i)

$$x = x_i'' + x_i'$$

where $x_i'' \in \mathfrak{N}_{k_i}$ and $|x_i'| \le 1/2^i$. Then $x_i'' - x_{i+1}'' = x_{i+1}' - x_i'$ and $|x_{i+1}' - x_i'| \le \frac{1}{2^{i+1}} + \frac{1}{2^i} = \frac{3}{2^{i+1}}$. Because $|x| \le |x_1''| + |x_1'| \le 1 + |x_1''|$ we have to estimate $|x_1''| \sim \sqrt{n}F(x_1'', \overline{y}_o)$. Again, we will abbreviate $F(z, \overline{y}_0)$ to $F(z)$. Note that F is a seminorm and $F(x) = 0$, so

$$\begin{aligned} |x_1''| &\simeq \sqrt{n}F(x_1'') = \sqrt{n}F(x_1'' - x) = \\ &= \sqrt{n}F\left(\sum_{i=1}^{t-1}(x_i'' - x_{i+1}'') + x_t'' - x\right) \le \\ &\le \sqrt{n}\sum_1^{t-1} F(x_i'' - x_{i+1}'') + \sqrt{n}F(x_t'' - x) \le \\ &\le (\mathrm{by}(4.1))\sum_1^{t-1}\sqrt{\frac{k_{i+1}}{k}}|x_i'' - x_{i+1}''| + \sqrt{n}F(x_t'' - x) \le \end{aligned}$$

(use Remark 2 of Lemma 3.1 to estimate the last term)

$$\le 3\sum_1^{t-1}\frac{\sqrt{k_{i+1}}e_{i+1}}{\sqrt{k}} + \sqrt{\frac{n}{k}}|x_t'| \, .$$

Choose now $t \sim \log\frac{n}{k}$. Then $\sqrt{\frac{n}{k}}|x_t'| \le 1$ and we can continue the above inequality

$$\lesssim \frac{S(K)}{\sqrt{k}}t + 1 \simeq \frac{S(K)\log\frac{n}{k}}{\sqrt{k}} \, . \qquad \square$$

Remark. Note an easy improvement of the inequality in Theorem 3: we can substitute $S(K)$ by

$$S_k(K) = \sup_{j \ge k}\sqrt{j}e_j \, .$$

References

[BLM] J. Bourgain, J. Lindenstrauss, V. Milman, Minkowski sums and symmetrizations, Springer-Verlag, Lecture Notes in Mathematics, 1317 (1988), 44-66.

[BTz] J. Bourgain, L. Tzafriri, Invertibility of large submatrices with applications to geometry of Banach spaces and harmonic analysis, Isr. J. Math. 57 (1987), 137-224.

[FLM] T. Figiel, J. Lindenstrauss, V. Milman, The dimension of almost spherical sections of convex bodies, Acta Math. 129 91977), 53-94.

[G] Y. Gordon, On Milman's inequality and random subspaces which escape through a mesh in $\mathbf{R}^n$, Springer-Verlag, Lecture Notes in Mathematics 1317 (1988), 84-106.

[H] W. Hoeffding, Probability inequalities for sums of bounded random variables, J. Amer. Stat. Assoc. 58 (1963), 13-30.

[JL] W.B. Johnson, L. Lindenstrauss, Extensions of Lipschitz mappings into a Hilbert space, Proc. Conf. in Honour of S.Katutani.

[K] B.S. Kašin, Sections of some finite dimensional sets and classes of smooth functions, Izv. ANSSSR, ser. mat. 41 (1977), 334-351 (Russian).

[LT] M. Ledoux, M. Talagrand, Isoperimetry and processes in probability in Banach spaces, Springer-Verlag (1990).

[Ma] Y. Makovoz, A simple proof of an inequality in the thoery of n-widths, Constructive Theory of Functions '87, Sofia (1988).

[M1] V. Milman, Geometrical inequalities and mixed volume in Local Theory of Banach Spaces, Asterisque 131 (1985), 373-400.

[M2] V. Milman, Volume approach and iteration procedures, in Local Theory of Normed Spaces, Springer-Verlag, Lecture Notes in Mathematics 1166 (1985), 99-105.

[M3] V. Milman, Random subspaces of proportional dimension of finite dimensional normed spaces; approach through the isoperimetric inequality, Springer-Verlag, Lecture Notes in Mathematics 1166 (1985), 106-115.

[M4] V. Milman, A new proof of the theorem of Dvoretzky on sections of convex bodies, Funkcional. Anal i Proložen 5 (1971), 28-37 (Russian).

[MP] V. Milman, G. Pisier, Banach spaces with a weak cotype 2 property, Israel J. Math. 54, No. 2 (1986), 139-158.

[MSch] V. Milman, G. Schechtman, Asymptotic theory of finite dimensional normed spaces, Springer-Verlag, Lecture Notes in Mathematics 1200 (1986), 156pp.

[PT] A. Pajor, N. Tomczak-Jaegermann, Subspaces of small codimension of finite dimensional Banach spaces, Proc. Amer. Math. Soc. 97 (1986), 637-642.

The $p^{\frac{1}{p}}$ in Pisier's Factorization Theorem

S.J. Montgomery-Smith

Abstract

We show that the constants in Pisier's factorization theorem for $(p,1)$-summing operators from $C(\Omega)$ cannot be improved.

* * * * * *

A theorem of Pisier (see [3]) states the following.

Theorem 1. *Let $T : C(\Omega) \to X$ be a bounded linear operator, where Ω is a compact Hausdorff topological space and X is a Banach space. Then the following are equivalent.*

i) T is $(p,1)$-summing.

ii) There is a constant $C < \infty$ and a Radon probability measure μ on Ω such that for all $f \in C(\Omega)$ we have

$$\|Tf\| \le C \, \|f\|_{L_1(\mu)}^{\frac{1}{p}} \, \|f\|_\infty^{1-\frac{1}{p}} . \tag{1}$$

iii) There is a constant $C < \infty$ and a Radon probability measure μ on Ω such that for all $f \in C(\Omega)$ we have

$$\|Tf\| \le C \, \|f\|_{L_{p,1}(\mu)} . \tag{2}$$

If we examine the proof of this theorem carefully, we can deduce the following. Let $\pi_{p,1}(T)$ denote the $(p,1)$-summing norm of T.

Theorem 2. *Let $T : C(\Omega) \to X$ be a bounded linear operator as in Theorem 1 that is $(p,1)$-summing. Then we can say the following.*

i) The set of C and μ that satisfy (1) coincide with the set of C and μ that satisfy (2).

ii) $\pi_{p,1}(T) \le C$ for all C satisfying (1) or (2).

iii) We can choose $C = p^{\frac{1}{p}} \, \pi_{p,1}(T)$ in (1) and (2).

Part (i) of this theorem is straight forward to demonstrate (we merely note that one needs the inequality: $\|f\|_{p,1} \le \|f\|_1^{\frac{1}{p}} \, \|f\|_\infty^{1-\frac{1}{p}}$). Part (ii) is also easy to verify. However, to show part (iii), the main part of Pisier's theorem, is much harder. Furthermore, the methods do not indicate whether the $p^{\frac{1}{p}}$ factor can be made smaller, or even completely removed (that is, replaced by 1).

The purpose of this paper is to show the rather surprizing fact that the $p^{\frac{1}{p}}$ factor cannot be reduced at all.

Theorem 3. *Given $\epsilon > 0$, there is an operator $T : C(\Omega) \to X$ such that for any Radon probability measure μ on Ω, if C is the least number satisfying (1) or (2), then $C \geq p^{\frac{1}{p}}\,\pi_{p,1}(T)\,(1-\epsilon)$.*

Construction. Let $1 \leq S \leq N$ be integers, and let Ω be the collection of S-subsets of $\{1, 2, \dots, N\}$. For each $1 \leq n \leq N$, let $\Omega_n = \{\omega \in \Omega : n \in \omega\}$. We note the following facts for later on:

$$|\Omega_n| = \binom{N-1}{S-1}, \tag{3}$$

$$\frac{1}{N}\sum_{n=1}^{N} \chi_{\Omega_n} = \frac{S}{N}\,\chi_{\Omega}. \tag{4}$$

We give Ω the discrete topology, and define a norm $\|\cdot\|_*$ on $C(\Omega)$ by

$$\|f\|_* = \sup_{1\leq n\leq N} \sum_{\omega\in\Omega_n} |f(\omega)|\,.$$

We let T be the canonical embedding

$$T : (C(\Omega), \|\cdot\|_\infty) \to (C(\Omega), \|\cdot\|_*).$$

Lemma 4. *For $1 \leq p < \infty$, the $(p,1)$-summing norm of T may be estimated by*

$$\pi_{p,1}(T) \leq \frac{N^{S-1+\frac{1}{p}}}{(S-1)!\,(Sp-p+1)^{\frac{1}{p}}}.$$

In order to show Lemma 4, we will need two more lemmas.

Lemma 5. *If $1 \leq p < \infty$, and $T : C(\Omega) \to X$ is a bounded linear operator, where X is a Banach space, then the $(p,1)$-summing norm of T may be calculated by the formula*

$$\pi_{p,1}(T) = \sup\left\{\left(\sum_{s=1}^{S} \|Tf_s\|^p\right)^{\frac{1}{p}}\right\},$$

where the supremum is over all sequences $f_1, f_2, \dots, f_S$ of disjoint elements of the unit ball of $C(\Omega)$.

Proof. See [2], Lemma 6 or [1], Proposition 14.4. □

Lemma 6. *If $1 \leq n \leq N$, then*

$$|\Omega_1 \cup \Omega_2 \cup \ldots \cup \Omega_n| \leq \frac{1}{(S-1)!} \sum_{m=1}^{n} (N-m)^{S-1}.$$

Proof. A simple counting argument shows that

$$|\Omega_1 \cup \Omega_2 \cup \ldots \cup \Omega_m \setminus \Omega_1 \cup \Omega_2 \cup \ldots \cup \Omega_{m-1}| = \binom{N-m}{S-1},$$

and this is bounded by $(N-m)^{S-1}/(S-1)!$. □

Proof of Lemma 4. By Lemma 5, it is easy to see that

$$\pi_{p,1}(T) = \sup \left\{ \left(\sum_{n=1}^{N} \|\chi_{B_n}\|_*^p \right)^{\frac{1}{p}} \right\} = \sup \left\{ \left(\sum_{n=1}^{N} |B_n|^p \right)^{\frac{1}{p}} \right\},$$

where the supremum is over disjoint sets $B_1, B_2, \ldots, B_N \subseteq \Omega$ such that $B_n \subseteq \Omega_n$ for each $1 \leq n \leq N$. Since $\Omega_1, \Omega_2, \ldots, \Omega_N$ interact with one another in a completely symmetric fashion, we may assume, without loss of generality, that $|B_1| \geq |B_2| \geq \ldots \geq |B_N|$.

Now

$$\sum_{n=1}^{N} |B_n|^p = \sum_{n=1}^{N} \left(\sum_{m=1}^{n} |B_m| \right) \left(|B_n|^{p-1} - |B_{n+1}|^{p-1} \right).$$

(We take $B_{N+1} = \emptyset$.) Since $|B_n|^{p-1} - |B_{n+1}|^{p-1} \geq 0$, and $B_1 \cup B_2 \cup \ldots \cup B_n \subseteq \Omega_1 \cup \Omega_2 \cup \ldots \cup \Omega_n$, we have, by Lemma 6, that

$$\begin{aligned} \sum_{n=1}^{N} |B_n|^p &\leq \frac{1}{(S-1)!} \sum_{n=1}^{N} \left(\sum_{m=1}^{n} (N-m)^{S-1} \right) \left(|B_n|^{p-1} - |B_{n+1}|^{p-1} \right) \\ &= \frac{1}{(S-1)!} \sum_{n=1}^{N} (N-n)^{S-1} |B_n|^{p-1}. \end{aligned}$$

Now, applying Hölder's inequality and dividing, we deduce

$$\left(\sum_{n=1}^{N} |B_n|^p \right)^{\frac{1}{p}} \leq \frac{1}{(S-1)!} \left(\sum_{n=1}^{N} (N-n)^{Sp-p} \right)^{\frac{1}{p}}.$$

Finally, we estimate the last quantity by an integral, and derive

$$\left(\sum_{n=1}^{N} |B_n|^p\right)^{\frac{1}{p}} \leq \frac{1}{(S-1)!}\left(\int_0^N x^{Sp-p}\,dx\right)^{\frac{1}{p}}$$
$$\leq \frac{N^{S-1+\frac{1}{p}}}{(S-1)!\,(Sp-p+1)^{\frac{1}{p}}},$$

as desired. □

Proof of Theorem 3. By the hypothesis on C, there is a probability measure μ on Ω such that inequality (1) holds. In particular, if we subsitute $f = \chi_{\Omega_n}$, we deduce that

$$|\Omega_n|^p = \|\chi_{\Omega_n}\|_*^p \leq C^p \int \chi_{\Omega_n}\,d\mu.$$

Hence

$$\frac{1}{N}\sum_{n=1}^{N} |\Omega_n|^p \leq C^p \int \frac{1}{N}\sum_{n=1}^{N} \chi_{\Omega_n}\,d\mu,$$

and so by equalities (3) and (4) we have

$$C \geq \frac{N^{\frac{1}{p}}\binom{N-1}{S-1}}{S^{\frac{1}{p}}}.$$

Thus, by Lemma 4, we deduce

$$C \geq \frac{N^{\frac{1}{p}}\binom{N-1}{S-1}(S-1)!\,(Sp-p+1)^{\frac{1}{p}}}{S^{\frac{1}{p}}N^{S-1+\frac{1}{p}}}\,\pi_{p,1}(T).$$

Choosing N much larger than S, we find that

$$C \geq \left(\frac{Sp-p+1}{S}\right)^{\frac{1}{p}}\,\pi_{p,1}(T)\,(1-\tfrac{1}{2}\epsilon).$$

Finally, choosing S large, we have the desired result, that is, $C \geq p^{\frac{1}{p}}\,\pi_{p,1}(T)\,(1-\epsilon)$. □

References

1. Jameson G.J.O.: Summing and Nuclear Norms in Banach Space Theory. London Math. Soc., Student Texts 8, 1987.
2. Maurey B.: Type et cotype dans les espaces munis de structures locales inconditionelles, Exposés 24–25. In: Seminaire Maurey-Schwartz 1973–74 (Ecole Polytechnique).
3. Pisier G.: Factorization of operators through $L_{p\infty}$ or L_{p1} and non-commutative generalizations. Math. Ann. **276** 105–136 (1986).

S.J. Montgomery-Smith,
Department of Mathematics,
University of Missouri at Columbia,
Columbia, Missouri 65211,
U.S.A.

Almost differentiability of convex functions in Banach spaces and determination of measures by their values on balls*

D. Preiss

Abstract. *We prove a new differentiability property of continuous convex functions on separable Banach spaces. As an application we give a new proof of the result of [2] that Borel measures on separable Banach spaces are determined by their values on balls.*

Let f be a convex continuous function defined on an open convex subset G of a real Banach space E. We recall that the subdifferential of f at a point $x \in E$ is the set $\partial f(x)$ of all (necessarily continuous) linear functionals x^* such that $f(x+h) - f(x) \geq \langle h, x^* \rangle$ for every $h \in G$. Using the fact that f is convex and continuous, we easily see that the set $T(f,x) := \{h \in E;\ f'(x,h) = -f'(x,-h)\}$ is a closed linear subspace of E. (Recall that $f'(x,h) = \lim_{t \searrow 0}(f(x+th) - f(x))/t$.) Moreover, the Hahn-Banach theorem immediately implies that $T(f,x) = \bigcap\{\mathrm{Ker}(x^* - y^*);\ x^*, y^* \in \partial f(x)\}$.

Clearly, the function f at the point x is Gateaux differentiable in the direction of the subspace $T(f,x)$ and is nondifferentiable in any direction not belonging to $T(f,x)$. Hence the codimension of $T(f,x)$ (which will be denoted as $\dim(\partial f(x))$ (and called the *dimension of the subdifferential of f at x*) measures how much f is Gateaux nondifferentiable at x. Clearly, Gateaux differentiability of f at x is equivalent to $\dim(\partial f(x)) = 0$.

For separable spaces, the classical result of Mazur according to which f is Gateaux differentiable at the points of a residual subset of G also implies

*Classification: Primary 46G05. Secondary 28C15, 46G12.

that for every $x \in G$

$$\partial f(x) = \bigcap_{r>0} w^*\text{-closed convex hull } \{f'(z);\ \|z - x\| < r,\ f'(z) \text{ exists}\}.$$

The proof of the Mazur theorem and other relevant information can be found in [1]. A more general form of the above corollary of the Mazur theorem is given in Proposition 1.1 below.

Motivated by an application to the question of determination of measures by their values on balls, we intend to replace in this statement "w^*-closed convex hull" by "w^*-closure". This needs, of course, a weakening of the requirement of Gateaux differentiability. (Just consider the absolute value function on $\mathbf{R}$.) However, it turns out that for our application, Gateaux differentiability in the direction of some finite codimensional linear subspace suffices. It is our aim to prove that, with these two replacements, the expression for the subdifferential remains true (see 1.4). This result is used to give a new proof of the result of D. Preiss and J. Tišer [2] that Borel measures on separable Banach spaces are uniquely determined by their values on balls.

Our main result is based on a new estimate of smallness of the sets $\{x \in G;\ \dim(\partial f(x)) > k\}$ for convex continuous functions f. (Theorem 1.3.) Let us remark that L. Zajíček [3] proved that a subset of G is of such a form if and only if it can be covered by countably many $k+1$ codimensional δ-convex (or $(c-c)$-convex) surfaces. (Since we will not use this fact, we refer the reader to [3] for the definition.) However, our result does not seem to follow directly from this interesting characterization of the smallness of these sets.

Except for the notation introduced above, we shall also denote by $B(x,r)$ and $B^0(x,r)$ the closed and open ball centered at x with radius r, respectively.

1. Almost differentiability of convex functions

1.1. Proposition. *Suppose that f is a convex continuous function defined on an open convex subset G of a real Banach space E and that, for each sufficiently small $r > 0$, D_r is a subset of $\bigcup\{\partial f(z);\ z \in B^0(x,r)\}$ such that the set $\{z \in B^0(x,r);\ D_r \cap \partial f(z) \neq \emptyset\}$ is dense in $B^0(x,r)$. Then*

$$\partial f(x) = \bigcap_{r>0} w^*\text{-closed convex hull } D_r.$$

Proof. Since the mapping $x \mapsto \partial f(x)$ has $w^\star$-compact convex values and is $w^\star$-upper semicontinuous,

$$\begin{aligned}\partial f(x) \supset & \bigcap_{r>0} w^\star\text{-closed convex hull} \bigcup\{\partial f(z);\ z \in B^0(x,r)\} \\ \supset & \bigcap_{r>0} w^\star\text{-closed convex hull } D_r.\end{aligned}$$

To prove the opposite inclusion, suppose, to the contrary, that $x^\star \in \partial f(x)\backslash M$ where $M = \bigcap_{r>0} w^\star$-closed convex hull D_r. Then the Hahn-Banach theorem provides us with $h \in E$ such that $||h|| = 1$ and $\langle h, x^\star\rangle > \sup\{\langle h, z^\star\rangle;\ z^\star \in M\}$. Let c, d be real numbers such that $\langle h, x^\star\rangle > c > d > \sup\{\langle h, z^\star\rangle;\ z^\star \in M\}$. For every $r > 0$ such that $B^0(x,r) \subset G$ and the set $\{z \in B^0(x,r);\ D_r \cap \partial f(z) \neq \emptyset\}$ is dense in $B^0(x,r)$ we find $\varepsilon \in (0, r/4)$ such that $|f(u) - f(x)| < (c-d)r/8$ and $|f(u+rh/4) - f(x+rh/4)| < (c-d)r/8$ whenever $u \in B^0(x,\varepsilon)$. Then for each $u \in B^0(x,\varepsilon)$ the function $g\colon t \in (-3r/4, 3r/4) \mapsto f(u+th)$ fulfils $g(r/4) - g(0) = f(u+rh/4) - f(u) \geq f(x+rh/4) - f(x) - (c-d)r/4 \geq (\langle h, x^\star\rangle - (c-d))r/4 > dr/4$. Since g is convex, it follows that $g(t) - g(s) > d(t-s)$ whenever $r/4 < s < t < 3r/4$. Consequently, $\langle h, z^\star\rangle \geq d$ whenever $z^\star \in \bigcup\{\partial f(z);\ ||z - (x + rh/2)|| < \varepsilon\}$. Since our assumption implies that $D_r \cap \bigcup\{\partial f(z);\ ||z - (x + rh/2)|| < \varepsilon\} \neq \emptyset$, there are $z_r^\star \in D_r$ such that $\langle h, z_r^\star\rangle \geq d$. Finally, using the fact that $\limsup_{r\searrow 0} ||z_r^\star|| < \infty$, we conclude that the set of all $w^\star$-accumulation points of $\{z_r^\star\}$ for $r \searrow 0$ is a nonempty subset of M. But, since each $z^\star \in M$ fulfils $\langle h, z^\star\rangle < d$, this is a contradiction.

1.2. Proposition. *Let f be a convex continuous function defined on an open convex subset G of a separable real Banach space E and let $x \in G$. Then the following two statements hold.*

(i) *The set of all $x^\star \in \partial f(x)$ such that $f'(x,h) > \langle h, x^\star\rangle$ for every $h \notin T(f,x)$ is norm dense in $\partial f(x)$.*

(ii) *If $\dim(\partial f(x)) < \infty$ then for any $x^\star$ with the property from* (i) *there is $\eta > 0$ such that $f(x+h) - f(x) \geq \langle h, x^\star\rangle + \eta \operatorname{dist}(h, T(f,x))$ for every $h \in E$.*

Proof. Since E is separable there is a sequence $z_1^\star, z_2^\star, \ldots \in \partial f(x)$ such that $T(f,x) = \bigcap_{i=1}^\infty \bigcap_{j=1}^\infty \operatorname{Ker}(z_i^\star - z_j^\star)$. Hence, to prove the first statement, it suffices to consider for any $x^\star \in \partial f(x)$ the sequence $x_q^\star := (1 - 2^{-q})x^\star + \sum_{i=1}^{\infty} 2^{-i-q} z_i^\star$.

To prove the second statement in case $\dim(\partial f(x)) > 0$ (if $\dim(\partial f(x)) = 0$ then $T(f,x) = E$ and the statement is obvious), we first use the facts that the function $h \mapsto f'(x,h) - \langle h, x^\star \rangle$ is convex, continuous, and positively homogeneous and that it equals zero on $T(f,x)$ to infer that it is of the form $g \circ \kappa$, where g is a convex, continuous, and positively homogeneous function on the factor space $E/T(f,x)$ and κ is the canonical projection. Since our condition upon $x^\star$ implies that g is positive off the origin, we deduce from the compactness of the unit sphere in $E/T(f,x)$ that there is $\eta > 0$ such that $g(u) \geq \eta \|u\|$ for every $u \in E/T(f,x)$. Hence $f(x+h) - f(x) - \langle h, x^\star \rangle \geq f'(x,h) - \langle h, x^\star \rangle \geq \eta \|\kappa(h)\| = \eta \operatorname{dist}(h, T(f,x))$ for every $h \in E$.

1.3. Theorem. *Suppose that f is a convex continuous function defined on an open convex subset G of a separable real Banach space E, $k = 0, 1, \ldots$, and that π is a linear map of E onto a real Banach space F such that $\dim(\operatorname{Ker}(\pi)) \leq k$. Then the π image of the set $\{x \in E;\ \dim(\partial f(x)) > k\}$ is a first category subset of F.*

Proof. Let S be a countable dense subset of E and let $\mathcal{V}$ be the family of all pairs $(V, e^\star)$ where V is a $k+1$ dimensional subspace of E generated by a finite subset A of S and $e^\star$ is a real valued linear functional on V attaining only rational values on A.

First we observe that for every $x \in E$ with $\dim(\partial f(x)) > k$ there is a pair $(V, e^\star) \in \mathcal{V}$ and $\eta > 0$ such that

$$f(x+h) - f(x) \geq \langle h, e^\star \rangle + \eta \|h\|$$

for every $h \in V$. Indeed, since the subspace $T(f,x)$ of E has codimension at least $k+1$, there is a $k+1$ dimensional subspace V of E spanned by a finite subset of S such that $V \cap T(f,x) = \{0\}$. Using first 1.2(i) and then 1.2(ii), we find $x^\star \in \partial f(x)$ and $\eta > 0$ such that $f(x+h) - f(x) \geq \langle h, x^\star \rangle + 2\eta \|h\|$ for every $h \in V$. Now it suffices to find a linearly independent subset A of $V \cap S$ and to define $e^\star$ so that its values on A are rational and sufficiently close to the values of $x^\star$.

Since the family $\mathcal{V}$ is countable, our statement will be proved by showing that for any $(V, e^\star) \in \mathcal{V}$ and for any $\eta > 0$ the π image of the set

$$M := \{x \in G;\ f(x+h) - f(x) \geq \langle h, e^\star \rangle + \eta \|h\| \text{ for every } h \in V\}$$

is a first category subset of F.

Let W be a subspace of V such that $V = W \oplus (V \cap \mathrm{Ker}(\pi))$ and let U be a closed subspace of E containing W such that $U \oplus \mathrm{Ker}(\pi) = E$. Also, let X be a subspace of $\mathrm{Ker}(\pi)$ such that $\mathrm{Ker}(\pi) = X \oplus (V \cap \mathrm{Ker}(\pi))$. Since the restriction of π to U is a linear isomorphism, we may identify F with U. The map π then becomes the projection of E onto U along $\mathrm{Ker}(\pi)$. We also observe that $\dim(X) < \dim(W)$, since $\dim(X) = \dim(\mathrm{Ker}(\pi)) - \dim(V \cap \mathrm{Ker}(\pi)) \le k - \dim(V \cap \mathrm{Ker}(\pi))$ and $\dim(W) = \dim(V) - \dim(V \cap \mathrm{Ker}(\pi)) = k + 1 - \dim(V \cap \mathrm{Ker}(\pi))$.

Since f is convex and continuous on G, each point of G has a neighbourhood on which f is Lipschitz. Consequently, it suffices to show that the π projection N of the set $M \cap B(x_0, r_0)$ is a nowhere dense subset of U whenever f is Lipschitz on $B(x_0, 3r_0) \subset G$ with Lipschitz constant, say, a. Also, since the spaces W and $V \cap \mathrm{Ker}(\pi)$ are finite dimensional and since $W \cap [V \cap \mathrm{Ker}(\pi)] = \{0\}$, there is a constant $b \in (0, \infty)$ such that $\|w\| \le b\|w + v\|$ whenever $w \in W$ and $v \in V \cap \mathrm{Ker}(\pi)$.

Whenever $u_j \to u$ is a convergent sequence of elements of N (considered as a subset of U), we find $z_j \in \mathrm{Ker}(\pi)$ such that $u_j + z_j \in M \cap B(x_0, r_0)$. From $E = U \oplus \mathrm{Ker}(\pi)$ we see that the sequence z_j is bounded. Since $\mathrm{Ker}(\pi)$ is finite dimensional, we may also assume that z_j converges to some element, say z, of $\mathrm{Ker}(\pi)$. But then the definition of M implies that $u + z \in M \cap B(x_0, r_0)$, which shows that $u \in N$. Consequently, N is a closed subset of U.

Finally, to prove that N has empty interior, we show that for each $w \in U$ the set $N \cap (w + W)$ has empty interior in $w + W$: Let $u_1, u_2 \in N \cap (w + W)$. We find $v_1, v_2 \in V \cap \mathrm{Ker}(\pi)$ and $x_1, x_2 \in X$ such that $u_1 + v_1 + x_1 \in M \cap B(x_0, r_0)$ and $u_2 + v_2 + x_2 \in M \cap B(x_0, r_0)$. Then the definition of M implies

$$f(u_1 + v_1 + x_1 + h) - f(u_1 + v_1 + x_1) \ge \langle h, e^* \rangle + \eta \|h\|$$

and

$$f(u_2 + v_2 + x_2 + h) - f(u_2 + v_2 + x_2) \ge \langle h, e^* \rangle + \eta \|h\|$$

for every $h \in V$. Since $u_2 - u_1 + v_2 - v_1 \in V$, we get by letting in the first of these inequalities $h = u_2 - u_1 + v_2 - v_1$ and in the second $h = u_1 - u_2 + v_1 - v_2$

$$f(x_1 + u_2 + v_2) - f(x_1 + u_1 + v_1) \ge \langle u_2 - u_1 + v_2 - v_1, e^* \rangle + \eta \|u_2 - u_1 + v_2 - v_1\|$$

and

$$f(x_2 + u_1 + v_1) - f(x_2 + u_2 + v_2) \ge \langle u_1 - u_2 + v_1 - v_2, e^* \rangle + \eta \|u_1 - u_2 + v_1 - v_2\|$$

Adding these two inequalities and using that the Lipschitz constant of f on $B(x_0, 3r_0)$ is at most a, we infer that

$$2\eta\|u_2 - u_1 + v_2 - v_1\| \leq 2a\|x_1 - x_2\|.$$

Since $u_2 - u_1$ and $v_2 - v_1$ belong to the spaces W and $V \cap \mathrm{Ker}(\pi)$, respectively, $\|u_2 - u_1\| \leq b\ \|u_2 - u_1 + v_2 - v_1\|$. Hence

$$\|u_2 - u_1\| \leq ab\|x_1 - x_2\|/\eta.$$

It follows that the set $N \cap (w + W)$ is the image of the set $D = \{x \in X;$ there are $u \in (w + W) \cap N$ and $v \in V \cap \mathrm{Ker}(\pi)$ such that $x + u + v \in M \cap B(x_0, r_0)\}$ under the map $\varphi: D \mapsto N \cap (w + W)$ defined by $\varphi(x) = u$ if and only if there is $v \in V \cap \mathrm{Ker}(\pi)$ such that $u + v + x \in M \cap B(x_0, r_0)$ and that φ is Lipschitz. Since $D \subset X$ and $\dim(X) < \dim(W)$, we conclude that the $\dim(W)$ dimensional measure of $N \cap (w + W)$ is zero, which clearly shows that $N \cap (w + W)$ has empty interior in $w + W$.

1.4. Corollary. *Let f be a convex continuous function defined on an open convex subset G of a separable real Banach space E and $x \in G$. Then*

$$\partial f(x) = \bigcap_{r>0} w^\star\text{–closure}\left[\bigcup\{\partial f(z);\ z \in B(x,r) \text{ and } \dim(\partial f(z)) < \infty\}\right].$$

Proof. The inclusion $\supset$ follows from 1.1. To prove the opposite inclusion, suppose that $x^\star \in \partial f(x)$, W is a finite dimensional subspace of E, $\varepsilon > 0$ and $r \in (0, \mathrm{dist}(x, E \setminus G))$. It suffices to find $y \in B(x, r)$ and $y^\star \in \partial f(y)$ such that $\dim(\partial f(y)) < \infty$ and $\sup\{|\langle h, y^\star - x^\star\rangle|;\ h \in W, \|h\| \leq 1\} < \varepsilon$. Let π denote the canonical projection of E onto the factor space E/W and let $\delta \in (0, r)$ be such that for every $w \in E/W$ with $\|w - \pi(x)\| < \delta$

$$\min\{f(z) - f(x) - \langle z - x, x^\star\rangle + \varepsilon\|z - x\|;\ z \in B(x,r) \cap \pi^{-1}(w)\} < \varepsilon r.$$

Using Theorem 1.3 we find a point $w \in E/W$ such that $\|w - \pi(x)\| < \delta$ and $\dim(\partial f(z)) \leq \dim(W)$ for every $z \in G \cap \pi^{-1}(w)$. Since W is finite dimensional, there is $y \in B(x, r) \cap \pi^{-1}(w)$ at which the function

$$z \in B(x,r) \cap \pi^{-1}(w) \mapsto f(z) - f(x) - \langle z - x, x^\star\rangle + \varepsilon\|z - x\|$$

attains its minimum. Moreover, our choice of δ implies that y belongs to $B^0(x,r)$. Hence $y+h \in B(x,r) \cap \pi^{-1}(w)$ if $h \in W$ has sufficiently small norm and so for such h

$$f(y+h) - f(y) \geq \langle h, x^\star \rangle - \varepsilon[\|y-x+h\| - \|y-x\|] \geq \langle h, x^\star \rangle - \varepsilon\|h\|.$$

Thus the Hahn-Banach theorem provides us with $y^\star \in \partial f(y)$ such that $\langle h, y^\star \rangle \geq \langle h, x^\star \rangle - \varepsilon\|h\|$ whenever $h \in W$ has sufficiently small norm. Consequently,

$$\sup\{|\langle h, y^\star - x^\star \rangle|;\ h \in W, \|h\| \leq 1\} \leq \varepsilon.$$

Since $\dim(\partial f(y)) \leq \dim(W) < \infty$, this proves our statement.

1.5. Corollary. *Let E be a separable real Banach space E and let Q denote the set of all $x^\star \in E^\star$ for which there are $x \in E$, $\eta > 0$, and a finite codimensional closed linear subspace H of* $\mathrm{Ker}(x^\star)$ *such that*

(i) $\|x\| = \|x^\star\| = \langle x, x^\star \rangle = 1$

(ii) $\lim_{t\to 0}(\|x+th\| + \|x-th\| - 2)/t = 0$ *for every $h \in H$.*

(iii) $\|x+h\| \geq 1 + \eta\,\mathrm{dist}(h,H)$ *for every $h \in$* $\mathrm{Ker}(x^\star)$.

Then Q is $w^\star$-dense in the unit sphere of $E^\star$.

Proof. Since the set of norm attaining unit functionals (which is the union of subdifferentials at nonzero points) is $w^\star$-dense in the unit sphere of $E^\star$, 1.4 and the positive homogeneity of the norm imply that the union of subdifferentials at those unit points x at which the dimension of the subdifferential of the norm is finite is also $w^\star$-dense in the unit sphere of $E^\star$. Whenever $y^\star$ belongs to the subdifferential at such a point x and $H = T(\|\cdot\|, x)$, then (i) and (ii) hold (with $x^\star$ replaced by $y^\star$). Finally, using 1.2(ii), we approximate $y^\star$ by another element $x^\star$ of the subdifferential of the norm at x so that (iii) holds.

2. Determination of measures

2.1. Theorem. *Whenever finite Borel measures μ and ν over a separable Banach space E agree on all balls in E, then they agree.*

To prove this theorem, we use the following lemma from [2].

2.2. Lemma. *Suppose that $\tilde{\mu}$ and $\tilde{\nu}$ are finite Borel measures over a finite dimensional Banach space F, $e^\star \in F^\star$, and $D \subset \{e \in F;\ \langle e, e^\star \rangle > 0\}$ is a nonempty open cone such that $\overline{D} \cap \operatorname{Ker}(e^\star) = \{0\}$ and such that $\tilde{\mu}$ and $\tilde{\nu}$ coincide on every translate of D.*

Then the images of $\tilde{\mu}$ and $\tilde{\nu}$ under $e^\star$ coincide.

Since measures on E are determined uniquely by their Fourier transform, it suffices to prove that the images of μ and ν under all functionals belonging to a $w^\star$-dense subset Q of the unit sphere of $E^\star$ coincide. We prove that this holds with the set Q defined in 1.5.

Let $x^\star \in Q$, $x \in E$, and let H be a subspace of $\operatorname{Ker}(x^\star)$ such that the properties 1.5(i)–(iii) hold. Then the set $C := \bigcup_{n=1}^{\infty} B^0(nx, n)$ is an open convex cone in E. Moreover, since the union is nondecreasing, μ and ν coincide on every translate of C. Let κ be the canonical projection of E onto the factor space $F := E/H$ and let $D := \kappa[C]$. Using the fact that $H \subset \operatorname{Ker}(x^\star)$, we write $x^\star = e^\star \circ \kappa$ where $e^\star \in F^\star$. Since 1.5(ii) implies that $C + H = C$, the image measures $\tilde{\mu} := \kappa[\mu]$ and $\tilde{\nu} := \kappa[\nu]$ coincide on every translate of D. Finally, 1.5(iii) implies that $D \subset \{e \in F;\ \langle e, e^\star \rangle > 0\}$ and $\overline{D} \cap \operatorname{Ker}(e^\star) = \{0\}$. Hence Lemma 2.2 implies that the images of $\tilde{\mu}$ and $\tilde{\nu}$ under $e^\star$ coincide, which, because of $x^\star = e^\star \circ \kappa$, shows that the images of μ and ν under $x^\star$ coincide.

References

[1] R. R. Phelps, *Convex Functions, Monotone Operators, and Differentiability*, Lecture Notes in Mathematics 1364, Springer-Verlag, Berlin · Heidelberg · New York, 1989.

[2] D. Preiss and J. Tišer, Measures on Banach spaces are determined by their values on balls. (To appear.)

[3] L. Zajíček, On the differentiation of convex functions in finite and infinite dimensional spaces. *Czechoslovak Math. J.* **29 (104)** (1979), 340–348.

WHEN E AND $E[E]$ ARE ISOMORPHIC

C.J. Read

Cambridge University, England

ABSTRACT

If E is a Banach space with 1-symmetric basic $(e_i)_{i=1}^{\infty}$, the Banach space $E[E]$ has an unconditional basis $(e_{ij})_{i,j=1}^{\infty}$ with norm $||\sum_{i,j} \lambda_{ij} e_{ij}|| = ||\sum_i e_i || \sum_j \lambda_{ij} e_j ||_E \; ||_E$.

If E is c_0 or $\ell_p(| \leq p < \infty)$ it is easy to see that $E[E]$ is isomorphic (indeed isometric) to E itself; however, this is not true in general, and at the recent conference in Austria, A.V. Bukvalov asked whether we could have $E[E]$ isomorphic to E **as a Banach space,** for any other space E with symmetric basis. Here we show that this can be done, by methods which are strongly reminiscent of the present author's first ever paper [3]. For completeness we give a sketch proof of a related fact, namely that if $E[E]$ is isomorphic to E **as a Banach lattice,** then E must be either c_0 or ℓ_p. I understand that isomorphisms between E and $E[E]$ are helpful in the theory of Sobolev spaces, and this was the reason for the question being raised.

§1. Introduction.

We shall use the following notation. E will denote a Banach space with 1-symmetric basis $(e_i)_{i=1}^{\infty}$; on the (not closed) linear span lin $\{e_i\}$ we may define the ℓ_p norms $(l \leq p < \infty)$ such that

$$||\sum_{i=1}^{N} \lambda_i e_i|| = \left(\sum_{i=1}^{N} |\lambda_i|^p\right)^{\frac{1}{p}}$$

and similarly the c_0 - norm $|| \sum_{i=1}^{N} \lambda_i e_i ||c_0 = \max_i |\lambda_i|$.

Usually the norm on E will satisfy $||\cdot||_{c_0} \leq ||\cdot|| \leq ||\cdot||_{\ell_1}$. Not unnaturally, we shall write ℓ_p (respectively c_0) for the completion of (lin $\{e_i\}, ||\cdot||_p$) (respectively, (lin $\{e_i\}, ||\cdot||_{c_0}$)). The space $E[E]$ is the space of all double series $x = \sum_{i,j} \lambda_{ij} e_{ij}$ such that

$$\|x\|_{E[E]} = \|\sum_i e_i \| \sum_j \lambda_{ij} e_j \|_E \|_E < \infty. \qquad (1.1.1.)$$

Note that $E[E]$ and $E[E] \oplus E$ are isomorphic (even as Banach lattices, given their natural lattice structures). $E[E]$ is just a countable number of copies of E summed in a certain symmetric way – adding on one more copy makes no odds up to isomorphism.

§1.2 Averaging projections

Let E be a Banach space with 1-symmetric basis $(e_i)_{i=1}^{\infty}$; let $\mathbf{N} = \bigcup_{i=1}^{\infty} \sigma_i$ be a partition of the natural numbers into countably many disjoint **finite** subsets σ_i. The **averaging projection** associated with this partition is the linear map $P : E \to E$ such that

$$P\left(\sum_{i=1}^{\infty} \lambda_i e_i\right) = \sum_{j=1}^{\infty} \left(\sum_{i\in\sigma_j} e_i\right)\left(\sum_{i\in\sigma_j} \lambda_i\right) / \overline{\overline{\sigma}}_j$$

where $\overline{\overline{\sigma}}_j$ denotes the number of elements in σ_j.

Now it is fairly clear that an "averaging projection" has norm 1; since for a vector x of finite length (i.e., a finite linear combination of the e_i's), Px is in the convex hull of various rearragements of x (rearrangements in the sense of the natural lattice structure on a Banach space with symmetric basis), and since the basis is 1-symmetric, these rearrangements have the same norms as x.

Not quite so obvious, (but nonetheless a standard application of the Pelcynski decomposition method) is the key result about averaging projections, namely that if P is an averaging projection on a space E with symmetric basis, then

$$E \cong E \oplus PE, \qquad (1.2.1)$$

where $\cong$ denotes Banach space isomorphism. This result will be found in [1], p.117, and it is the one "external quote" necessary for a complete understanding of this paper.

Now if E has a 1-symmetric (or even just a 1-unconditional) basis, then $E[E]$ also has a 1-uncoditional basis $(e_{ij})_{i,j=1}^{\infty}$. It is our task, in the next section, to give a sketch proof of the result alluded to in the Abstract, namely that if E and $E[E]$ are isomorphic **as Banach lattices,** then E must be either ℓ_p or c_0.

§2. Lattice isomorphisms.

Now if $E[E]$ is isomorphic to E as a Banach lattice, then the isomorphism must send "atoms to atoms", i.e.,

$$e_i \to e_{\phi(i),\psi(i)}$$

where

$$\Psi : n \to (\phi(n), \psi(n))$$

in some bijection $\mathbf{N} \to \mathbf{N} \times N$. Because E has a symmetric basis, it doesn't matter which bijection we choose. So any bijection Ψ will do; now take any normalised block basis $(u_i)_{i=1}^{\infty}$ of E, and for all n we can pick a Ψ such that the vectors $u_1 \dots u_n$ are mapped onto n distinct "rows" of $E[E]$, and hence the (finite length) basic sequence $u_1 \dots u_n$ is uniformly equivalent to the subsequence $e_1 \dots e_n$ of the standard basis of E. So the whole sequence (u_i) is uniformly equivalent to (e_i), and E has "perfectly homogeneous" basis (all block bases are uniformly equivalent to it). Therefore by Zippin [5], E is c_0 or ℓ_p.

We now prove the main result of this paper, that E can be isomorphic to $E[E]$ as a Banach space, without E being isomorphic to c_0 or ℓ_p.

§3. Proof of the main result.

We wish to construct a "symmetric norm" $||.||$ on a vector space V with basis $(e_i)_{i=1}^{\infty}$, such that the completion E of this normed space will satisfy $E \cong E[E]$ as a Banach space, though the isomorphism will not be a lattice isomorphism. Let us begin by defining some very simple symmetric norms on V.

Defintion 3.1 For each $n \in \mathbf{N}$ let $||.||^{(n)}$ be the norm on V such that

$$||\sum_{i=1}^{N} \lambda_i e_i||^{(n)} = \left(\sup_{\pi \in S(\mathbf{N})} \sum_{i=1}^{n} |\lambda_{\pi(i)}| \right) \cdot \frac{\log n}{n}; \qquad (3.1.1)$$

where $S(\mathbf{N})$ is the group of permutations on $\mathbf{N}$.

This norm just takes the sum of the n largest absolute values of the coefficients λ_i, and then scales the answer by a factor $(\log n)/n$, so

$$||e_1 + \cdots + e_n||^{(n)} = \log n.$$

[This scaling is done for ease in constructing a symmetric norm which obviously isn't equivalent to the c_0 norm, but increases more slowly than any ℓ_p norm]

Note that

$$\frac{\log n}{n}||.||_{c_0} \leq ||.||^{(n)} \leq \frac{\log n}{n}.||.||_{\ell_1}.$$

We shall not be using all of the $||.||^{(n)}$'s in our construction, so for the sake of being definite, let us write

$$n_i = 10^{10^{10^i}}$$

and define

$$||.||_i = ||.||^{(n_i)}.$$

Then define a map $J : \ell_1 \to \ell_1$ by

$$J(\tilde{a}) = (||\tilde{a}||_i)_{i=1}^{\infty}$$

So $J(\tilde{a})$ is a positive sequence in ℓ_1 consisting of the sequence of norms $||\tilde{a}||_1, ||\tilde{a}||_2, ||\tilde{a}||_3, \ldots$

We know that

$$||\tilde{a}||_i = ||\tilde{a}||^{(n_i)} \leq \frac{\log n_i}{n_i}.||\tilde{a}||_{\ell_1}$$

hence

$$||J(\tilde{a})||_{\ell_1} \leq ||\tilde{a}||_{\ell_1} \left(\sum_{i=1}^{\infty} \frac{\log n_i}{n_i} \right) \leq \frac{1}{4}||\tilde{a}||_{\ell_1}$$

say.

Recall that we have chosen ourselves a bijection $\Psi : \mathbf{N} \to \mathbf{N} \times \mathbf{N}$. We now use this map to define some more symmetric norms, in the following way. For $x \in V$ let

$$\begin{aligned} ||x||_{E_0} &= ||x||_{c_0}, \\ ||x||_{E_1} &= ||x||_{c_0} + ||\hat{\Psi} Jx||_{E_0[E_0]}, \end{aligned}$$

where $\hat{\Psi}$ is the linear map sending $e_i \to e_{\phi(i),\psi(i)}$;

$$||x||_{E_2} = ||x||_{c_0} + ||\hat{\Psi} Jx||_{E_1[E_1]}$$

$$\vdots$$

$$||x||_{E_n} = ||x||_{c_0} + ||\hat{\Psi} Jx||_{E_{n-1}[E_{n-1}]}.$$

Then an elementary piece of induction tells us that $||x||_{E_i}$ is an increasing function of i; furthermore it is a bounded function of i, since if for all x we have

$$||x||_{E_i} \leq \rho.||x||_{\ell_1}$$

then

$$||x||_{E_i[E_i]} \leq \rho^2.||x||_{\ell_1[\ell_1]}$$

therefore

$$\begin{aligned}||x||_{E_i+1} &= ||x||_{c_0} + ||\hat{\Psi} Jx||_{E_i[E_i]}\\ &\leq ||x||_{\ell_1} + \rho^2 ||Jx||_{\ell_1}\\ &\leq (1 + \frac{1}{4}\rho^2)||x||_{\ell_1}.\end{aligned}$$

Hence, for all i, $||x||_{E_i} \leq 2||x||_{\ell_1}$.

So we may define $||x||_E = \lim_i ||x||_{E_i}$ and $||x||_E = ||x||_{c_0} + ||\hat{\Psi} Jx||_{E[E]}$ (3.2)

Now let $\mathbf{N} = \bigcup_1^\infty \sigma_i$ be a partition of the natural numbers into sets σ_i of size $\overline{\overline{\sigma_i}} = n_i$. Let π be the averaging projection on E associated with the partition $(\sigma_i)_1^\infty$. Then the image of π is the closed linear span of the vectors $x_i = \sum_{j \in \sigma_i} e_j$. We define a unique linear map $\Phi : \text{lin}\{x_i\} \to E[E]$ by requiring that

$$\Phi(x_i) = (\log n_i).\hat{\Psi}(e_i) = (\log n_i).e_{\phi(i),\psi(i)}.$$

This brings us to our first main theorem:

Theorem 3: Φ is continuous and extends to an isomorphism of $Im\,\pi$ with $E[E]$.

Proof: It is clear (because $\Psi : \mathbf{N} \to \mathbf{N} \times \mathbf{N}$ is a bijection) that Φ has dense range in $E[E]$. We must therefore establish that both Φ and Φ^{-1} are bounded linear maps where defined.

To begin with, we note by (3.1.1) that $||x_i||^{(n_i)} = \log n_i$.

So if $x = \sum_{i=1}^N \lambda_i x_i \in \text{lin}\{x_i\}$, since the x_i are disjointly supported and $||.||^{(n_i)}$ is a symmetric norm, we have

$$J(x) \geq \sum_{i=1}^N \log n_i.|\lambda_i|.e_i$$

where "$\geq$" is in the sense of the lattice ordering on ℓ_1. So,

$$\hat{\Psi} J(x) \geq \sum_{i=1}^N \log n_i.|\lambda_i| \hat{\Psi}(e_i)$$

and so

$$\|\hat{\Psi} Jx\|_{E[E]} \geq \|\sum_{i=1}^{N} \log n_i.|\lambda_i|.\hat{\Psi}(e_i)\|_{E[E]}$$

$$= \|\sum_{i=1}^{N} (\log n_i).\lambda_i.\hat{\Psi}(e_i)\|_{E[E]}$$

(for $E[E]$ has a natural unconditional structure)

$$= \|\Phi(x)\| \qquad (1)$$

However, by (1.2) we have

$$\|x\|_E = \|x\|_{c_0} + \|\hat{\Psi} Jx\|_{E[E]} \qquad (2)$$

so (1) and (2) tell us that $\|x\| \geq \|\Phi(x)\|$, *i.e.* Φ is a contraction.

Getting a bound on Φ^{-1} is not quite so simple, but is broadly similar. The idea is that if $x = \sum \lambda_i x_i = \sum \alpha_i e_i$ say then $J(x)$ is roughly equal to $\sum \log n_i.|\lambda_i|.e_i$.

To implement our idea, write $|||x||| = \max_i\{|\lambda_i|.\log n_i\} = \|\Phi(x)\|_{c_0}$. Then, $|||x||| \leq \|\Phi(x)\|$. Now we know that

$$J(x) \geq \sum (\log n_i).|\lambda_i|.e_i;$$

however, the coefficient of e_i in $J(x)$ is

$$\|x\|_i \leq \sum_{j=1}^{\infty} \|\lambda_j x_j\|_i = (\log n_i).|\lambda_i| + \sum_{j\neq i} \|\lambda_j x_j\|_i$$

$$= \log n_i.|\lambda_i| + \sum_{j=1}^{i-1} \frac{n_j \log n_i}{n_i}.|\lambda_j| + \sum_{j=i+1}^{\infty} \log n_i.|\lambda_j|$$

$$\leq \log n_i.|\lambda_i| + |||x|||.\left(\sum_{j=1}^{i-1} \frac{n_j \log n_i}{n_i \log n_j} + \sum_{j=i+1}^{\infty} \frac{\log n_i}{\log n_j}\right)$$

$$\leq \log n_i.|\lambda_i| + |||x|||.2^{-i}$$

(say). Hence, $J(x) \leq (\sum_i \log n_i.|\lambda_i|.e_i) + |||x|||.\sum 2^{-i} e_i$;

$$\|\hat{\Psi} Jx\|_{E[E]} \leq \|\sum_i \log n_i.|\lambda_i|.e_i\|_{E[E]} + |||x|||.\|\sum_i 2^{-i}\hat{\Psi} e_i\|_{E[E]}$$

$$= \|\Phi(x)\| + |||x|||.\|\sum 2^{-i}\hat{\Psi} e_i\|_{E[E]}$$

$$\leq \|\Phi(x)\| + 4|||x|||.\|\sum 2^{-i} e_i\|_{\ell_1}$$

(since $||.||_E \leq 2.||.||_{\ell_1}$),

$$=||\Phi(x)|| + 4|||x|||$$
$$\leq 5||\Phi(x)||.$$

Therefore,

$$||x||_E = ||x||_{c_0} + ||\hat{\Psi}Jx||_{E[E]}$$
$$\leq ||x||_{c_0} + 5||\Phi(x)|| \leq |||x||| + 5||\Phi(x)||$$
$$\leq 6||\Phi(x)||.$$

Thus, $||\Phi^{-1}|| \leq 6$. This is proves Theorem 3.

But Theorem 3 gives us our result. For (1.2.1) tells us that

$$E \cong E \oplus Im\pi \quad \text{(since } \pi \text{ is an averaging projection)},$$

and Theorem 3 adjusts this to $E \cong E \oplus E[E]$.

But we have already made the trivial observation that

$$E[E] \cong E[E] \oplus E.$$

Therefore $E \cong E[E]$, (incidentally it's then obvious that E and $E[E[E]]$ are isomorphic, and so on). Furhermore we do not have E isomorphic to c_0 or to any ℓ_p, because for selected values of n we have

$$||e_1 + \cdots + e_n||_E = O(\log n),$$

which of course doesn't happen in c_0 or ℓ_p.

Thus the question of Bukvalov is answered, and we find that we can have E and $E[E]$ isomorphic as Banach spaces without $E \cong \ell_p$ or c_o. Note also that, by our earlier arguments, this is an example of Banach space isomorphism without Banach lattice isomorphism.

Note: The referee pointed out that, as well as c_0 and ℓ_p, the universal space U_1 of Pelcynski also has the property that $U_1 \cong U_1[U_1]$. For it is known to have a symmetric basis, and $U_1[U_1]$ obviously contains a complemented U_1 so it is another universal space with unconditional basis, therefore it is U_1. For a discussion of U_1 see Pelcynski [2] or Singer [4], pp. 547–550.

I would like to thank Walter Schachermayer for organising a jolly good conference in Strobl, which I found very stimulating – this is my claim, at any rate, and here at least is a paper in support of it! But he knows, and we all know, that we had a very good time.

4. References

[1] J. Lindenstrauss and L. Tzafiri, *Classical Banach Spaces,* Vol. 1, Springer, (1977).

[2] A. Pelcynski, *Universal bases,* Studia Math. 32, 247–268, (1969).

[3] C.J. Read, *A Banach space with, up to equivalence, precisely two symmetric bases,* Israel J. Math., Vol. 40, No. 1, (1981).

[4] I. Singer, *Bases in Banach spaces I,* Springer, (1970).

[5] M. Zippin, *On perfectly homogeneous bases in Banach spaces,* Israel J. Math. 4, 265–272, (1966).

A note on Gaussian measure of translates of balls

Michel Talagrand(*)

Abstract. Consider a centered Gaussian measure μ on a separable Banach space X. W. Lindé [2] proved that if B is the unit ball of X, the function $x \to \mu(B) - \mu(x+B)$ is Gâteaux-differentiable at zero, and asked whether $\mu(B) - \mu(x+B) \leq C\|x\|^2$. We show that this is not the case; actually, if $X = \ell_p, 1 \leq p \leq 2$, and μ is "diagonal", and not supported by a finite dimensional space, then no estimate $\mu(B) - \mu(B+x) = o(\|x\|^p)$ can hold.

Denote by $(e_n)_{n\geq 1}$ the canonical basis of ℓ_p, and by B its unit ball. Throughout the paper, we assume $1 \leq p \leq 2$, and we denote by (g_n) an i.i.d. sequence of $N(0,1)$ random variables. Consider a sequence $a_n > 0$, such that $\sum_{n\geq 1} a_n^p < \infty$, and the law μ on ℓ_p of $\sum_{n\geq 1} a_n g_n e_n$.

Proposition. For some constant c depending on μ and p only, we have for all k

$$\mu(B) - \mu(B + a_k e_k) \geq c\, a_k^p = c\|a_k e_k\|^p. \tag{1}$$

Condition (1) implies that we cannot have an estimate $\mu(B) - \mu(B+x) = o(\|x\|^p)$. In particular, for $p = 1$, the map $x \to \mu(B) - \mu(B+x)$ is not Frechet differentiable at zero. This complements the results of [3], where it is shown that

$$\mu(B) - \mu(B+x) = O(\|x\|^p).$$

Proof. We fix k, and we denote by μ_k the law of $\sum_{n\neq k} a_n g_n e_n$. We set, for $v \geq 0$

$$h_k(v) = \mu_k(vB).$$

(*) Work partially supported by an N.S.F. grant.

Essential to our approach is the fact, consequence of a well known result of C. Borell [1], that $\log h_k$ is concave. For $0 \le t \le 1$, we set

$$f(t) = h_k((1-t^p)^{1/p})$$

so that

$$f'(t) = t^{p-1}(1-t^p)^{-1+1/p} h_k'((1-t^p)^{1/p}). \tag{2}$$

By Fubini's theorem, we have

$$\begin{aligned} \mu(ue_k + B) &= \frac{1}{a_k\sqrt{2\pi}} \int_{u-1}^{u+1} f(t-u) e^{-t^2/2a_k^2} dt \\ &= \frac{1}{a_k\sqrt{2\pi}} \int_{-1}^{1} f(t) e^{-(t+u)^2/2a_k^2} dt. \end{aligned} \tag{3}$$

We set $\Phi(t) = \frac{1}{\sqrt{2\pi}} \int_{-\infty}^{t} e^{-v^2/2}\, dv = \frac{1}{a_k\sqrt{2\pi}} \int_{-\infty}^{a_k t} e^{-v^2/2a_k^2}\, dv$.

Integrating (3) by parts, we get

$$\mu(ue_k + B) = \int_{-1}^{1} -f'(t)\Phi((t+u)/a_k)\, dt,$$

so that

$$\mu(B) - \mu(ue_k + B) = \int_{-1}^{1} -f'(t)[\Phi(t/a_k) - \Phi((t+u)/a_k)]\, dt.$$

Since $f(-t) = f(t)$ and $\Phi(x) - \Phi(z) = -\Phi(-x) + \Phi(-z)$, we get

$$\mu(B) - \mu(ue_k + B) = \int_0^1 -f'(t)\Psi(t/a_k, u/a_k)\, dt \tag{4}$$

where

$$\Psi(t,u) = 2\Phi(t) - \Phi(t+u) - \Phi(t-u).$$

We apply (2), (4), to get, after change of variable,

$$\mu(B) - \mu(B - a_k e_k) = a_k^p \int_0^{1/a_k} h_k'((1-a_k^p t^p)^{1/p})(1-a_k^p t^p)^{-1+1/p} \theta(t)\, dt \tag{5}$$

where $\theta(t) = t^{p-1}\Psi(t,1) > 0$.

Since $\log h_k$ is concave, for $v > 1$ we have

$$\frac{h_k'(u)}{h_k(u)} \geq \frac{\log h_k(v) - \log h_k(u)}{v - u}$$

so that, for $1 \geq u \geq 1/2$,

$$h_k'(u) \geq \frac{1}{v-u} h_k\left(\frac{1}{2}\right) \log \frac{h_k(v)}{h_k(u)} \geq \frac{1}{v} h_k\left(\frac{1}{2}\right) \log \frac{h_k(v)}{h_k(1)}. \tag{6}$$

Obviously for all k and t, $h_k(t) = \mu_k(tB) \geq \mu(tB)$. There is no loss of generality to assume that the sequence (a_n) decreases. Since $\mu(B)$ is clearly a decreasing function of the sequence (a_n) (for the componentwise order) we have $\mu_k(B) \leq \mu_1(B)$. If we choose v large enough that $\mu(vB) > \mu_1(B)$, we get from (6) that for $0 \leq u \leq 1/2$,

$$h_k'(u) \geq \frac{1}{v} \mu\left(\frac{1}{2} B\right) \log \frac{\mu(vB)}{\mu_1(B)}.$$

Together with (5), this implies the result. □

Acknowledgement. The problem investigated in this paper is motivated by questions of W. Lindé.

References

[1] C. Borell. The Brunn-Minkowski in Gauss Space, Invent. Math. 30, 207-216, 1975.

[2] W. Lindé, Gaussian measure of translated balls in a Banach space, to appear in Teor. Veroj. i. Primenen.

[3] M. Ryznar and T. Żak, "The measure of a translated ball in uniformly convex space", manuscript, 1989.

[4] T. Żak, On the difference of Gaussian measure of two balls in Hilbert spaces, Probability Theory on Vector Spaces IV, Lecture Notes in Math 1391, p. 401-405, Springer Verlag, 1989.

Equipe d'Analyse - Tour 46
U.R.A. au C.N.R.S N° 754
Université Paris VI
4 Pl. Jussieu
75230 Paris cedex 05

Department of Mathematics
The Ohio State University
231 West 18th Avenue
Columbus, Ohio 43210-1174

Sublattices of $M(X)$ isometric to $M[0,1]$

Lutz W. Weis
Louisiana State University
Baton Rouge, Louisiana 70803

1. Introduction

Let X, Y always denote infinite Polish spaces and $M(X)$ stands for the Banach space of Radon measures on X with the variation norm. While the isometric type of sublattices of $L_1(X,\mu)$ for some $\mu \in M(X)$ is completely understood (they are either l_1, $L_1[0,1]$, or $l_1 \oplus L_1[0,1]$, see e.g. [S] III Prop. 11.2) the structure of sublattices of $M(X)$ is much more complicated. Interesting examples of such sublattices are the Henkin-measures on the unit sphere of $\mathbb{C}^n$ for $n > 1$ (see [Rn], Chap. IX), Rajchman measures on the unit circle (see [Ke], Chap. IX), invariant mesures of a family of measurable tranformations of X (see [Ph], Chap. X) and, more generally, the invariant measures of a H-sufficient statistic in the sense of Dynkin (see [Dy], [Ma]). The first two examples are actually bands in $M(X)$ and there is a very nice characterization of such bands in terms of the compact subsets of X that they annihilate due to Mokobodski ([Ke], Chap. IX.1). The last two examples are usually true sublattices of $M(X)$ isometric to $M(0,1)$.

In this note we characterize sublattices L of the latter kind, (i.e. L is isometric to $M(0,1)$) in terms of the existence of strongly affine projections, the w^*-Radon-Nikodym-property, martingale compactness, a choquet-type integral representation theorem and finally in terms of the embedding of their unit sphere into $M(X)$ (see section 2 for precise statements).

These characterizations are derived from general integral representation theorems of Bourgin-Edgar ([Bo], [Ro]) and results on orthogonal kernels by Mauldin, Preiss and v. Weizsäcker ([Ma]), but in some cases we can give (in our lattice setting) more direct and easier proofs than the ones known in the literature.

Let us recall some definitions. Let F be a separable Banach space and U a bounded convex subset of F^*. U has the *w^*-RNP* if for every probability space $(\Omega,\mathcal{F},P)$ and bounded linear operator $T: L_1(\Omega, P) \longrightarrow F^*$ with $Tf \in U$ for all f in the positive sphere of $L_1(P)$ there is a w^*-measurable function

$$\phi : \Omega \longrightarrow U \qquad \text{with} \qquad Tf = w^* - \int \phi f dP$$

It is well known that every w^*-closed U has the w^*-RNP and for a line-closed, w^*-Borel w^*-measure convex U the w^*-RNP is equivalent to the w^*-martingale compactness, i.e. every U-valued martingale (f_n) with f_n finite-valued for each n, w^*-converges almost every where to a U-valued function f. (see e.g. [Gh], Chap. I).

Recall that U is *w^*-measure convex* if the w^*-barycenter of every w^*-Radon measure supported on U belongs to U.

A function $t : U \to F^*$ is *strongly affine* if it is universally measurable for the w^*-topology and for every probability measure P on U with barycenter $x \in U$ we have that $t(x)$ is the barycenter of the image measure $P \circ t^{-1}$ on $t(U)$. (see [D-M] Chap. X, Sect. 3, No. 50).

The following notion arises in mathematical statistics (see [Dy]). Let $(X, \mathcal{B})$ be a measure space and U a family of measures on $(X, \mathcal{B})$. A sub-σ-algebra Σ of $\mathcal{A}$ is called *H-sufficient* for U if there is a stochastic kernel $(\mu_x)_{x \in X}$ with $\mu_x \in U$ such that for all $\mathcal{B}$-measurable $f : X \to \mathbb{R}$ we have for all $\mu \in U$

$$E^\mu(f|\Sigma)(x) = \int f d\mu_x \qquad \mu - a.e.,$$

i.e. $g(x) = \int f d\mu_x$ is a common conditional disbribution of f given Σ for all $\mu \in U$.

Throughout we refer to the $\sigma\big(M(X), C_b(X)\big)$-topology on $M(X)$ as the w^*-topology which on the positive unit ball of $M(X)$ is induced by the $\sigma(F^*, F)$-topology, where F is the separable Banach space of functions on X uniformly continuous with respect to a totally bounded metric on X.

2. The Result

L denotes a fixed, norm-closed, non separable sublattice of $M(X)$, the space of Radon-measures on a Polish space X. $P(X)$ is the set of probability measures on X and

$$U = \{\lambda \in L : \|\lambda\| = 1, \quad \lambda \geq 0\}$$

$$E = \{\lambda \in U : \quad \lambda \text{ an atome of } L\}.$$

Notice that E is the set of extreme points of U.

$$M(\mu) = \{\nu \in M(X) : \sigma \text{ is } \mu\text{-absolutey continuous}\}$$

Theorem. Let L, U, E as above and assume that U is w^*-measure convex and w^*-analytic and E is w^*-Borel. Then the following four conditions are equivalent:

a) There is a strongly affine order isometry J of $M(0,1)$ onto L.

b) For every $\mu \in U$, there is a unique w^*-probability measure on E with μ as its w^*-barycenter.

c) U has the w^*-Radon-Nikodym property (or, equivalently, is martingale-compact.)

d) For every $\mu \in L_+$ there is a w^*-measurable kernel $x \in X \longrightarrow \mu_x \in E$ such that the positive projection $P : M(\mu) \longrightarrow M(\mu) \cap L$ is given by

$$P(\lambda) = w^* - \int \mu_x d\lambda \qquad \text{for} \quad \lambda \in M(\mu).$$

For the following conditions e), f), g) we always have e) $\Leftrightarrow$ f) $\Rightarrow$ g) $\Rightarrow$ a) and if we assume Martin's axiom all of the conditions are equivalent.

e) There is a positive, contractive projection P of $M(X)$ onto L, which is also strongly affine.

f) There is a countably generated σ-algebra Σ of universally measurable subsets of X such that Σ is H-sufficient for U.

g) There are w^*-universally measurable, w^*measure convex sets M_n so that

$$P(X) \backslash U = \bigcup_n M_n.$$

Remarks: 1) The implication c) $\Rightarrow$ b) follows from Edgar's choquet-type theorem for simplices with w^*-RNP ([Bo]) and its extentions (see e.g. [Ro]). We take advantage of our more specialized lattice setting to give a different and simpler proof below.

2) Of course there is always a positive projection $P : M(\mu) \longrightarrow M(\mu) \cap L$ which — as a conditional expectation operator — is given by a kernel of measures (μ_x) on X. So the essential requirement in d) is that μ_y can be chosen in U. Similarly, there is always a positive projection $P : M(X) \longrightarrow L$ ([S] III, Prop. 1.1.). Therefore the main point in e) is that P is in addition strongly affine. A typical projection without the latter property is the projection $P : M(X) \longrightarrow M(\mu)$ given by Lebesgue's decomposition. Notice that the unit ball of $M(\mu)$ has no extreme points.

3) The set theoretic assumptions enter through the separation theorems of Mauldin, Preiss and v. Weizäcker in [Ma]. They only use the weaker (but less well known) assumption of "existence of medial limits" (see e.g. [D-M], Chap. X, Sect. 3, No. 55).

4) The condition "$\mu_x \in U$" in the definition of H-sufficiency, is essential; if we replace 'H-sufficiency' in g) by the usual notion of 'sufficiency', then condition f) would be satisfied for $L = M(X, \mu)$ and Σ the Borel sets in X.

Also, the counterexamples in [Ma] show, that in general we cannot choose Σ as a σ-algebra of Borel subsets of X.

5) While d) $\Rightarrow$ e) may very well depend on set theoretic assumptions we suspect that a) $\Rightarrow$ g) holds in general. This is certainly true if U is a w^*-G_δ set since then we can apply the results of Ghoussoub and Maurey on w^*-H_δ-embeddings ([Gh], Chap. I). Condition g) is of course inspired by their "H_δ-embeddings".

6) The assumptions that U is w^*-measure convex and E is w^*-Borel cannot be dropped completely. Of course a) implies both of these properties but some of the other conditions do not. E.g., if Y is an uncountable, universal zero set in X, then the space L of Radon measures concentrated on Y satisfies b) to d) but not a) although E is universally measur-

able and U is measure convex. Also, the space L of purely atomic measures on X satisfies b) to d) but U is not measure convex.

7) The requirement in a) that J is strongly affine cannot be dropped. Just consider an isometry that maps the atomic measures on $[0,1]$ onto the atomic measures on $[0,\frac{1}{2}]$ and the diffuse measures on $[0,1]$ onto the diffuse measures on $[\frac{1}{2},1]$.

Corollary: A band $L \subset M(X)$ satisfies the assumptions of the theorem and one of the conditions a) to b) if and only if L is of the form $L = M(Y)$ for a Borel subset $Y \subset X$.

Proof: Since L is a band, E must consist of point measure. Now choose $Y \subset X$ such that $E = \{\delta_x : x \in Y\}$. ∎

The following classical exampe illustrates the conditions of the theorem.

Example: X is a Polish space and $\mathcal{B}$ the Borel sets of X. Let $\phi_n : X \to X$ be a sequence of Borel measurable mappings and L be the set of measures which are ϕ_n-invariant for all n; i.e.

$$L = \{\mu \in M(X) : \mu(\phi_n^{-1}(A)) = \mu(A) \quad \text{for all} A \in B,\ n \in \mathbb{N}\}$$

There are elementary arguments that show that L is a sublattice of $M(X)$ and the extreme points E of U are precisely the ergodic measures ([Ph], Sect. 10). Also, U is w^*-measure convex and

$$P(X)\backslash U = \bigcup_{i,j,k} M^{\pm}_{i,j,k}$$

$$M^{+}_{i,j,k} = \left\{\mu \in P(X) : \mu(\phi_i^{-1}(A_j)) \le \mu(A_j) - \frac{1}{k}\right\}$$

$$M^{-}_{i,j,k} = \left\{\mu \in P(X) : \mu(\phi_i(A_j)) \ge \mu(A_j) + \frac{1}{k}\right\}$$

where A_j is an enumeration of a countable subalgebra $\mathcal{B}_0$ of $\mathcal{B}$ which separates the points of X. It is easily checked that all $M^{\pm}_{i,j,k}$'s are w^*-Borel subsets of $P(X)$ and w^*-measure convex. Now g) $\Rightarrow$ b) of the theorem gives the usual representation of invariant measures as integrals of ergodic measures ([Dy] Sect. 6, [Ph]). The classical proofs show directly that the σ-algebra

$$\Sigma = \{A \in B : \mu(\phi_n^{-1} A \triangle A) = 0 \text{ for all } \mu \in L \text{ and } n \in \mathbb{N}\}$$

of $\{\phi_n\}$-invariant sets is sufficient as in condition f) or, if $\phi_n = \phi^n$, that

$$Q_n\mu = \frac{1}{n}\sum_{k=0}^{n-1} \mu\circ(\phi_n)^{-1} : M(X) \to M(X)$$

has a weak limit which defines a projection of $M(X)$ onto L as in e).

3. Proofs

a) $\Rightarrow$ c) Let $T : L_1(\Omega, P) \to L$ be an operator that maps the unit sphere of $L_1(\Omega, P)$ into U. By [Du], VI.8.6, there is w^*-measurable function $\phi : \Omega \to P[0,1]$ with

$$J^{-1}\circ T(f) = w^* - \int \phi f dP \quad \text{for} \quad f \in L_1(P).$$

Since J is strongly affine we get

$$T(f) = w^* - \int (J\circ\phi) f dP.$$

c) $\Rightarrow$ d) We may identify $M(\mu)$ with $L_1(X, \mu)$ and $M(\mu) \cap L$ with $L_1(X, \Sigma, \mu)$ where Σ is a sub-σ-algebra of the Borel sets of X, generated by a Borel measurable function $\varphi : X \longrightarrow \mathbb{R}$. The positive projection $P : L_1(X, \mu) \longrightarrow L_1(\Sigma, \mu)$ is then the conditional expectation operator $E(\cdot|\Sigma)$ and it is well known that there is a measurable kernel $x \in X \longrightarrow \mu_x \in M(X)$ of probability measures with

$$Pf(x) = \int f(y) d\mu_x(y) \qquad \text{for} \quad f \in L_1(X, \mu)$$

and

$$\mu_x\Big(\varphi^{-1}(\varphi(x))\Big) = 1 \qquad \text{for } \mu\text{-almost all } x \in X.$$

See e.g. [D-M] Chap. III, No. 70–72. One can check that this implies

$$P\lambda = w^*\text{-}\int \mu_x d\lambda(x) \qquad \text{for} \quad \lambda \in M(\mu).$$

Indeed, using the standard properties of conditional expectations we get for $f \in C_b(X)$ and $\lambda \in M(\mu)$ with $d\lambda = g d\mu$, $g \in L_1(\mu)$

$$\begin{aligned} P\lambda(f) &= \int P(g) \cdot f d\mu = \int E(g|\Sigma) f d\mu \\ &= \int g \cdot E(f|\Sigma) d\mu = \int \int f d\mu_x g(x) d\mu \\ &= \left(\int \mu_x d\lambda \right)(f). \end{aligned}$$

It remains to show that $\mu_x \in E$ for μ-almost all $x \in X$. First of all, we can assume that $\mu_x \in U$. Indeed, since U has the w^*-RNP there is a kernel $\mu'_x \in U$ for all $x \in X$ such that for the operator $P : M(\mu) \longrightarrow M(X)$ we have

$$P\lambda = w^*\text{-}\int \mu'_x d\lambda(x) \qquad \text{for} \quad \lambda \in M(\mu).$$

Since this kernel representation is unique up to a μ-nullset we have $\mu_x = \mu'_x \in U$ μ-a.e. Now assume that there is a Borel set

$$A \subset \{x \in X : \quad \mu_x \notin E\}$$

with positive μ-measure.

$$\widetilde{A} = \{(\lambda, x) \in L \times A : O \leq \lambda \leq \mu_x, \quad O \neq \lambda, \quad \lambda \neq \mu_x\}$$

is an analytic subset of the Polish space $U_{M(X)} \times X$ and by applying the von Neumann selection theorem ([D-M], Chap. III, No. 81) to the coordinate projection $U_L \times A \longrightarrow A$ restricted to $\widetilde{A}$ we find a w^*-universally measurable kernel $x \in A \longrightarrow \lambda_x \in U$ with $\lambda_x \neq O$, $\lambda_x \neq \mu_x$ and $O \leq \lambda_x \leq \mu_x$. By Lusin's theorem there is an $A_0 \subset A$ with $\mu(A_0) > O$ such that $x \in A_0 \longrightarrow \lambda_x$ is w^*-measurable and we define

$$\mu'_x = \begin{cases} \lambda_x(X)^{-1}\lambda_x & \text{for } x \in A_0 \\ \mu_x & \text{for } x \in X \backslash A_0 \end{cases}$$

the w^*-measurable kernel $x \in X \longrightarrow \mu'_x \in U$ still satisfies $\mu'_x\left(\varphi^{-1}(\varphi(x))\right) = 1$ for μ-almost all $x \in X$. This implies again that $P = E(\cdot|\Sigma)$ on $M(\mu)$, or

$$P\lambda = \int \mu'_x d\lambda(x) \qquad \text{for} \quad \lambda \in M(\mu).$$

The uniqueness of this kernel representation gives the contradiction that $\mu_x = \mu'_x$ for μ-almost all $x \in X$.

d) $\Rightarrow$ b) Given $\mu \in U$, we choose a kernel $x \in X \longrightarrow \mu_x \in E$ as in d). As in c) $\Rightarrow$ d) we see that for a Borel function f on X

$$E^{\mu}(f|\Sigma) = \int f d\mu_x$$

where Σ is the sub-σ-algebra generated by the map $\phi : X \to E$, $x \to \mu_x$. Now, if $Q = \mu \circ \phi^{-1}$, then for every $\mu_A = \mu(A)^{-1}\mu(A \cap \cdot)$, $A \in \Sigma$, we have

$$(*) \qquad \mu_A = P(\mu_A) = \int_X \mu_x d\mu_A(x) = \frac{1}{Q(\phi(A))} \int_{\phi(A)} \lambda dQ(\lambda).$$

For $A = X$ we see that μ is the barycenter of Q on E.

As a preparation for the proof of uniqueness, we choose an increasing sequence of σ-algebras each generated by a finite set $\mathcal{B}_n$ of atoms such that $\bigcup \sigma(\mathcal{B}_n)$ generates all Borel subsets of E. Then, for $\mathcal{A}_n = \{\phi^{-1}(B) : B \in \mathcal{B}_n\}$, the union of the increasing σ-algebras $\sigma(\mathcal{A}_n)$ generates Σ and we get from (*) that $Q(B)^{-1} \cdot b(Q|_B) = \mu_{\phi^{-1}(B)}$ for all $B \in \mathcal{B}_n$.

In particular, the ranges of

$$F_n(\nu) = \sum_{B \in \mathcal{B}_n} Q(B)^{-1} b(Q|_B) \chi_B(\nu) = \sum_{A \in \mathcal{A}_n} \mu_A \chi_A(\phi^{-1}(\nu))$$

(where terms with $Q(B) = 0$ are omitted) are precisely the atoms of $L_1(\sigma(\mathcal{A}_n), \mu)$. Let $Q_n = Q \circ F_n^{-1}$ be the distribution of this $(\mathcal{A}_n, Q)$-martingale on U.

Now let R be another probability measure on E with barycenter μ. We form the martingale

$$G_n(\nu) = \sum_{B \in \mathcal{B}_n} R(B)^{-1} b(R|_B) \chi_B(\nu)$$

with respect to $(\mathcal{A}_n, R)$ and denote by $R_n = R \circ G_n^{-1}$ its distribution on U. We are going to show that in the Choquet order of measures on U R_n is "almost" smaller than Q_{m_n} for an appropriate n_m. Indeed, since $b(R) = \mu$ and U is measure-convex, G_n has its values in $M(\mu) \cap L = L_1(\Sigma, \mu)$ and for every n we can find an $m = m(n)$ such that for $\nu_B = E\big(R(B)^{-1} b(R|_B) | \mathcal{A}_{m(n)}\big)$ we have

$$\|R(B)^{-1} b(R|_B) - \nu_B\| \le \frac{1}{n} \quad \text{for all} \quad B \in \mathcal{B}_n$$

and

$$\sum_{B \in \mathcal{B}_n} R(B) \nu_B = \mu.$$

If $\nu_B = \sum\limits_{A \in \mathcal{A}_{m(n)}} \alpha_A^B \mu_A$ we get

$$\sum_{A \in \mathcal{A}_{m(n)}} \sum_{B \in \mathcal{B}_n} R(B) \alpha_A^B \mu_A = \mu = \sum_{A \in \mathcal{A}_{m(n)}} Q(A) \mu_A$$

and therefore $Q(A) = \sum\limits_{B \in \mathcal{B}_n} R(B) \alpha_A^B$ and $\sum\limits_{A \in \mathcal{A}_{m(n)}} \alpha_A^B = 1$. Put $\widetilde{R}_n = \sum_{B \in \mathcal{B}_n} R(B) \delta_{\nu_B}$. In order to show that $\widetilde{R}_n \prec Q_{m(n)}$ in the Choquet order, we choose a continuous convex $\psi : U \to \mathbb{R}$. Then

$$\int_n \psi d\widetilde{R}_n = \sum_{B \in \mathcal{B}_n} R(B) \psi(\nu_B) \le \sum_{B \in \mathcal{B}_n} \sum_{A \in \mathcal{A}_{m(n)}} \alpha_A^B \psi(\mu_A)$$

$$= \sum_{A \in \mathcal{A}_{m(n)}} Q(A) \psi(\mu_A) = \int_U \psi dQ_n.$$

For the martingale F_n with respect to $(\mathcal{A}_n, Q)$ we have $F_n(\nu) \to \nu$ in the topology of U for Q-almost all $\nu \in E$. Therefore, $Q_n \to Q$ weakly on U. Similarly, since

$G_n(\nu) \to \nu$ R-a.e. on U implies that $R_n \to R$ and $\tilde{R}_n \to R$ weakly on U. Since $R_n \prec Q_{n(m)}$ we get for $n \to \infty$ that $R \prec Q$ in the Choquet order of U.

But since R and Q are supported by the extreme points E of U, it follows that $R = Q$ (see e.g. [Ed], theorem 2.2).

b) $\Rightarrow$ a) Since L is non-separable, b) implies that E is a uncountable Borel subset of the Polish space $P(X)$. Then there is a Borel isomorphism $j : [0,1] \to E$ ([D-M], Chap. III, No. 80). Define $J : M_+[0,1] \to L_+$ by taking $J(\lambda)$ as the w^*-barycenter of the measure $\lambda \circ j^{-1}$ on E. By b) this J is bijective and standard lattice arguments show that $\lambda \perp \nu$ implies $J\lambda \perp J\nu$ for $\lambda, \nu \in M_+[0,1]$. Now $J(\lambda) = J(\lambda^+) - J(\lambda^-)$ extends J to a lattice isometry of $M[0,1]$ onto L. $J : M[0,1]) \to M(X)$ is w^*-Borel measurable and strongly affine. Indeed, for a probability measure on $P([0,1])$ with barycenter $\mu \in P[0,1]$ and a Borel function f on X and $\tilde{f} : P(X) \to \mathbb{R}$, $\tilde{f}(\nu) = \int f d\nu$, we get

$$\begin{aligned} P \circ J^{-1}(\tilde{f}) &= \int (\tilde{f}) \circ J dP = \int J\nu(f) dP(\nu) \\ &= \int \left(\int j(y)(f) d\nu(y) \right) dP(\nu) \\ &= \int j(y)(f) d\mu(y) = J\mu(f) \end{aligned}$$

i.e. $J(\mu)$ is the barycenter of $P \circ J^{-1}$.

a) $\Rightarrow$ e) Assuming a), $y \in [0,1] \longrightarrow \tilde{\mu}_y := J(\delta_y)$ is an orthogonality preserving kernel in the sense of [Ma]. For such a kernel it is shown in [Ma], theorem 4.3, that — assuming the existence of medial limits — there is a universally measurable $\varphi : X \longrightarrow [0,1]$ such that $\tilde{\mu}_y(\varphi^{-1}(y)) = 1$ for all $y \in [0,1]$.

Define $\mu_x = \tilde{\mu}_{\varphi(x)}$ for all $x \in X$ and for $\lambda \in M(X)$

$$(+) \qquad P\lambda = \int \mu_x d\lambda(x) \qquad \text{for} \quad \lambda \in M(X).$$

Since U is measure convex we have $P(P(X)) \subset U$ and it is easily checked that

$P\lambda = \lambda$ for $\lambda \in L$. The same argument that showed in b) $\Rightarrow$ a) that J is strongly affine also shows that P is strongly affine, i.e. P is the required projection.

e) $\Rightarrow$ f) Given P we consider the universally measurable stochastic kernel $\mu_x = P\delta_x \in U$, $x \in X$. For $\lambda \in P(X)$ think of $\widehat{\lambda}$ as the induced measure on $\widehat{X} = \{\delta_x : x \in X\}$ with barycenter λ. Since P is strongly affine we get that

$$P\lambda = w^* - \int \mu_y d\lambda.$$

Let Σ be the smallest sub-σ-algebra of the universally measureable sets of X that makes the map $x \in X \longrightarrow \mu_x \in P(X)$ measurable.

For a fixed $\mu \in U$ we again identify $M(\mu)$ with $L_1(X,\mu)$ and $L_1(X,\mu) \cap L$ with $L_1(\Sigma_0,\mu)$ for an appropriate sub-σ-algebra Σ_0 of Borel sets of X. Then P induces on $L_1(\Sigma_0,\mu)$ the conditional expectation operator $E(\cdot|\Sigma)$. Observe that for all $f \in L_1(X_1\mu)$ and bounded Borel function g on X :

$$\begin{aligned}\int E(f|\Sigma_0)\cdot g d\mu &= \int f \cdot E(g|\Sigma_0) d\mu \\ &= P(g\cdot\mu)(f) = \int \mu_x(f) g(x) d\mu(x)\end{aligned}$$

Hence $E(f|\Sigma_0) = \int f d\mu_x \qquad \mu$-a.e.

Since functions of the form $x \to \mu_x(f)$, $f \in C_b(X)$ generate the σ-algebra Σ and each of them equals μ-a.e. the Σ_0-measurable function $E(f|\Sigma_0)$ it follows that modulo μ-zero sets Σ_0 and Σ are identical and that Σ is a H-sufficient σ-algebra for U with statistic $(\mu_x)_{x\in X}$.

f) $\Rightarrow$ e) By f) there is a sub-σ-algebra Σ of universally measurable sets of X and a universally measurable stochastic kernel $(\mu_x)_{x\in X}$ with $\mu_x \in U$ such that for all Borel functions f on X and all $\mu \in U$ we have

$$(++) \qquad E^{\mu}(f|\Sigma)(x) = \int f d\mu_x, \qquad \mu\text{-a.e.}$$

Define $P\lambda = w^*\text{-}\int \mu_y d\lambda$ for $\lambda \in M(X)$. Since U is measure convex, the image of P must be contained in L. By (++) (and the same arguments used in e) $\Rightarrow$ f)) P is a projection on every $M(\mu)$ for al $\mu \in U$. Hence P is a strongly affine projection onto L.

e) $\Rightarrow$ g) A measure $\mu \in M_+(X)$, $\|\mu\| \leq 1$, belongs to U if and only if $P\mu = \mu$ or $P\mu(B_n) = \mu(B_n)$ for all n, where $\{B_n\}$ is a countable subalgebra of the Borel sets that generates all Borel sets. Therefore

$$P(X)\backslash U = \bigcup_{i,j}(M_{ij} \cup N_{ij})$$

where

$$M_{ij} = \left\{\lambda \in P(X) : P\lambda(B_i) \leq \lambda(B_i) - \frac{1}{j}\right\}$$

$$N_{ij} = \left\{\lambda \in P(X) : P\lambda(B_i) \geq \lambda(B_i) + \frac{1}{j}\right\}$$

Since P is a w^*-universally measurable kernel operator it follows that M_{ij} and N_{ij} are all w^*-universally measurable and w^*-measure-convex. Indeed, if Q is a probability measure on M_{ij} with barycenter μ, may use that P is strongly affine and obtain

$$P\mu(B_i) = \int_{M_{ij}} P\lambda(B_i)dQ(\lambda) \leq \int_{M_{ij}} \lambda(B_i)dQ(\lambda) - \frac{1}{j} = \mu(B_i) - \frac{1}{j}$$

g) $\Rightarrow$ a) uses an idea of Edgar and Wheeler (see also [Gh] Theorem I.3).

Let $T : L_1(\Omega, P) \longrightarrow M(X)$ be a bounded linear operator which maps the unit sphere of $L_1(P)$ into U. Since $P(X)$ has the w^*-RNP there is a w^*-measurable $\phi : \Omega \longrightarrow P(X)$ with

$$Tf = w^* - \int \phi(t)f(t)dP(t), \qquad \text{for} \quad f \in L_1(P)$$

We want to show that ϕ takes its values in U P-almost surely. By assumption it is enough to show that for all i, j

$$P(\phi \in M_{ij}) = 0, \qquad P(\phi \in N_{ij}) = 0$$

Otherwise, there is a M_{ij} (or N_{ij}) so that $A = \phi^{-1}(M_{ij})$ has positive P-measure. For the function $f = P(A)^{-1}\chi_A$ we have on one hand $Tf \in U$, on the other hand

$$\begin{aligned} Tf &= w^* - \int \phi(t)f(t)dP(t) = P(A)^{-1}\int_A \phi(t)dP(t) \\ &= \int_{M_{i,j}} \nu dQ(\nu) \in M_{i,j} \end{aligned}$$

where $Q = (f \circ P) \circ \phi^{-1}$ is the distribution of $\phi : (A, f \cdot P) \longrightarrow M_{ij}$ and we used the w^*-measure convexity of M_{ij}. This contradiction shows that $P(A) = 0$. ∎

References

1) [Bo] R. D. Bourgin and G. E. Edgar : Non-compact simplices in Banach spaces with the Randon Nikodym Property, J. Funct. anal. 23 (1976), 162–176.

2) [D-M] C. Dellacherie and P. Meyer : Probabilités et potentiel, Herman, Paris.

3) [Du] N. Dunford and J. Schwartz : Linear Operators I, Interscience Publishers, New York (1958).

4) [Dy] E. B. Dynkin : Sufficient statistics and extreme points, Ann. of Prob. 6, 1978, 705–730.

5) [Ed] G. A. Edgar : On the Radon-Nikodym Property and Martingale Convergence, in : Vector Spaces Measures and Applications II, Lecture Notes in Mathematics 654, p. 62-76, Springer Verlag, Berlin-Heidelberg-New York, 1978.

6) [Gh] N. Ghoussoub and B. Maurey : H_δ-embeddings in Hilbert space and optimization on G_δ-sets. Memoirs of the AMS 349, Providence, 1986.

7) [Ke] A. S. Kechris and A. Louveau : Discriptive Set theory and the Structure of Sets of Uniqueness, Cambridge University Press, 1987.

8) [Ma] R.D. Mauldin, D. Preiss and H. v. Weizsäcker : Orthogonal Transition kernels, Ann. of Prob. 11, 1983, 970–988.

9) [Ph] R. P. Phelps : Lectures on Choquet's theorem, D. von Nostrand Comp., Princeton-New York, 1966.

10) [Ro] H. Rosenthal : Sub-simplices of convex sets and some characterizations of simplices with the RNP, in :Banach Space theory, Comtemporary Mathematics 85 (1987), 447-465.

11) [Ru] W. Rudin : Function theory in the unit ball of C^n, Springer Verlag, New York-Heidelbert-Berlin, 1980.

12) [S] H. H. Schaefer : Banach lattices and Positive Operators, Springer Verlag, Berlin-Heidelberg-New York, 1974.

Printed in the United States
By Bookmasters